Advanced Composites in Bridge Construction and Repair

Related titles:

Eco-efficient construction and building materials
(ISBN 978-0-85709-767-5)

Rehabilitation of metallic civil infrastructure using fiber-reinforced polymer (FRP) composites
(ISBN 978-0-85709-653-1)

Transport properties of concrete
(ISBN 978-1-78242-306-5)

Woodhead Publishing Series in Civil and Structural Engineering:
Number 50

Advanced Composites in Bridge Construction and Repair

Edited by
Yail J. Kim

ELSEVIER

AMSTERDAM • BOSTON • CAMBRIDGE • HEIDELBERG • LONDON
NEW YORK • OXFORD • PARIS • SAN DIEGO
SAN FRANCISCO • SINGAPORE • SYDNEY • TOKYO
Woodhead Publishing is an imprint of Elsevier

WP WOODHEAD PUBLISHING

Woodhead Publishing is an imprint of Elsevier
80 High Street, Sawston, Cambridge, CB22 3HJ, UK
225 Wyman Street, Waltham, MA 02451, USA
Langford Lane, Kidlington, OX5 1GB, UK

British Library Cataloguing-in-Publication Data
A catalogue record for this book is available from the British Library.

Library of Congress Control Number: 2014933286

ISBN 978-0-85709-694-4 (print)
ISBN 978-0-85709-701-9 (online)

For information on all Woodhead Publishing publications visit our website at http://store.elsevier.com/

Typeset by Newgen Knowledge Works Pvt Ltd, India

Contents

Contributor contact details

(* = main contact)

Editor

Y. J. Kim
Department of Civil Engineering
University of Colorado Denver
1200 Larimer Street
Denver, CO 80217, USA

E-mail: jimmy.kim@ucdenver.edu

Chapter 1

B. Wan
Department of Civil, Construction and Environmental Engineering
Marquette University
211 Engineering Hall
1637 W. Wisconsin Ave
Milwaukee, WI 53233, USA

E-mail: baolin.wan@marquette.edu

Chapter 2

Y. J. Kim*
Department of Civil Engineering
University of Colorado Denver
1200 Larimer Street
Denver, CO 80217, USA

E-mail: jimmy.kim@ucdenver.edu

M. F. Green
Department of Civil Engineering
Queen's University
58 University Avenue
Kingston, ON, K7L 3N6, Canada

E-mail: greenm@civil.queensu.ca

R. Gordon Wight
Department of Civil Engineering
Royal Military College of Canada
Kingston, ON, Canada

E-mail: gordon.wight@rmc.ca

Chapter 3

C. Carloni
University of Hartford
200 Bloomfield Avenue, Room UT 302 I
West Hartford, CT 06117, USA

E-mail: carloni@hartford.edu

Chapter 4

P. Z. Qiao* and W. Fan
Department of Civil and Environmental Engineering
Washington State University
Sloan 117, Spokane Street
Pullman, WA 99164-2910, USA

E-mail: qiao@wsu.edu

Chapter 5

J.G. Dai* and Y.L. Bai
Department of Civil and Environmental Engineering
The Hong Kong Polytechnic University
Hung Hom
Kowloon, Hong Kong, China

E-mail: cejgdai@polyu.edu.hk

Chapter 6

H. S. Ji
Department of Civil and Railroad Engineering
Daewon University College
274 Daehak-Ro
Jecheon-Si
Chungbuk-Do, 390-702, Republic of Korea

E-mail: hsji@mail.daewon.ac.kr

W. Song and Z. J. Ma*
Department of Civil and Environmental Engineering
University of Tennessee at Knoxville
313 J. D. Tickle
Knoxville, TN 37996-2313, USA

E-mail: zma2@utk.edu

Chapter 7

M. Li
University of Houston
Department of Civil and Environmental Engineering
N107 Engineering Building 1
4800 Calhoun Road
Houston, TX 77204-4003, USA

E-mail: moli@uh.edu

Chapter 8

W. Xiong
Department of Bridge Engineering
School of Transportation
Southeast University
2# Sipailou, Xuanwu District
Nanjing, 210096, China

E-mail: wxiong@seu.edu.cn

C. S. Cai
Department of Civil and Environmental Engineering
Louisiana State University
Baton Rouge, LA 70803, USA

E-mail: cscai@lsu.edu

R. C. Xiao
Department of Bridge Engineering
Tongji University
1239 Siping Road
Shanghai, 200092, China

E-mail: xiaorc@tongji.edu.cn

Chapter 9

M. E. Williams
Diagnostics Group
Walter P. Moore and Associates, Inc.
1301 McKinney, Suite 1100
Houston, TX 77010, USA

E-mail: mwilliams@walterpmoore.com

Chapter 10

Y. Kitane
Department of Civil Engineering
Nagoya University
Furo-cho, Chikusa-ku
Nagoya, Aichi 464-8603, Japan

E-mail: ykitane@civil.nagoya-u.ac.jp

A. J. Aref*
Department of Civil, Structural and Environmental Engineering
University at Buffalo – The State University of New York
235 Ketter Hall
Buffalo, NY 14260, USA

E-mail: aaref@buffalo.edu

Woodhead Publishing Series in Civil and Structural Engineering

1 **Finite element techniques in structural mechanics**
C. T. F. Ross

2 **Finite element programs in structural engineering and continuum mechanics**
C. T. F. Ross

3 **Macro-engineering**
F. P. Davidson, E. G. Frankl and C. L. Meador

4 **Macro-engineering and the earth**
U. W. Kitzinger and E. G. Frankel

5 **Strengthening of reinforced concrete structures**
Edited by L. C. Hollaway and M. Leeming

6 **Analysis of engineering structures**
B. Bedenik and C. B. Besant

7 **Mechanics of solids**
C. T. F. Ross

8 **Plasticity for engineers**
C. R. Calladine

9 **Elastic beams and frames**
J. D. Renton

10 **Introduction to structures**
W. R. Spillers

11 **Applied elasticity**
J. D. Renton

12 **Durability of engineering structures**
J. Bijen

13 **Advanced polymer composites for structural applications in construction**
Edited by L. C. Hollaway

14 **Corrosion in reinforced concrete structures**
Edited by H. Böhni

15 **The deformation and processing of structural materials**
Edited by Z. X. Guo
16 **Inspection and monitoring techniques for bridges and civil structures**
Edited by G. Fu
17 **Advanced civil infrastructure materials**
Edited by H. Wu
18 **Analysis and design of plated structures Volume 1: Stability**
Edited by E. Shanmugam and C. M. Wang
19 **Analysis and design of plated structures Volume 2: Dynamics**
Edited by E. Shanmugam and C. M. Wang
20 **Multiscale materials modelling**
Edited by Z. X. Guo
21 **Durability of concrete and cement composites**
Edited by C. L. Page and M. M. Page
22 **Durability of composites for civil structural applications**
Edited by V. M. Karbhari
23 **Design and optimization of metal structures**
J. Farkas and K. Jarmai
24 **Developments in the formulation and reinforcement of concrete**
Edited by S. Mindess
25 **Strengthening and rehabilitation of civil infrastructures using fibre-reinforced polymer (FRP) composites**
Edited by L. C. Hollaway and J. C. Teng
26 **Condition assessment of aged structures**
Edited by J. K. Paik and R. M. Melchers
27 **Sustainability of construction materials**
J. Khatib
28 **Structural dynamics of earthquake engineering**
S. Rajasekaran
29 **Geopolymers: Structures, processing, properties and industrial applications**
Edited by J. L. Provis and J. S. J. van Deventer
30 **Structural health monitoring of civil infrastructure systems**
Edited by V. M. Karbhari and F. Ansari
31 **Architectural glass to resist seismic and extreme climatic events**
Edited by R. A. Behr
32 **Failure, distress and repair of concrete structures**
Edited by N. Delatte
33 **Blast protection of civil infrastructures and vehicles using composites**
Edited by N. Uddin
34 **Non-destructive evaluation of reinforced concrete structures Volume 1: Deterioration processes**
Edited by C. Maierhofer, H.-W. Reinhardt and G. Dobmann

35 **Non-destructive evaluation of reinforced concrete structures Volume 2: Non-destructive testing methods**
Edited by C. Maierhofer, H.-W. Reinhardt and G. Dobmann
36 **Service life estimation and extension of civil engineering structures**
Edited by V. M. Karbhari and L. S. Lee
37 **Building decorative materials**
Edited by Y. Li and S. Ren
38 **Building materials in civil engineering**
Edited by H. Zhang
39 **Polymer modified bitumen**
Edited by T. McNally
40 **Understanding the rheology of concrete**
Edited by N. Roussel
41 **Toxicity of building materials**
Edited by F. Pacheco-Torgal, S. Jalali and A. Fucic
42 **Eco-efficient concrete**
Edited by F. Pacheco-Torgal, S. Jalali, J. Labrincha and V. M. John
43 **Nanotechnology in eco-efficient construction**
Edited by F. Pacheco-Torgal, M. V. Diamanti, A. Nazari and C. Goran-Granqvist
44 **Handbook of seismic risk analysis and management of civil infrastructure systems**
Edited by F. Tesfamariam and K. Goda
45 **Developments in fiber-reinforced polymer (FRP) composites for civil engineering**
Edited by N. Uddin
46 **Advanced fibre-reinforced polymer (FRP) composites for structural applications**
Edited by J. Bai
47 **Handbook of recycled concrete and demolition waste**
Edited by F. Pacheco-Torgal, V. W. Y. Tam, J. A. Labrincha, Y. Ding and J. de Brito
48 **Understanding the tensile properties of concrete**
Edited by J. Weerheijm
49 **Eco-efficient construction and building materials: Life cycle assessment (LCA), eco-labelling and case studies**
Edited by F. Pacheco-Torgal, L. F. Cabeza, J. Labrincha and A. de Magalhães
50 **Advanced composites in bridge construction and repair**
Edited by Y. J. Kim
51 **Rehabilitation of metallic civil infrastructure using fiber-reinforced polymer (FRP) composites**
Edited by V. Karbhari

52 **Rehabilitation of pipelines using fiber-reinforced polymer (FRP) composites**
Edited by V. Karbhari
53 **Transport properties of concrete: Measurement and applications**
P. A. Claisse

Preface

Advanced composites such as fiber-reinforced polymer (FRP) are promising materials for civil infrastructure application. Despite their relatively short history in research and implementation, significant advances have been made over the last two decades. Federal and municipal agencies now accept these non-conventional construction materials. Practitioners are eager to use them for their projects to improve the sustainability of constructed structural members. The objective of this book, *Advanced Composites in Bridge Construction and Repair*, is to identify current scientific challenges facing the infrastructure community, and to address them for the benefit of the general public, technically and socioeconomically. A total of ten chapters have rigorously been reviewed and selected for publication. The authors of each chapter are considered experts and leaders in the research community. Technical contents include use of advanced composites for bridge construction and performance monitoring, prestressed FRP composites, bond behavior of the composite–concrete interface, non-conventional composite honeycomb bridge decks and all composite superstructures, and repair of deteriorated bridges. The emphasis of this book is placed on fundamental research issues and practical application on site. *Advanced Composites in Bridge Construction and Repair* provides critical information to practicing engineers, government officials, and academics. There is no doubt that the compilation of the state-of-the-art technologies will advance our knowledge-oriented society, in particular constructed civil infrastructure. The editor gratefully acknowledges Ms Emily Cole and Mr Francis Dodds at Woodhead Publishing for their tireless professional service.

Yail J. Kim
Denver, Colorado, USA

Part I

General issues

1
Using fiber-reinforced polymer (FRP) composites in bridge construction and monitoring their performance: an overview

B. WAN, Marquette University, USA

DOI: 10.1533/9780857097019.1.3

Abstract: The superstructure and substructure of bridges can be made from all fiber-reinforced polymer (FRP), or from FRP composited with concrete or steel. The different applications of FRP composite materials in bridges are discussed. Common nondestructive evaluation/testing (NDE/NDT) methods are reviewed for their potential application in monitoring FRP bridges. Smart FRP bars and cables can be used to monitor the bridge, as well as help the bridge resist loads. A case study of monitoring a bridge with an FRP stay-in-place formwork deck is presented. A short commentary on likely future trends in FRP bridges and monitoring techniques is provided. Sources of further information related to FRP bridges and monitoring are listed at the end of this chapter.

Key words: fiber-reinforced polymer (FRP) bridge, FRP repaired and strengthened bridge, nondestructive evaluation/testing (NDE/NDT), smart FRP bridge, monitoring FRP bridge.

1.1 Introduction

Because of their high strength and low density, fiber-reinforced polymer (FRP) composite materials have been widely used in the aerospace and automotive industries since the middle of the last century. FRP composites do not corrode in concrete in the way steel does. Goldsworthy (1954) predicted that FRP composites could be used as reinforcements in concrete and as structural members subject to corrosive environments. Glass fiber reinforced polymer (GFRP) rods were used first as reinforcements for concrete buildings. Since the late 1980s, FRP rebars have been used more extensively in concrete structures, especially in highway bridge decks, because of their resistance to corrosion (Bank, 2006). For the same reason, FRP composites have been used more and more widely when repairing and retrofitting deteriorated bridge superstructures, to reinforce bridge decks, girders and piles, and when replacing structural members (e.g. decks). FRP materials are also increasingly being used in bridges because new manufacturing techniques have reduced their cost (Telang *et al.*, 2006). Although FRP materials have been used in the manufacture of missiles, planes, cars, boats, tanks, sporting goods, etc. for over half a century, their use in building bridges is still relatively recent. How well FRP composites in bridges will perform in the long term is one of our major concerns. Therefore, a monitoring program is normally set up to examine the behavior of FRP composites in demonstration bridges constructed around the world.

In this chapter, we consider how the superstructure and substructure of bridges can be made up of all FRP, or of FRP composited with concrete or steel. We discuss the different applications of FRP composites in bridge construction. Then we look at how bridges constructed using FRP composites are monitored by visual inspection and experimental nondestructive evaluation (NDE) techniques. Such evaluation can reveal damage such as blistering, voids, discoloration, fiber exposure, cracks, and scratches. We review common NDE techniques to see if they are suitable for monitoring FRP bridges, and we present a case study of a bridge with an FRP stay-in-place (SIP) formwork deck. This chapter also comments briefly on likely future trends, and includes sources of further information and advice.

1.2 Fiber-reinforced polymer (FRP) composites for bridge construction

FRP composite materials have been widely used to repair deteriorated bridges and to retrofit bridges that do not meet updated code requirements, and especially to retrofit columns to improve their seismic load capacity. They are also used in new construction as reinforcement in concrete decks, SIP formworks, pure FRP bridge decks, composite columns/piers, cables for suspension bridges, pedestrian bridges, and more. In fact, FRP composite materials can be used to replace steel in almost any component in bridge construction.

1.2.1 FRP pedestrian bridges

Because of their light weight and corrosion resistance properties, FRP profiles have been increasingly used in the decks and superstructure members of bridges since the mid 1970s (Bank, 2006). FRP was first used in short-span pedestrian bridges, the first FRP pedestrian bridge being built by the Israelis in 1975 (Tang and Podolny, 1998). Since then, hundreds of FRP footbridges have been built worldwide. The main structural type of these bridges is the truss. Cable-stayed and other traditional styles are also used (Feng, 2012).

Because of the low modulus of FRP composite materials, the stiffness of a pure FRP deck is relatively low, and thus deflection control is an issue. Relatively small amounts of concrete can be used to increase the stiffness of an FRP footbridge deck, keeping the total weight low. Bank *et al.* (2010) used pultruded GFRP planks with a cement board or concrete topping slabs to form a 75 mm deep hollow section to be used in place of traditional timber decking in pedestrian bridges. Neto and La Rovere (2010) developed a footbridge deck system consisting of a fiber reinforced concrete (FRC) top layer supported by GFRP wide-flange pultruded profiles, with foam blocks filling the voids to keep the bottom flat.

Researchers also tried combining FRP materials with other newly developed construction materials to improve the performance of pedestrian bridges. Mendes *et al.* (2011) designed and analyzed a 12 m long single span pedestrian bridge with two GFRP I-shape profiles and a thin steel fiber reinforced self-compacting concrete (SFRSCC) deck. Both steel anchors and epoxy resin were used to ensure the composite action between the two materials. The SFRSCC deck increases the flexural stiffness of the whole structure, resists compression stress, and provides better crack control. The GFRP I-shape resists tensile stress in this bridge.

The structural style can also be modified to counteract the disadvantage of the low modulus of FRP composite materials. Caron *et al.* (2009) proposed a self-stressed footbridge in which energy stored elastically in the bent bow generates the required stress to maintain the stability of the whole structure. This self-stressed FRP footbridge takes advantage of the low stiffness of the pultruded FRP profile.

1.2.2 FRP vehicular bridges

Using knowledge and experience gained from building FRP pedestrian bridges, engineers went on to build all FRP vehicular bridges. The first FRP vehicular bridge was built in Beijing, China in 1982. It has six hand-laminated GFRP sandwich girders, spans 20.7 m, and is 9.2 m wide (Ye *et al.*, 2003). In the early 1990s, many FRP deck systems were developed and evaluated in laboratories and in demonstration projects supported by federal or local departments of transportation and FRP manufacturers. Many of these FRP bridges were built with short spans over secondary or rural roads where traffic is light. The first US all-composite vehicular bridge went into public service on 4 December 1996 in Russell, Kansas (Tang and Podolny, 1998). In 1998, the US Federal Highway Administration (FHWA) set up the Innovative Bridge Research and Construction (IBRC) Program to develop cost-effective innovative material applications in highway bridges (FHWA, 2003). This program boosted the use of FRP materials in highway bridges in the US. Many FRP vehicular bridges have also been constructed in Europe, China, and Japan over the past two decades.

Replacing deteriorated reinforced concrete decks with FRP decks is an attractive option for increasing the live load capacity of old bridges because FRP decks are much lighter than concrete decks. For example, the non-composite concrete deck of the Bentley Creek Bridge in New York was replaced with an FRP deck in 1999. Changing the weight and stiffness changes the natural frequencies of such a bridge (Hag-Elsafi *et al.*, 2012). FRP decks are also composited with steel or concrete girders and trusses in new bridges (Turner *et al.*, 2003; Liu *et al.*, 2008).

FRP deck configurations can be cellular, honeycomb, or hybrid/sandwich. Cellular decks are manufactured by combining several pultruded FRP profiles to create different geometries. Although the weight of cellular decks is very low compared to other deck formats, their stiffness is relatively low and may cause buckling. An FRP honeycomb panel consists of top and bottom face sheets and a sinusoidal core (Davalos *et al.*, 2012), whereas the compression zone of an FRP-concrete hybrid deck often uses concrete (Aref *et al.*, 2005; Schaumann *et al.*, 2009). A sandwich panel consists of upper and lower layers and a core, with the bottom layer usually made of FRP, and the top layer and the core made of FRP and/or traditional materials. Many creative FRP sandwich decks have been invented. For instance, Schaumann *et al.* (2008) produced a sandwich deck panel consisting of three layers: an FRP sheet with T-upstands for the bottom skin, lightweight concrete for the core, and a thin layer of ultra-high-performance reinforced concrete for the top layer. Ji *et al.* (2009) proposed a composite sandwich panel consisting of a wrapped hybrid core of GFRP grid and multiple steel box cells with upper and lower GFRP layers.

1.2.3 FRP reinforced concrete bridge decks

FRP composite materials were first used to reinforce concrete structures in the 1950s (Rubinsky and Rubinsky, 1954). Corrosion of steel reinforcement is a major problem for bridge decks, worsened by increased traffic and deicing salts. As FRP composite materials are corrosion resistant, they are a good alternative to steel reinforcement for concrete bridge construction. Because of cost, GFRP reinforcement is more attractive than carbon FRP (CFRP) or aramid FRP (AFRP) for bridge decks, and has been used in many bridge decks in North America (El-salakawy *et al.*, 2005; Bank, 2006).

One of the barriers to using FRP in bridge decks is the initial material cost. Bouguerra *et al.* (2011) tested different parameters of eight full-scale concrete decks including slab thickness, concrete compressive strength, reinforcement ratio, and reinforcement type, and found that all slabs failed in punching shear. Punching shear was still the critical failure mode for concrete slabs reinforced with FRP grids when the span increased (Brunton *et al.*, 2012). The common slab-on-girder superstructure used for highway bridges may also cause punching shear failure in a bridge deck due to the lateral restraint of the supporting girders and the continuity of the slab (El-Ragaby *et al.*, 2007). Such lateral restraint, however, can produce compressive membrane action in the FRP reinforced concrete slab, which benefits the load-bearing capacities and deflection control of the bridge. Zheng *et al.* (2012) found that the reinforcement percentages did not affect the ultimate capacity of the GFRP-reinforced concrete bridge deck with the compressive membrane action. Therefore, the current high reinforcement requirement based on flexural design is very conservative.

The punching shear failure mode may lead to a reduction in FRP reinforcement required in bridge decks and thereby reduces the initial cost of GFRP-reinforced concrete bridge decks (Bouguerra *et al.*, 2011). Berg *et al.* (2006) compared the cost of a two-span highway bridge deck with an FRP-SIP formwork, deformed FRP bars, and a prefabricated pultruded FRP reinforcing grid, and concluded that an FRP reinforced bridge deck might be cost-effective even though its material costs were 60% higher than those of a steel-reinforced bridge deck. The saving of labor costs and construction, and maintenance due to potential long-term durability, can therefore overcome the initial high material cost.

Fatigue life and durability are also major concerns when using GFRP reinforcement in bridge decks. El-Ragaby *et al.* (2007) found that the bridge decks with GFRP reinforcement had a better fatigue performance and longer fatigue life than traditional steel-reinforced concrete decks. But laboratory tests showed that when GFRP is exposed to an alkaline solution, which simulates the concrete pore water solution, its mechanical properties (i.e., elastic modulus, tensile, shear, and bond strengths) decrease significantly (Bank and Russell, 1995; Sen *et al.*, 2002). So, although several bridges had used GFRP as reinforcement at the end of the 1990s, the designers of these bridges were questioning the durability of the GFRP bars (Bradberry, 2001). However, ISIS Canada studied several in-service bridges with GFRP-reinforced concrete decks and found no degradation of the GFRP in the concrete of structures exposed to natural environmental conditions for durations of 5–8 years (Mufti *et al.*, 2005). Therefore, the deterioration of a GFRP-reinforced concrete deck in field is not as serious as results from laboratory tests had indicated.

1.2.4 FRP stay-in-place formworks for bridge construction

FRP-SIP formworks serve as formworks for fresh concrete during construction and as reinforcement during service. They can be used as open forms for bridge decks, and as closed forms for bridge columns. The major advantages of FRP-SIP formworks are that they are lightweight, do not have to be removed after concrete is set, and provide crack control (Bank *et al.*, 2007; Oliva *et al.*, 2008; Hanus *et al.*, 2009). However, moisture may be trapped on the interface between the FRP formwork and the concrete, so increasing freeze–thaw damage (Wan *et al.*, 2007; Hanus *et al.*, 2009).

FRP-SIP formworks can be used to build reinforcement-free concrete decks if the gaps between the flange edges of the precast concrete girders are small (Bank *et al.*, 2007). Although no reinforcement is provided, such a slab is normally made by FRC for crack control. Bank *et al.* (2009) proposed a design specification for FRP-SIP formworks for concrete bridge deck construction. It includes a permanent FRP-SIP formwork system, a two-way mat of FRP I-shaped pultruded bars, and three-piece orthogonally placed cross-rods.

Since the FRP-SIP formwork serves as reinforcement during service, the bond between the FRP-SIP form and the concrete is very important for developing composite action, and it is made up of both mechanical and chemical bonding. Aggregates can be bonded with epoxy to the FRP plank before the concrete is poured to improve the mechanical bond (Bank *et al.*, 2007). Sand coating improves the connection between the FRP form and the concrete, and increases the flexural capacity of the composite deck (Cho *et al.*, 2010; He *et al.*, 2012). Using penetrating bars inserted into the webs of the FRP-SIP formwork and perpendicular to the major fiber direction of FRP formwork can also improve mechanical bonding (He *et al.*, 2012). Wet bonding, where the FRP form is coated with a layer of adhesive resin and fresh concrete is poured before the resin is cured, can be used to improve chemical bonding between the FRP-SIP formwork and the concrete (Li *et al.*, 2010).

Concrete-filled FRP tubes have been studied and used as bridge piers and girders (Son and Fam, 2008). During construction, FRP tubes act as a permanent formwork and, when in service, they can protect the concrete from aggressive environments, provide confinement, and act as shear and/or flexural reinforcement (Mohamed and Radhouane, 2010). Because of the confinement effect, using an FRP-SIP formwork for a bridge column can improve its deformation capacity significantly (Ozbakkaloglu and Saatcioglu, 2006, 2007), thereby improving its seismic performance. FRP-SIP formworks can also be used for bridge girders to create a hybrid FRP/reinforced concrete bridge superstructure (Chen *et al.*, 2009).

1.2.5 FRP cable-stayed bridges

A cable-stayed bridge is one where the bridge deck is stayed by cables attached to towers. It is a very efficient means of achieving a long span and aesthetically pleasing. The span of steel cable-stayed bridges has now exceeded 1000 m, with the Sutong Bridge spanning the Yangtze River in China with a main span of 1088 m, being the longest cable-stayed bridge in the world until 2012. Currently, the longest cable-stayed bridge in the world is Russky Bridge in Russia, which has a main span of 1104 m and was completed in July 2012. With a longer span, the self-weight of traditional steel and concrete limits the advantage of a cable-stayed structure, and steel cables also develop sag effect and corrosion problems.

As a high strength and lightweight material, FRP has the potential to span further. Moreover, it does not corrode and, therefore, it is logical to use FRP in a cable-stayed bridge to achieve a super long span. Meier (1987) proposed building a CFRP cable-stayed all FRP bridge to cross the Strait of Gibraltar with a maximum span of 8400 m. The first FRP cable-stayed bridge to be constructed, however, was Links Leader Bridge in Scotland, which was constructed in 1992 with a main span of 63 m (Khalifa *et al.*, 1996). After that, several such bridges for pedestrians and vehicles were constructed in other countries, including Herning Bridge (Denmark, 1999), I-5/Gilman Bridge (USA, 2002) (Scalea, 2000), and a pedestrian bridge at Jiangsu University (China, 2004) (Wu and Wang, 2008). Wang and Wu (2010) evaluated the suitability of different FRP and hybrid FRP cables for 1000–10 000 m super-long-span cable-stayed bridges. They concluded that basalt FRP (BFRP), hybrid basalt and steel wire FRP (B/SFRP) 20% and 30% cables have pronounced integrated advantages compared with steel cables within a span range of 3000 m, and that hybrid B/CFRP 25% and 50% cables exhibit cost advantages over CFRP cables, with promising applications for longer spans (Wang and Wu, 2010).

1.2.6 Retrofitting existing bridges

Existing bridges may have deteriorated due to:

- environmental factors, as with the corrosion of steel rebars in bridge superstructures
- vehicle collision
- design and construction errors resulting in lower load capacity than planned.

Deteriorated bridges have to be repaired and, due to the updating of design codes, some may no longer satisfy current requirements (especially seismic resistance) and must also be strengthened. FRP materials have exceptional stiffness- and strength-to-weight ratios, and do not require heavy equipment to be installed. Because of this, external bonding of FRP plates or sheets has become a popular method for strengthening or retrofitting reinforced concrete bridges and other structures (Bakis *et al.*, 2002; Teng *et al.*, 2002). The epoxy normally used in FRP strengthening can be cured and acts quickly. As it reduces the time that traffic is blocked or detoured, FRP strengthening is a very attractive solution.

For flexural strengthening, there are three major methods of bonding FRP composite materials to the soffit of the bridge girders and the bottom of the decks:

1. Using an adhesive to bond premanufactured FRP strips to the surface of bridge members.
2. Hand lay-up, in which dry fiber fabrics or sheets are saturated with liquid adhesives in field and bonded to the bridge members. A number of layers of FRPs may be used in hand lay-up applications.
3. Near surface mounting (NSM), where grooves are cut along the surface of the concrete members. Premanufactured FRP strips or small-diameter FRP bars are inserted into the grooves and bonded adhesively.

FRPs can also be wrapped around concrete columns or piers to increase their capacity and ductility. After the Prieta earthquake in 1989, FRP and steel jackets became popular and routine methods of retrofitting concrete bridge piers/columns in the US (Tang and Podolny, 1998; Bank, 2006). Compared with steel jackets, FRP composite materials are lightweight and corrosion free, making them more attractive in most cases. As well as increasing the ductility of bridge columns (Moran and Pantelides, 2002) and repairing piers or piles damaged by steel reinforcement corrosion (Pantazopoulou *et al.*, 2001; Winters *et al.*, 2008), FRP wrap is used as a barrier to prevent chloride penetration further corroding steel reinforcement (Klaiber *et al.*, 2004).

1.3 Monitoring problems in bridges using FRP composites

Although FRP composite materials have a lot of advantages, as mentioned in the previous sections, they also have some disadvantages compared with steel and concrete used in bridge construction:

- Besides its high initial cost, pultruded FRP has lower stiffness and, therefore, deflection control is necessary in FRP bridges.
- Most FRP deck systems are enclosed, making it difficult to detect problems by visual inspection.
- Compared with traditional materials used in bridge construction, there are relatively few long-term performance data on FRPs, and not enough for

bridge owners and engineers to use FRPs as confidently as they use steel and concrete. Therefore, a monitoring program is normally required after the construction of FRP bridges (Turner *et al.*, 2003; Foley *et al.*, 2010).

1.3.1 Potential problems of FRP bridges in service

In this section, we consider potential problems detected in FRP bridges in service.

An FRP composite material used in a bridge in field may deteriorate or become damaged due to moisture absorption, alkali solution in concrete, temperature cycling, freeze–thaw, creep and relaxation, fatigue, UV radiation, or fire (Karbhari *et al.*, 2003). The following damage may be observed:

- blistering
- voids
- discoloration
- wrinkling
- fiber exposure
- cracks
- delamination.

After moisture penetrates FRP components, local hydrostatic pressure between the layers and stresses due to freeze–thaw cycles may delaminate the layers and cause bubbles on the surface i.e., *blistering*.

Voids are submerged gaps within the laminates of a deck, and are usually invisible.

A long period of exposure to UV lights, heat, and the chemical environment may cause FRP *discoloration*.

Excessive stretching or shearing of the fabric during wet-out may cause *wrinkling*.

Improper handling may cause *fiber exposure* and scratches on the surface of FRP components. If it is not repaired, the damage would make the fibers susceptible to moisture and contamination, leading to further laminate deterioration in the area.

Vehicle and debris impact can cause *cracks* and result in the material separating through the entire thickness of the laminate (Telang, 2006).

Moisture trapped between the FRP-SIP formwork and the concrete, and its corresponding freeze–thaw cycles, may cause the loss of shear strength between the FRP formwork and the concrete, and *delamination* (Wan *et al.*, 2007; Foley *et al.*, 2010). FRP repaired and retrofitted concrete bridge flexural members also suffer delamination problems (Wan *et al.*, 2006).

FRP decks can be connected to steel girders by shear studs, clips, or bolts. For shear stud connections, the studs are installed onto the steel girders through stud pocket openings, and then the stud pockets are grouted to obtain composite action between the FRP and the steel girder (Turner *et al.*, 2003). For clip connections, the FRP deck is vertically clamped to the girder flange with steel side clips, and for bolt connections, threaded studs, nuts, and washers anchor the FRP deck to steel girder (Telang, 2006). If the girder is a precast concrete girder, a set of bent bars is cast as shear studs, and then inserted into the stud pocket openings in the FRP deck to be grouted. Good deck-to-girder connections are critical to the composite action between the deck and girder, and the integrity of the entire structure in service. Damage to these connections will cause gaps or relative movement between the deck and girder, and the bridge will no longer be safe for pedestrians and vehicles. However, these connections, apart from the bolted connection, are hidden in grout and are difficult to inspect.

It is possible to connect FRP deck-to-deck using interference fit with shear keys, tongue-and-groove connections, or butt joints with shear-splice plate strips (Telang, 2006). These connections are relatively easy to inspect, but problems can also develop in these areas as a result of improper installation, large vehicle loads, and environmental exposures. Loosened deck-to-deck joints may cause the decks to move vertically relative to each other, causing an uneven surface on the bridge, and cracks in the overlay and surface.

1.4 Common nondestructive evaluation/testing (NDE/NDT) methods for bridges using FRP composites

FRP bridges can be inspected and monitored by common NDE/NDT methods that have been used widely by bridge maintenance engineers and by engineers in the aerospace and ship industries that have a longer history of using FRP composite materials. Some simple but efficient methods, such as visual inspection, chain dragging, and tap testing, do not need special equipment. Some methods, such as thermal, acoustic, and ultrasonic testing, radiography, modal analysis, and *in situ* load testing, are more complex and do require specific equipment (Telang, 2006).

Visual inspection is the simplest method and is widely used to check FRP bridges. The inspector uses the naked eye with the help of simple tools, such as a flash light, measuring tape, or mirrors, to rapidly detect defects and problems on the bridge. Cracks, delamination, discoloring, voids, and deformations detected by visual inspection may be further investigated by more sophisticated methods if needed. The major disadvantage of visual inspection is that it can only detect surface problems and cannot quantify the extent of the damage.

In *tap testing*, the inspector taps the FRP surface with a coin, hammer, or electronic tapping unit and detects delamination or voids by listening to the distinctive changes in sound. Although it does not require expensive equipment or extensive training, tap testing can efficiently identify most common problems in FRP bridges in conjunction with visual inspection. Void in the bottom of an FRP deck was found in an FRP slab bridge by tapping the underside of the bridge surface with a rubber mallet (Alampalli, 2006). Visual inspection and *chain dragging* were also performed on this bridge to check for delamination between the polymer-concrete wearing surface and the FRP structure (Alampalli, 2006). Similarly to tap testing, in chain dragging a steel chain is dragged along the bridge deck and an inspector listens to the changes in acoustic response to locate potential delamination (Guthrie *et al.*, 2006). Rosenboom *et al.* (2009) found that visual inspection and tapping (by a steel rod) were very effective in inspecting bonding between FRP and concrete for FRP-strengthened prestressed concrete girders.

The theory behind infrared thermography (IRT) is that subsurface defects affect heat flow in a material (Rens *et al.*, 2005). An ambient or artificial heat source is used to heat one side of the bridge deck, and an infrared camera is used to measure the temperature variation of the other side, to detect discontinuities inside the tested area, such as delamination, debonding, impact damage, moisture, and voids. IRT imaging has been successfully used to evaluate the quality of the bond between FRP and concrete in an FRP-strengthened reinforced concrete bridge (Hag-Elsafi *et al.*, 2004; Taillade *et al.*, 2012). Digital IRT was used to detect subsurface defects in FRP bridge decks and the debonding between the wearing surface and underlying deck (Halabe *et al.*, 2007). The temperature inside a bridge can also be measured by thermal sensors. Teng *et al.* (2003) installed thermal sensors on GFRP-wrapped concrete columns of a highway bridge to monitor the temperature difference between the FRP and substrate concrete column over 2 years and detect possible debonding.

The principle behind *acoustic testing* is that a structure under certain load levels produces acoustic sounds or emissions (AE) due to deformation, cracks initiating and growing, cracks opening and closing, fibers breaking, and delamination in composite materials (Telang, 2006). Testing is carried out with AE piezoelectric sensors, couplant, multi-channel data acquisition hardware, and a fully integrated analysis and data acquisition software system. Gostautas *et al.* (2005) successfully used AE to evaluate six full-scale GFRP bridge deck panels with different cross-sections.

In *ultrasonic testing*, high-frequency sound waves are sent into the material being tested, and the change of speed of the reflected waves is detected. Pulse velocities can be correlated to material quality or bond quality to detect discontinuity (Hellier, 2001). An ultrasonic testing system includes a transducer, pulser, receiver/amplifier, and a display screen. Ultrasonic testing has been successfully used to detect debonding damage in FRP-strengthened concrete members (Bastianini *et al.*, 2001) and an FRP-strengthened cast iron bridge (Bastianini *et al.*, 2004). Mahmouda *et al.* (2010) used surface acoustic waves (SAWs) to evaluate the interface deterioration of CFRP-strengthened concrete specimens, and proposed using this method to monitor the health of concrete bridge structures repaired or retrofitted with FRP composite materials. The operator of ultrasonic testing should have a high level of expertise to properly conduct the test and interpret the data, and be certified by the American Society of Nondestructive Testing (ASNT) (Telang, 2006).

Radiography uses X-rays or gamma rays to penetrate the deck and 'see' defects. Defects in an FRP deck, such as delamination, differ from healthy parts in the amount of penetrating radiation they absorb. The radiographic film or computer screen of the test system can clearly indicate where there are defects.

If an FRP bridge has deteriorated, its overall stiffness will change, and therefore its dynamic response will differ from that of the original structure. Using this theory to evaluate the condition of a bridge structure is known as *modal analysis*. To perform a modal analysis, the bridge is instrumented with an array of accelerometers, and then prescribed dynamic load tests are performed to extract modal parameters with selected frequencies and mode shapes (Telang, 2006). How much the modal parameters differ from the baseline parameters indicates how much the FRP bridge has deteriorated or been damaged since its original construction. Modal analysis was used to characterize the dynamic behavior of the first FRP composite bridge superstructure built in New York State (Alampalli, 2006). Guan *et al.* (2007) installed a distributed array of accelerometers on an FRP highway bridge to collect structural vibration data and carried out a modal analysis to determine the long-term performance of the bridge.

For *in situ* load testing, a known live load with predefined patterns is applied to a bridge instrumented with strain gages, accelerometers, and displacement sensors. The strain and deflection of the bridge are recorded and analyzed to evaluate its performance. The load test can be performed over a certain time interval (e.g., 1 or 2 years) to discover any change in bridge response and determine any deterioration or damage during this time. Repeated *in situ* load tests can help to monitor the long-term performance of a bridge.

Because most of the FRP bridges built over the past two decades were built for demonstration purposes and technical verification, they all had short- and/or long-term monitoring programs, including *in situ* load testing, set up for them (Shenton and Chajes, 1999; Alampalli and Kunin, 2001; Turn *et al.*, 2003; Hag-Elsafi *et al.*, 2004; Reising *et al.*, 2004; Shahrooz and Boy, 2004; Zhou *et al.*, 2005; Alampalli, 2006; Foley *et al.*, 2010). The case study in

Section 1.5 describes one such program, which monitors a highway bridge using an FRP-SIP formwork and an FRP grid in a concrete deck.

Radiography, modal analysis, and load testing require special and normally expensive equipment, compared to visual inspection, tap testing and chain dragging, thermography, and acoustic testing. They also require a relatively high level of expertise to operate and interpret the data, and are time consuming. Therefore, they are only used for detailed assessments of potentially damaged or defective areas already identified using simpler methods (e.g. by visual inspection, chain dragging, or tap testing). More detailed discussion of common NDE/NDT methods can be found in many books and manuals covering this topic for regular and FRP bridges (Hellier, 2001; Telang, 2006).

1.4.1 Smart FRP bridges

This section introduces the recently developed smart FRP bars and smart FRP cables that integrate FRPs and fiber optic sensors used to monitor FRP bridge components. The use of strain gages to measure the strains of bridge components is a well-developed technology widely used in FRP bridge monitoring projects (Shenton and Chajes, 1999; Alampalli and Kunin, 2001; Turn *et al.*, 2003; Foley *et al.*, 2010). However, strain gages can be damaged in field by environmental exposure or vandalized. Instead, optical fiber Bragg grating (OFBG) sensors can measure the strain and temperature of a component by monitoring the Bragg light spectrum of the grating reflection and transmission inside the optic fibers. OFBGs have a small diameter and are lightweight, and therefore have only a very small effect on the components tested when they are installed. OFBG sensors are not affected by electromagnetic interference and can be easily made into distributed sensor networks, and have a long service life. Therefore, they are attractive for long-term monitoring of bridges (Zou *et al.*, 2003).

Two different fiber optic sensors were installed in an FRP reinforced concrete highway bridge in Canada to create a smart bridge that continuously sent bridge information to the engineers' office (Shehata and Rizkalla, 1999). More recently, OFBG sensors were installed on Hong Kong's Tsing Ma Bridge and successfully monitored the performance of the bridge in service (Chan *et al.*, 2006). Jiang *et al.* (2010) also embedded global and local fiber optic sensors in FRP materials to create a smart FRP system to strengthen reinforced concrete and steel members.

As the bare OFBG sensor is fragile, it is easily damaged if directly installed on structural members. For this reason, some researchers integrated OFBG sensors into FRP bars or cables to create smart FRP bars or smart FRP cables (Kalamkarov *et al.*, 1999, 2005; Zhang *et al.*, 2006; Lu and Xia, 2007;

Li *et al.*, 2009, 2011, 2012). To produce OFBG FRP smart bars, a bare OFBG sensor is put in the center of the die for FRP bars or cables during pultrusion, which are then cut to the required length for structural applications (Zhou *et al.*, 2003). The bond between the smart FRP bar and the concrete is critical in transferring the strain between them. In order to improve this bond, surface deformations are normally added to the FRP bars and sand used to roughen their surface (Zhou *et al.*, 2003; Kalamkarov *et al.*, 2005).

To make a smart cable for cable-stayed bridges, a steel strand was opened; then its middle wire was taken out and replaced by an FRP smart bar; finally, the rest of the wires and the FRP smart bar were re-twisted to make a new smart strand (Li *et al.*, 2011). Smart bars and cables with OFBG can be used as regular bars and cables, but they also act as sensors measuring strain and temperature inside structural members. Wang *et al.* (2009) installed smart GFRP bars with fiber optic sensors in reinforced concrete beams strengthened by NSM FRP bars to achieve the dual purposes of structural strengthening and strain monitoring.

1.5 Case study: monitoring a bridge with an FRP composite stay-in-place (SIP) formwork and an FRP composite reinforced concrete deck

As part of the FHWA IBRC Program, the state of Wisconsin constructed a highway bridge with decks using an FRP-SIP formwork and FRP grillage reinforcement in 2003 (Foley *et al.*, 2005, 2010). The bridge superstructure is composed of traditional precast concrete I-girders acting compositely with the bridge deck. The bridge is Bridge B-20-133 on US Highway 151 in Waupun, Wisconsin, and it is a typical precast, pretensioned two-span continuous superstructure system as shown in Fig. 1.1. A 5-year program was set up to monitor its long-term performance with two *in situ* load tests at 2-year intervals and field inspections.

Electrical resistance strain gages are often used to record the strains in different components of a bridge during *in situ* load testing (Turner *et al.*, 2003). As mentioned above, they have some drawbacks when used for long-term monitoring, for example, being vulnerable to environmental degradation, damage, and vandalism. Therefore, portable and removable strain sensors were developed and used to measure normal strains in the bridge deck and girders of this monitoring project (Schneeman, 2006; Foley *et al.*, 2010).

The portable strain sensor includes a material carrier with low modulus of elasticity for the strain gage and a conventional electronic resistance strain gage as shown in Fig. 1.2a and 1.2b. Two holes were drilled at each end of the portable strain sensor through which mechanical anchors were inserted to mount the sensor on the concrete girder or deck as shown in Figs 1.2c and

1.1 South side of US 151 overpass bridge, B-20-133. (*Source*: Courtesy of Dr Christopher Foley.)

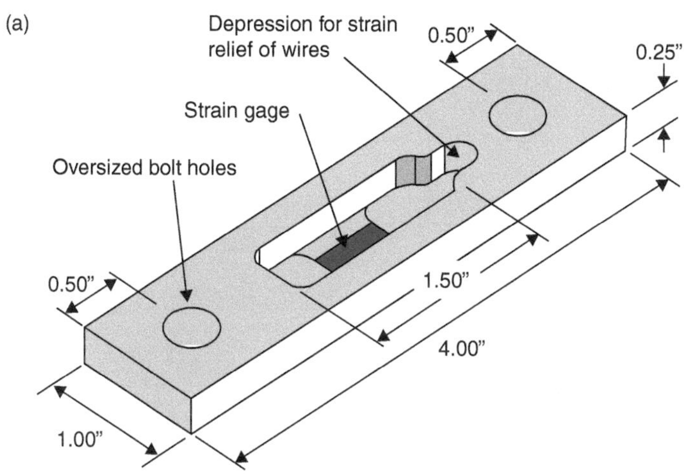

1.2 Portable strain sensor: (a) configuration of the portable strain sensor; (b) constructed strain sensors without connection tabs or protective coating; (c) field installation of the strain sensor to concrete; and (d) portable strain transducer mounted to lower flange of precast 54 W girder. (*Source*: Schneeman, 2006; Foley *et al.*, 2010; Courtesy of Dr Christopher Foley.)

(*Continued*)

(b)
(c)
(d)

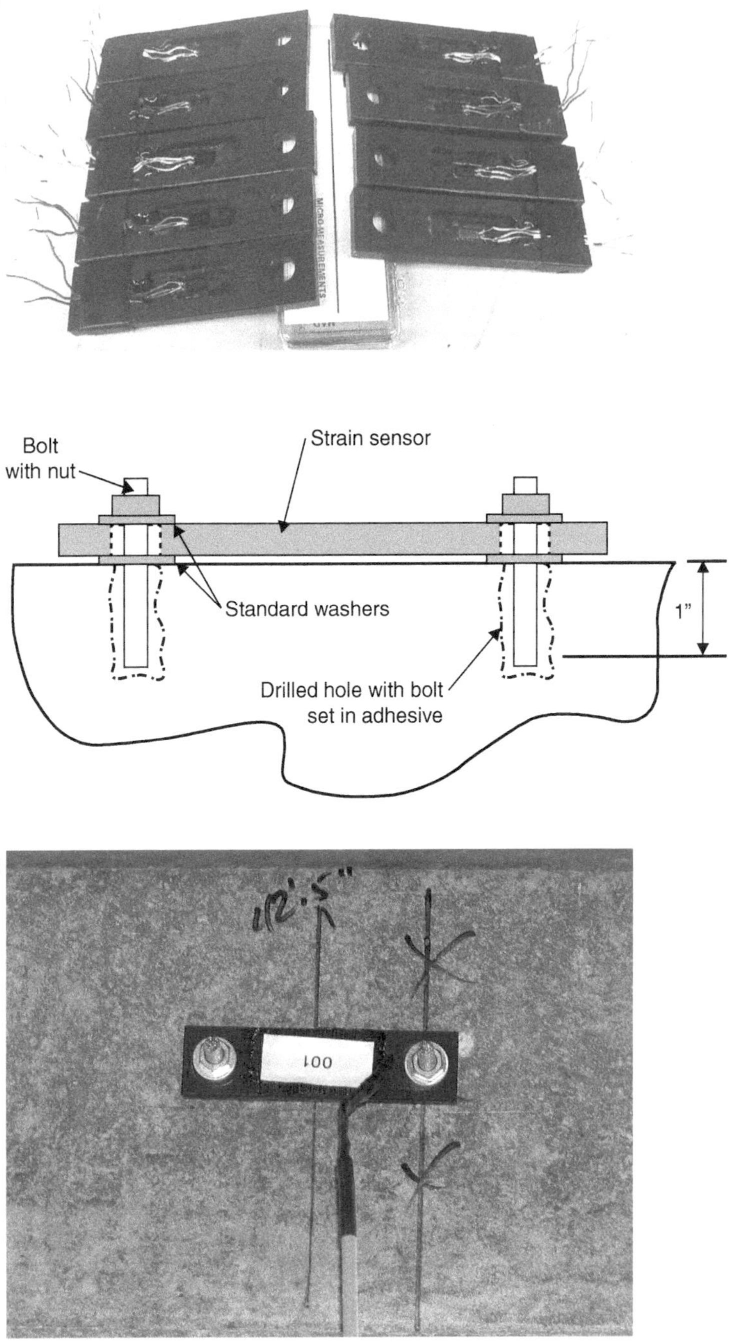

1.2 Continued

1.2d. For each sensor, a quick connect/disconnect tab was wired to the strain gage. The mechanical anchors and wires were installed permanently on the bridges. Before load testing, strain sensors were installed at designated locations on the anchors and connected to a data acquisition system via the quick connect/disconnect tabs. All strain sensors were tested and calibrated in the laboratory before they were used on the bridge. Details of the testing and calibration of the portable strain sensors can be found in Schneeman (2006).

The removable strain sensors were used to measure the strains of the bridge deck and commercially available draw wire transducers (DWTs) were used to measure the deflection of the bridge deck relative to the bridge girders. A schematic illustrating the instrumentation is shown in Fig. 1.3. The instrumentation in this bridge structure was focused on measuring the distribution of wheel loads within the bridge deck and the deflection of the bridge deck relative to the girders.

Two load tests were conducted on this bridge in July 2007 and July 2009. A weighted dump truck was driven slowly onto the bridge and stopped at designated locations to test the deck strain and deflection. Two loading path protocols were used to generate large strains in the exterior and interior spans, and the data acquisition system recorded the strains and deflections when the truck was on the bridge. Details of the loading test process can be found in Foley *et al.* (2010) and Wan and Foley (2010). Peak deck deflections measured in the first and second load tests were similar, on the order of 0.381 mm. Therefore, it can be concluded that there has been no significant change in the flexure load transfer mechanism in the bridge deck over the 2-year period between the two load tests.

The strain measurements from the July 2007 load test were used to estimate the bridge deck width that resists wheel loading, and a standard specification design procedure for bridge deck analysis (AASHTO, 2002) was also used to calculate the live load bending moment when the bridge was subjected to the truck load. Compared with the results calculated from measured strains, the American Association of State Highway and Transportation Officials (AASHTO) specification (2002) methodology is conservative in estimating the bending moment per foot of width that would be seen in the deck. The moment strip width was also computed using the measured strain data (Foley *et al.*, 2008) and compared with the results calculated using the AASHTO (2006) load and resistance factor design (LRFD) procedure. This shows that LRFD specifications are also conservative in analyzing this innovative FRP-SIP bridge deck.

Moisture accumulation at the interface between the FRP-SIP formwork and the concrete in this FRP-SIP bridge deck was assessed using a digital hygrometer inserted into holes drilled from the bottom of the deck to

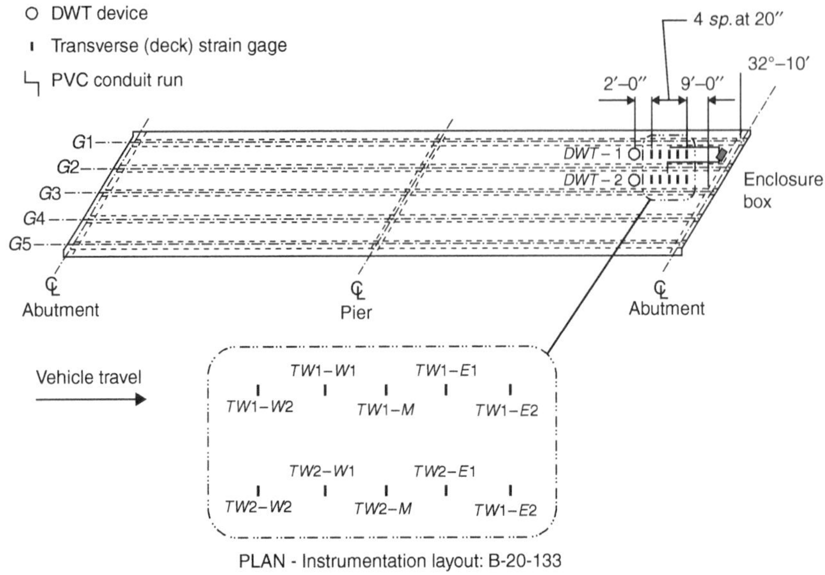

1.3 Instrumentation layout for B-20-133 (1 in. = 25.4 mm). (*Source*: Foley *et al.*, 2010; Courtesy of Dr Christopher Foley.)

different depths. No moisture was found when the probe holes were drilled. Therefore, no moisture had accumulated when this test was conducted in 2009 when the bridge had been in service for 5 years. However, the humidity at the interface of the FRP-SIP formwork and the concrete was higher than at other locations, indicating a tendency for moisture to accumulate at the interface. Experimental work relating to this monitoring program has shown that freeze–thaw cycling and the presence of water could be detrimental to the interfacial shear strength between the FRP-SIP formwork and the concrete (Wan *et al.*, 2007). Therefore, it is recommended that relative humidity is monitored at the interface for long-term evaluation of bridge decks with FRP-SIP formwork.

This bridge with its FRP-SIP deck and its sister bridge with a traditional steel-reinforced concrete deck were visually inspected in October 2005, July 2007, and July 2009. The visual inspections examined all superstructure elements including the abutments, piers, deck surface, deck soffit, and parapets. They revealed that both bridges were in very good condition.

In the first visual inspection, scaled crack maps were drawn for both bridges as shown in Fig. 1.4. The figure shows extensive cracking in both bridge decks and that most of the cracks were concentrated in the negative moment regions above the central piers. A numerical simulation indicated that the traffic load combined with concrete shrinkage caused transverse

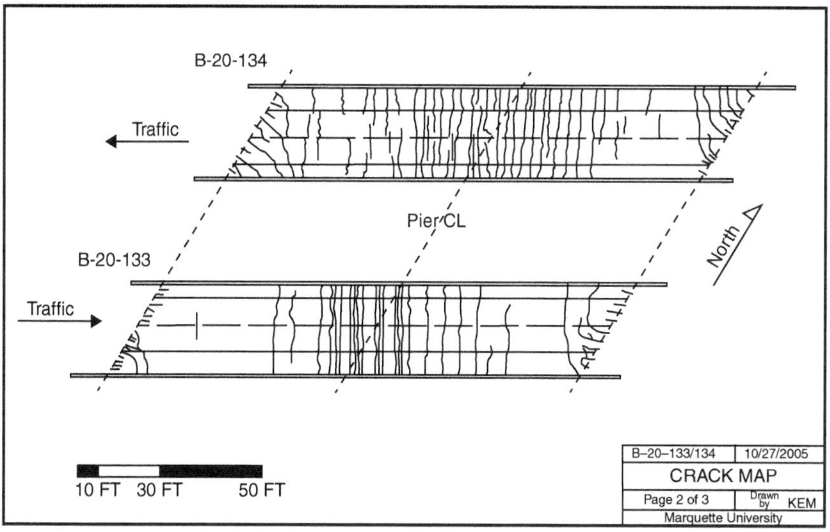

1.4 Crack map of B-20-133/134 (B-20-133 is the bridge with FRP-SIP form deck, and B-20-134 is the bridge with traditional steel reinforced concrete deck). (*Source*: Martin, 2006; Foley *et al.*, 2010; Courtesy of Dr Christopher Foley.)

cracking in the deck over the interior piers (Foley *et al.*, 2010; Wan *et al.*, 2010). The bridge with the FRP-SIP deck (Bridge B-20-133) appeared to have less cracking at the mid span location between the abutment and central pier than the bridge with the traditional steel-reinforced concrete deck (Bridge B-20-134). This may be a result of the SIP-FRP formwork restraining the shrinkage of the deck, as well as the tight spacing of the FRP grillage. The cracking at acute corners in skewed superstructures resulted from concrete shrinkage in the bridge deck. The visual inspection did not detect any cracking at the overhanging portion of the decks for either bridge. However, hairline cracks in the bridge decks had propagated to and through the parapet, with efflorescence showing on the underside of the overhang of each bridge deck. Because of the FRP-SIP formwork, it was impossible to check the condition of the concrete deck from the underside by visual inspection.

1.6 Future trends

The unique properties of high specific strength, high specific stiffness, and corrosive resistance have made FRP composite materials acceptable for all kinds of bridge construction, as described. Following two decades of field application and research, manufacturers, researchers, bridge engineers, owners, and government agencies are putting a lot of effort into increasing the application of FRP in bridges. To regulate design, construction, and inspection, different countries have published their own specifications and codes. As we gain greater understanding of their long-term performance, more bridges will be constructed using FRP composite materials. It is expected for more FRP to be used in the following three areas of bridge construction:

- FRP pedestrian bridges
- FRP reinforced concrete bridge decks
- super-long-span all FRP bridges.

An *FRP pedestrian bridge* can be manufactured in a factory and then installed on site within several hours. Because it is light compared with a steel and concrete pedestrian bridge, the FRP pedestrian bridge does not require heavy equipment for transporting and installing. Its corrosion resistance also reduces its life cycle cost. Therefore, an FRP pedestrian bridge will be seen as a very attractive option when an old pedestrian bridge needs to be replaced or a new one needs to be built.

Contrary to results from accelerated tests in the laboratory, GFRP reinforcement did not show any deterioration in *concrete decks* after several years of service in field (Mufti *et al.*, 2005). This has given bridge engineers and bridge owners the confidence to increase its application in areas where steel reinforcement corrosion is a serious problem, such as traditional steel-reinforced concrete bridge decks.

As the economy has developed, *long-span bridges* have been required worldwide to connect islands or cross straits. The high material strength of CFRP and the excellent structural style of a cable-stayed bridge enable an all FRP bridge to span much further than a bridge constructed using steel and concrete.

Basic and simple methods of bridge testing, such as visual inspection and tap testing, will be the regular NDE/NDT methods for FRP bridges. When needed, thermography, load testing, and other test methods, as discussed in the previous section, will be used. It is believed that more new FRP bridges will be constructed as 'smart' bridges, with various sensors integrated into the decks, girders, cables, and other important components to continuously monitor the performance of each bridge. The data of these smart bridges will be transmitted through wired or wireless networks to an office where they will be analyzed. Some prototype smart systems have already been established and installed on FRP bridges (Shehata and Rizkalla, 1999; Guan and Karbhari, 2006). In future, a complete system of software and hardware will be developed to automatically analyze technical data transmitted from the bridge. Once the critical parameter, such as the strain value at a predefined location, exceeds a certain threshold, the system will send an alert to the maintenance engineers to take action. Using such a system, engineers will be able to fix problems promptly when they first occur, avoiding further damage.

1.7 Sources of further information and advice

In recent years, a significant number of specifications and codes have been published to guide engineers when designing FRP bridges. In the USA, the American Concrete Institute (ACI) has published a series of guides on using FRP composite materials in concrete structures. All highway bridges in the US including FRP bridges should satisfy the requirements of the AASHTO *Standard Specifications for Highway Bridges*. AASHTO has also published *Guide Specifications for Design of FRP Pedestrian Bridges*. Canada, Japan, Europe, and China have published guide specifications for FRP applications in infrastructure.

The International Institute for FRP in Construction (IIFC) is the only international professional organization dedicated to the use of fiber reinforced composite materials in civil infrastructure (http://www.iifc-hq.org/). The IIFC hosts two official conferences in alternating years in which state-of-the-art research on FRP applications for infrastructures is presented and discussed. The FHWA maintains an FRP bridge library (http://www.fhwa.dot.gov/bridge/frp/frppaper.cfm) in which reports, design and construction guides, research papers, and other related information on FRP bridges in the US are collected. Significant journals publish research papers on FRP bridge practice and inspection, such as those listed in the reference section of this chapter.

For inspection and monitoring, *Handbook of Nondestructive Evaluations* by Chuck Hellier (2001) covers the majority of NDE methods. FRP bridge inspectors should also follow the federal and state published manuals for inspection of bridges. The National Cooperative Highway Research Program (NCHRP) report 564: *Field inspection of in-service FRP bridge decks* (Telang *et al.*, 2006) gives guidance on inspection and assessment of in-service FRP bridge decks.

1.8 References

AASHTO (2002) *Standard Specifications for Highway Bridges, Customary Units*, 17th Edition, American Association of State Highway and Transportation Officials, Washington, DC.

AASHTO (2006) *AASHTO LRFD Bridge Design Specifications Including 2006 Interim Revisions, Customary U.S. Units*, 3rd Edition, American Association of State Highway and Transportation Officials, Washington, DC.

Alampalli S. (2006) 'Field performance of an FRP slab bridge', *Composite Structures*, Vol. **72**, pp. 494–502.

Alampalli, S. and Kunin, J. (2001) *Load Testing of an FRP Bridge Deck on a Truss Bridge*, Transportation Research and Development Bureau, New York State Department of Transportation, Report No. FHWA/NY/SR-01/137.

Aref, A.J., Kitan, Y. and Lee, G.C. (2005) 'Analysis of hybrid FRP concrete multi-cell bridge superstructures', *Composite Structures*, Vol. **69**, pp. 346–359.

Bakis, C.E., Bank, L.C., Brown, V.L., Cosenza, E., Davalos, J.F. and Lesko, J.J. (2002) 'Fiber-reinforced polymer composites for construction state-of-the-art review', *ASCE Journal of Composite for Construction*, Vol. **6**, No. 2, pp. 73–87.

Bank, L. (2006) *Composites for Construction: Structural Design with FRP Materials*, Hoboken, New Jersey, John Wiley & Sons, Inc.

Bank, L.C., Malla, A.P., Oliva, M.G., Russell, J.S., Bentur, A. and Shapira, A. (2009) 'A model specification for fiber reinforced non-participating permanent formwork panels for concrete bridge deck construction', *Construction and Building Materials*, Vol. **23**, pp. 2664–2677.

Bank, L.C., Oliva, M.G., Bae, H., Barker, J.W. and Yoo, S. (2007) 'Pultruded FRP plank as formwork and reinforcement for concrete members', *Advances in Structural Engineering*, Vol. **10**, No. 5, pp. 525–535.

Bank, L.C., Oliva, M.G., Bae, H. and Bindrich, B.V. (2010) 'Hybrid concrete and pultruded-plank slabs for highway and pedestrian bridges', *Construction and Building Materials*, Vol. **24**, pp. 552–558.

Bank, L.C. and Russell, G.T. (1995) 'Accelerated test methods to determine the long-term behavior of FRP composite structures: environmental effects', *Journal of Reinforced Plastics and Composites*, Vol. **14**, pp. 559–587.

Bastianini, F., Di Tommaso, A. and Pascale, G. (2001) 'Ultrasonic nondestructive assessment of bonding defects in composite structural strengthening', *Composite Structures*, Vol. **5**, pp. 463–467.

Bastianini, F., Ceriolo, L., Di Tommaso, A. and Zaffaroni, G. (2004) 'Mechanical and nondestructive testing to verify the effectiveness of composite strengthening on historical cast iron bridge in Venice, Italy', *Journal of Materials in Civil Engineering*, Vol. **16**, pp. 407–413.

Berg, A.C., Bank, L.C., Oliva, M.G. and Russell, J.S. (2006) 'Construction and cost analysis of an FRP reinforced concrete bridge deck', *Construction and Building Materials*, Vol. **20**, pp. 515–526.

Bouguerra, K., Ahmed, E.A., El-Gamal, S. and Benmokrane, B. (2011) 'Testing of full-scale concrete bridge deck slabs reinforced with fiber-reinforced polymer (FRP) bars', *Construction and Building Materials*, Vol. **25**, pp. 3956–3965.

Bradberry (2001) 'Concrete bridge decks reinforced with fiber-reinforced polymer bars', *Transportation Research Record*, Vol. **1770**, pp. 94–104.

Brunton, J.J., Bank, L.C. and Oliva, M.G. (2012) 'Punching shear failure in double-layer pultruded FRP grid reinforced concrete bridge decks', *Advances in Structural Engineering,* Vol. **15**, No. 4, pp. 601–613.

Caron, J.F., Julich, S. and Baverel O. (2009) 'Self-stressed bowstring footbridge in FRP', *Composite Structures*, Vol. **89**, pp. 489–496.

Chan, T., Yu, L. and Tam, H. (2006) 'Fiber Bragg grating sensors for structural health monitoring of Tsing Ma bridge: background and experimental observation', *Engineering Structures*, Vol. **28**, pp. 648–659.

Chen, Y., Ziehl, P.H. and Harrison, K.W. (2009) 'Experimental characterization and optimization of hybrid FRP/RC bridge superstructure system', *Journal of Bridge Engineering*, Vol. **14**, pp. 45–54.

Cho, J., Cho, K, Park, S.Y., Kim, S.T. and Kim, B. (2010) 'Bond characteristics of coarse sand coated interface between stay-in-place fibre-reinforced polymer formwork and concrete based on shear and tension tests', *Canadian Journal of Civil Engineering*, Vol. **37**, pp. 706–718.

Davalos, J.F., Chen, A. and Zou, B. (2012) 'Performance of a scaled FRP deck-on-steel girder bridge model with partial degree of composite action', *Engineering Structures*, Vol. **40**, pp. 51–63.

El-Ragaby, A., El-Salakawy, E. and Benmokrane, B. (2007) 'Fatigue analysis of concrete bridge deck slabs reinforced with E-glass/vinyl ester FRP reinforcing bars', *Composites: Part B*, Vol. **38**, pp. 703–711.

El-Salakawy, E.F., Benmokrane, B., El-Ragaby, A. and Nadeau, D. (2005) 'Field investigation on the first bridge deck slab reinforced with glass FRP bars constructed in Canada', *ASCE Journal of Composites for Construction*, Vol. **9**, No. 6, pp. 470–479.

Feng, P. (2012) 'All FRP and FRP-concrete hybrid components for bridges: experiments, theories and case study', *The Third Asia-Pacific Conference on FRP in Structures (APFIS 2012)*, Hokkaido University, Sapporo, Japan, 2–4 February, CD-ROM.

FHWA (2003) *Innovative Bridge Research and Construction Program*, U.S. Department of Transportation, Federal Highway Administration, Washington, D.C., http://ibrc.fhwa.dot.gov.

Foley, C.M., Wan, B. and Liu, J. (2008) 'Wheel load distribution in concrete bridge decks with FRP stay-in-place forms', *ACI SP-257: FRP Stay-In-Place Forms for Concrete Structures*, CD-ROM, SP-257-3.

Foley, C.M., Wan, B., Schneeman, C., Barnes, K., Komp, J., Liu, J. and Smith, A. (2010) *In-Situ Monitoring and Testing of IBRC Bridges in Wisconsin*, Wisconsin Highway Research Program, report No. WHRP 10-09.

Goldsworthy, B. (1954) 'The continuous extrusion of RP', *Proceedings of the 9th SPI RPD Conference*, Chicago, 3–5 February, Section 13.

Gostautas, R.S., Ramirez, G., Peterman, R.J. and Meggers, D. (2005) 'Acoustic emission monitoring and analysis of glass fiber-reinforced composites bridge decks', *Journal of Bridge Engineering*, Vol. **10**, No. 6, pp. 713–721.

Guan, H. and Karbhari, V.M. (2006) 'Web-based structural health monitoring of an FRP composite bridge', *Computer-Aided Civil and Infrastructure Engineering*, Vol. **21**, pp. 39–56.

Guan, H., Karbhari, V.M. and Sikorsky, C.S. (2007) 'Long-term structural health monitoring system for a FRP composite highway bridge structure', *Journal of Intelligent Material Systems and Structures*, Vol. **18**, pp. 809–823.

Guthrie, W.S., Frost, S.L., Birdsall, A.W., Linford, E.T., Ross, L.A., Crane, R.A. and Eggert, D.L. (2006) 'Effect of stay-in-place metal forms on performance of concrete bridge decks', *Transportation Research Board Annual Meeting*, Washington D.C., CD-ROM.

Hag-Elsafi, O., Alampalli, S. and Kunin, J. (2004) 'In-service evaluation of a reinforced concrete T-beam bridge FRP strengthening system', *Composite Structures*, Vol. **64**, pp. 179–188.

Hag-Elsafi, O., Albers, W.F. and Alampalli, S. (2012) 'Dynamic analysis of the Bentley Creek Bridge with FRP deck', *Journal of Bridge Engineering*, Vol. **17**, pp.318–333.

Halabe, U.B., Vasudeven, A., Klinkhachorn, P. and Ganagarao, H.V.S. (2007) 'Detection of subsurface defects in fiber reinforced polymer composite bridge decks using digital infrared thermography', *Nondestructive Testing and Evaluation*, Vol. **22**, pp. 155–175.

Hanus, J.P., Bank, L.C. and Oliva, M.G. (2009) 'Combined loading of a bridge deck reinforced with a structural FRP stay-in-place form', *Construction and Building Materials*, Vol. **23**, pp. 1605–1619.

He, J., Liu, Y., Chen, A. and Dai, L. (2012) 'Experimental investigation of movable hybrid GFRP and concrete bridge deck', *Construction and Building Materials*, Vol. **26**, pp. 49–64.

Hellier, C. (2001) *Handbook of Nondestructive Evaluation*, New York, McGraw-Hill.

Ji, H.S., Son, B.J. and Ma, J. (2009) 'Evaluation of composite sandwich bridge decks with hybrid FRP-steel core', *Journal of Bridge Engineering*, Vol. **14**, pp. 36–44.

Jiang, G., Dawood, M., Peters, K. and Rizkalla, S. (2010) 'Global and local fiber optic sensors for health monitoring of civil engineering infrastructure retrofit with FRP materials', *Structural Health Monitoring*, Vol. **9**, pp. 309–322.

Kalamkarov, A.L., Fitzgerald, S.B., MacDonald, D.O. and Georgiades, A.V. (1999) 'On the processing and evaluation of pultruded smart composites', *Composites: Part B*, Vol. **30**, pp. 753–763.

Kalamkarova, A.L., Sahaa, G., Rokkama, S., Newhookb, J. and Georgiades, A. (2005) 'Strain and deformation monitoring in infrastructure using embedded smart FRP reinforcements', *Composites: Part B*, Vol. **36**, pp. 455–467.

Karbhari, V.M., Chin, J.W., Hunston, D., Benmokrane, B., Juska, T., Morgan, R., Lesko, J.J., Sorathia, U. and Reynaud, D. (2003) 'Durability gap analysis for fiber-reinforced polymer composites in civil infrastructure', *Journal of Composites for Construction*, Vol. **7**, No. 3, pp. 238–247.

Khalifa, M.A., Hodhodt, O.A. and Zaki, M.A.(1996) 'Analysis and design methodology for an FRP cable-stayed pedestrian bridge', *Composites Part B: Engineering*, Vol. **27**, pp. 307–317.

Klaiber, F.W., Wipf, T.J. and Kash, E.J. (2004) *Effective Structural Concrete Repair: Use of FRP to Prevent Chloride Penetration in Bridge Columns, Iowa Project TR-428*, Iowa Department of Transportation.

Li, H., Ou, J.P. and Zhou, Z. (2009) 'Application of optical fibre Bragg grating sensing technology-based smart stay cables', *Optics and Lasers in Engineering*, Vol. **47**, No. 10, pp. 1077–1084.

Li, L., Shao, Y. and Wu, Z. (2010) 'Durability of wet bond of hybrid laminates to cast-in-place concrete', *Journal of Composites for Construction*, Vol. **14**, pp. 209–216.

Li, H., Zhou, Z. and Ou, J.P. (2011) 'Development and sensing properties study of FRP–FBG smart stay cable for bridge health monitoring applications', *Measurement*, Vol. **44**, pp. 722–729.

Li, H., Zhou, Z. and Ou, J.P. (2012) 'Dynamic behavior monitoring and damage evaluation for arch bridge suspender using GFRP optical fiber Bragg grating sensors', *Optics & Laser Technology*, Vol. **44**, pp. 1031–1038.

Liu, Z., Cousins, T.E., Lesko, J.J. and Sotelino, E.D. (2008), 'Design recommendations for a FRP bridge deck supported on steel superstructure', *Journal of Composites for Construction*, Vol. **12**, pp. 660–668.

Lu, S. and Xia, H. (2007) 'Strengthen and real-time monitoring of RC beam using 'intelligent' CFRP with embedded FBG sensors', *Journal of Construction and Building Materials*, Vol. **21**, pp. 1839–1845.

Mahmouda, A. M., Ammara, H.H., Mukdadi, O.M., Ray, I., Imani, F.S., Chen, A. and Davalos, J.F. (2010) 'Non-destructive ultrasonic evaluation of CFRP–concrete specimens subjected to accelerated aging conditions', *NDT&E International*, Vol. **43**, pp. 635–641.

Martin, K.E. (2006) *Impact of Environmental Effects on, and Condition Assessment of, IBRC Bridge Decks in Wisconsin*, M.S. Thesis, Marquette University, Milwaukee, Wisconsin.

Mendes, P.J.D., Barros, J.A.O., Sena-Cruz, J. M. and Taheri, M. (2011) 'Development of a pedestrian bridge with GFRP profiles and fiber reinforced self-compacting concrete deck', *Composite Structures*, Vol. **93**, pp. 2969–2982.

Mohamed, H. and Radhouane, M. (2010) 'Axial load capacity of concrete-filled FRP tube columns: experimental versus theoretical predictions', *Journal of Composites for Construction*, Vol. **14**, pp. 231–243.

Moran, D.A. and Pantelides, C.P. (2002) 'Variable strain ductility ratio for fiber-reinforced polymer-confined concrete', *Journal of Composites for Construction*, Vol. **6**, No. 4, pp. 224–232.

Mufti, A. Onofrei, M., Benmokrane, B., Banthia, N., Boulfiza, M., Newhook, J., Bakht, B., Tadros, G. and Brett, P. (2005) 'Durability of GFRP reinforced concrete in field structures', *The 7th International Symposium on Fiber Reinforced Polymer Reinforcement for Concrete Structures (FRPRCS-7)*, New Orleans, Louisiana, USA, 7–10 November, pp. 1361–1378.

Neto, A.B.S. and La Rovere H.L. (2010) 'Composite concrete/GFRP slabs for footbridge deck systems', *Composite Structures*, Vol. **92**, pp. 2554–2564.

Oliva, M., Bae, H., Bank, L. and Russell, J. (2008) 'FRP stay-in-place formwork for floor and deck construction', *ACI SP257 FRP Stay-in-Place Forms for Concrete Structures*, CD-ROM.

Ozbakkaloglu, T. and Saatcioglu, M. (2006) 'Seismic behavior of high-strength concrete columns confined by fiber-reinforced polymer tubes', *Journal of Composites for Construction*, Vol. **10**, pp. 538–549.

Ozbakkaloglu, T. and Saatcioglu, M. (2007) 'Seismic performance of square high-strength concrete columns in FRP stay-in-place formwork', *Journal of Structural Engineering*, Vol. **133**, pp. 44–56.

Pantazopoulou, S.J., Bonacci, J.F., Sheikh, S., Thomas, M.D.A. and Hearn, N. (2001) 'Repair of corrosion-damaged columns with FRP wraps', *Journal of Composites for Construction*, Vol. **5**, No. 1, pp. 3–11.

Reising, R.M.W., Shahrooz, B.M., Hunt, V.J., Neumann, A.R. and Helmicki, A.J. (2004) 'Performance comparison of four fiber-reinforced polymer deck panels', *Journal of Composites for Construction*, Vol. **8**, No. 3, pp. 265–274.

Rens, K.L., Nogueira, C.L. and Transue, D.J. (2005) 'Bridge management and nondestructive evaluation', *Journal of Performance of Constructed Facilities*, Vol. **19**, No. 1, pp. 3–16.

Rosenboom, O., Walter, C. and Rizkalla, S. (2009) 'Strengthening of prestressed concrete girders with composites: installation, design and inspection', *Construction and Building Materials*, Vol. **23**, pp. 1495–1507.

Rubinsky, I. A. and Rubinsky, A. (1954) 'An investigation into the use of fiberglass for prestressed concrete', *Magazine of Concrete Research*, Vol. **6**, No. 17, pp. 71–78.

Schaumann, E., Vallée, T. and Keller, T. (2008) 'Direct load transmission in hybrid FRP and lightweight concrete sandwich bridge deck', *Composites: Part A*, Vol. **39**, pp. 478–487.

Schaumann, E., Vallée, T. and Keller, T. (2009) 'Shear resistance of lightweight concrete core of fiber-reinforced polymer concrete sandwich structure', *ACI Materials Journal*, Vol. **106**, No. 1, pp. 33–41.

Schneeman, C.L. (2006) *Development and Evaluation of a Removable and Portable Strain Sensor for Short-Term Live Loading of Bridge Structures*, M.S. Thesis, Marquette University, Milwaukee, WI.

Sen, R., Mullins, G. and Salem, T. (2002) 'Durability of E-glass/vinylester reinforcement in alkaline solution', *ASI Structural Journal*, Vol. **99**, No. 3, pp. 369–375.

Shahrooz, B.M. and Boy, S (2004) 'Retrofit of a three-span slab bridge with fiber reinforced polymer systems – testing and rating', *Journal of Composites for Construction*, Vol. **8**, No. 3, pp. 241–247.

Shehata, E. and Rizkalla, S. (1999) 'Intelligent sensing for innovative bridges', *Journal of Intelligent Material Systems and Structures*, Vol. **10**, pp. 304–313.

Shenton, H.W. and Chajes, M. J. (1999) 'Long-term health monitoring of an advanced polymer composite bridge', *Proceedings of the SPIE – The International Society for Optical Engineering*, Vol. **3671**, pp. 143–151.

Son, J.K. and Fam, A. (2008) 'Finite element modeling of hollow and concrete-filled fiber composite tubes in flexure: Model development, verification and investigation of tube parameters', *Engineering Structures*, Vol. **30**, pp. 2656–2666.

Taillade, F., Quiertant, M., Benzarti, K., Dumoulin, J. and Aubagnac, C. (2012) 'Nondestructive evaluation of FRP strengthening systems bonded on RC structures using pulsed stimulated infrared thermography', *Infrared Thermography*, Raghu V. Prakash (Ed.), InTech, Available from: http://www.intechopen.com/books/infrared-thermography/pulsed-stimulated-infrared-thermography-applied-to-a-nondestructive-evaluation-of-frp-strengthening-.

Tang, B. and Podolny, W. (1998) 'A successful beginning for fiber reinforced polymer (FRP) composite materials in bridge applications', *FHWA Proceedings, International Conference on Corrosion and Rehabilitation of Reinforced Concrete Structures*, 7–11 December, 1998, Orlando, FL.

Telang, N.M., Dumlao, C., Mehrabi, A.B., Ciolko, A.T. and Gutierrez, J. (2006) *NCHRP Report 564: Field Inspection of In-Service FRP Bridge Decks*, Washington, D.C., Transportation Research Board of the National Academies.

Teng, J.G., Chen, J.F., Smith, S.T. and Lam, L. (2002) *FRP Strengthened RC Structures*, UK, John Wiley and Sons.

Teng, M.H., Sotelino, E.D. and Chen, W.F. (2003) 'Performance evaluation of reinforced concrete bridge columns wrapped with fiber reinforced polymers', *Journal of Composites for Construction*, Vol. **7**, No. 2, pp. 83–92.

Turner, M.K., Harries, K.A., Petrou, M.F., Rizos, D. and Wan, B. (2003) *In-Situ Evaluation of Demonstration GFRP Bridge Deck System Installed on South Carolina Route S655*, University of South Carolina, Department of Civil and Environmental Engineering Report No. ST 03–02.

U Meier, D. (1987) 'Proposal for a carbon fiber reinforced composite bridge across the strait of Gibraltar at its narrowest site', *Proceedings of the Institution of*

Mechanical Engineers, Part B: Management and Engineering Manufacture, Vol. **201**, No. 2, p73–78.

Wan, B., Foley, C.M. and Martin, K. (2007) 'Freeze-thaw cycling effects on shear transfer between FRP stay-in-place formwork and concrete', *The Third International Conference on Durability & Field Applications of Fiber Reinforced Polymer (FRP) Composites for Construction*, 23–27 May, Quebec City, Quebec, Canada, pp. 227–234.

Wan, B. and Foley, C.M. (2010) '*In-situ* load tests and FE modeling of concrete bridge with FRP stay-in-place forms', *The 5th International Conference on FRP Composites in Civil Engineering (CICE 2010)*, 27–29 September, Beijing, China, pp. 207–210.

Wan, B., Petrou, M.F. and Harries, K.A. (2006) 'Effect of the presence of water on the durability of bond between CFRP and concrete', *Journal of Reinforced Plastics and Composites*, Vol. **25**, No. 8, pp. 875–890.

Wang, B., Teng, J.G., De Lorenzis, L., Zhou, L., Ou, J., Jin, W. and Lau, K.T. (2009) 'Strain monitoring of RC members strengthened with smart NSM FRP bars', *Construction and Building Materials*, Vol. **23**, pp. 1698–1711.

Wang, X. and Wu, Z.S. (2010) 'Evaluation of FRP and hybrid FRP cables for super long-span cable-stayed bridges', *Composite Structures*, Vol. **92**, pp. 2582–2590.

Winters, D., Mullins, G., Sen, R., Schrader, A. and Stokes, M. (2008) 'Bond enhancement for FRP pile repair in tidal waters', *Journal of Composites for Construction*, Vol. **12**, No. 3, pp. 334–343.

Wu, Z.S. and Wang, X. (2008) 'Investigation on a thousand-meter scale cable-stayed bridge with fibre composite cables', *The Fourth International Conference on FRP Composites in Civil Engineering (CICE2008)*, 22–24 July, Zurich, Switzerland, CD-ROM.

Ye, L.P., Feng, P., Zhang, K., Lin, L., Hong, W.H. and Yue Q.R. (2003) 'FRP in civil engineering in China: research and applications', *Proceedings of Fiber Reinforced Polymer: Reinforcement for Concrete Structures (FRPRCS-6)*, 8–10 July, Singapore, pp. 1401–1412.

Zhang, Z.C., Zhou, Z. and Wang C. (2006) 'Long-term monitoring FBG-based cable load sensor', *SPIE Proceedings Vol. 6167, Smart Structures and Materials 2006: Smart Sensor Monitoring Systems and Applications*, February 27 – March 1, 2006, San Diego, California, USA, CD-ROM, Paper No. 55, 7 pages.

Zheng, Y., Yu, G. and Pan, Y. (2012), 'Investigation of ultimate strengths of concrete bridge deck slabs reinforced with GFRP bars', *Construction and Building Materials*, Vol. **28**, pp. 482–492.

Zhou, A., Coleman, J.T., Temeles, A.B., Lesko, J.J. and Cousins, T.E. (2005) 'Laboratory and field performance of cellular fiber-reinforced polymer composite bridge deck systems', *Journal of Composites for Construction*, Vol. **9**, No. 5, pp. 458–467.

Zhou, Z., Graver, T.W. and Hsu, L. (2003), 'Techniques of advanced FBG sensors: fabrication, demodulations, encapsulation and their applications in the structural health monitoring of bridges', *Pacific Scientific Review*, Vol. **5**, pp. 116–121.

2

Prestressed fiber-reinforced polymer (FRP) composites for concrete structures in flexure: fundamentals to applications

Y. J. KIM, University of Colorado, Denver, USA, M. F. GREEN, Queen's University, Canada and R. GORDON WIGHT, Royal Military College of Canada, Canada

DOI: 10.1533/9780857097019.1.30

Abstract: This chapter presents a comprehensive overview of prestressed fiber-reinforced polymer (FRP) composites for concrete structures. Of interest are the prestressing methods and flexural responses of such structures. This state-of-the-art review provides a synthesis of all existing prestressing applications for reinforcing or strengthening concrete members, including bonded and unbonded FRP composites. The review examines published codes and design manuals with their limitations, prestressing operation, failure modes, design methods, bond characteristics, and deformability (or ductility) of various prestressed FRP applications. Current research needs are discussed to further improve an understanding of the behavior and application of prestressed FRP composites for concrete structures.

Key words: application methods, bonding, ductility, fiber-reinforced polymer (FRP), prestressed concrete, reviews.

2.1 Introduction

Prestressed concrete structures may suffer from damage of steel strands, including insufficient load-carrying capacity and degraded serviceability (e.g., large deflection and excessive cracks). Corrosion of steel strands may be the most significant contribution to the initiation and acceleration of damage in prestressed concrete members. To overcome this critical problem, various efforts have been made: for example, galvanized or epoxy-coated steel, and polymer concrete or low permeability concrete to reduce pores of the concrete (Toutanji and Saafi, 1999). None of these methods, however, effectively resolved corrosion problems.

Recent advances in materials technologies have resulted in an alternative reinforcing material, FRP composites. This chapter presents a synthesis of recent advances in prestressed FRP composites for concrete members subjected to flexural load, from fundamentals to applications. Focus is on the practical implementation of prestressed FRPs to better transfer the emerging technologies to practicing engineers and research professionals, including various prestressing operations and corresponding flexural responses of FRP-prestressed members. Current research needs to fill the gap between theory and practice are also discussed.

2.2 Types and characteristics of fiber-reinforced polymer (FRP) composites

Typical FRP composites consist of unidirectional fibers and a matrix resin. The fibers provide the load-carrying capacity of the composite. The matrix resin binds the fibers and transfers applied stresses between the fibers. Typical matrix resins include polyester, vinylester, and epoxy (Dolan, 1999). Two types of FRP reinforcement are currently available for concrete structures, namely FRP bars and laminates. A typical FRP tendon consists of 40% resin and 60% fiber (Dolan, 1999), and FRP laminates have 20–35% resin and 65–80% fiber (Meier, 1995; Wight *et al.*, 2001; El-Hacha *et al.*, 2003). Fiber content controls the mechanical properties of FRP composites. It should be noted that, for FRP-sheet applications, volume–fraction ratios may slightly change depending upon wet-lay-up (i.e., dry fibers and wet resins are incorporated on site and cured). In this case, the contribution of resin to the load-carrying capacity of FRP composites can be ignored (MBrace, 1998).

For infrastructure applications, three types of FRP composites are primarily used: aramid FRP (AFRP), glass FRP (GFRP), and carbon FRP (CFRP) (Bakis *et al.*, 2002; Teng *et al.*, 2003). The advantages of these FRP composites are non-corrosive characteristics, light weight, high tensile strength, low relaxation, good fatigue resistance, reduced maintenance costs, and ease of handling (Grace and Abdel-Sayed, 2000; Zhang *et al.*, 2000; Bakis *et al.*, 2002; Ghallab and Beeby, 2005; Youakim and Karbhari, 2007; Kim and Heffernan, 2008).

The benefits of prestressed FRP composites include the efficient use of FRP materials, reduced deflection, controlled cracks, active load-carrying mechanism, restored prestress losses in a prestressed concrete beam, improved cracking and yield loads of strengthened beams, and stress redistribution (Meier, 1995; Wight *et al.*, 2001; Kim *et al.*, 2007a, 2008). The most notable benefits of FRP composites for prestressed concrete applications are their non-corrosive characteristics and high tensile strength. Taking into account the high tensile strength of FRP composites (e.g., 4–10 times greater than yield strength of reinforcing steel and comparable to prestressing steel strands, as shown in Table 2.1), a prestress may be applied to efficiently use the material strength. The most remarkable difference between concrete

Table 2.1 Various FRP products for prestressing applications reported by the manufacturers

Typical property	Steel		CFRP					AFRP		
Type	Rebar	Tendon	Tendon		Sheet/laminate			Tendon		
Product	Generic	Generic	Leadline	CFCC	MBrace (CF-130)	Replark	Tyfo UC	Arapree	Parafil	Technora
Density (g/cm^3)	7.85	7.85	1.5	1.5	300 g/m^2	200 g/m^2	1.81	1.25	1.44	1.3
Longitudinal tensile strength (MPa)	400*	1860	2550	2100	3800	1480	2510	1500	1900	2100
Longitudinal tensile modulus (GPa)	200	195	150	137	227	144	139	64	130	54
Poisson's ratio	0.3	0.3	0.27	–	–	0.3	–	0.38	–	0.35
Ultimate strain (%)	0.2*	>3.5	1.5	1.57	1.67	1.4	1.8	2.4	1.7	3.8
Longitudinal coefficient of thermal expansion (CTE) ($\times 10^{-6}$/°C)	12	12	–0.9	0.5	–0.38	0.7×10^{-4}	–	–2	–	-3

*Yield property.

beams with prestressed FRP composites and steel-prestressed concrete beams is the absence of yield load, because of the linear behavior of FRP composites until failure occurs. The failure of the beams with prestressed FRP composites may, thus, be substantiated by either concrete crushing or FRP rupture.

AFRP materials are sensitive to the exposure of ultraviolet rays and thus they may be appropriate for internal prestressing application (McKay and Erki, 1993). AFRP composites show good fatigue resistance (Meier, 1995; Pisani, 1998); however, their time-dependent behavior is an important consideration for prestressing applications, taking into account their large relaxation characteristics and creep (McKay and Erki, 1993; Zhang *et al.*, 2000). For example, typical relaxation losses of AFRP tendons are 6–18% in a 100 year design life (Dolan, 1989) which are significantly greater than those of steel strands, having approximately 3% of relaxation losses in 100 years. GFRP composites are limitedly used in prestressing application because of their poor resistance to creep (ACI, 2004; ISIS, 2007; Youakim and Karbhari, 2007). GFRP tendons show a typical strength decrease of 30% in 100 years (Pisani, 1998). It should be noted that an additional strength reduction of GFRP tendons may occur if the tendons are embedded in concrete because of their inadequate performance in an alkaline environment (Dolan, 1999). CFRP composites exhibit high strength (over 2100 MPa), low density (typically 1.5 g/cm^3), and high temperature resistance, as shown in Table 2.1. CFRP materials also include excellent time-dependent behavior (e.g., no creep (Pisani, 1998)). However, material costs of CFRP are higher than those of GFRP and AFRP.

2.3 Using FRP composites in structures: design and applications

Despite the advantages of FRP composites and potential for cost-saving solutions, the application of prestressed FRP composites in practice is still in an early stage. This is attributed to the fact that practicing engineers and designers may not be familiar with such emerging prestressing technologies. Recently published code and design manuals reflect the significance of prestressed FRP composites for concrete structures.

fib Task Group 9.3 (*fib*, 2001) regards the application of prestressed FRP reinforcement as a special technique and addresses a number of advantages. The manual, however, does not specifically provide design guidelines for use of prestressed FRP composites. ACI 440.4R-04 (ACI, 2004) and ISIS Canada (ISIS, 2007) provide comprehensive design considerations for concrete structures using prestressed FRP composites, including flexural and shear design as well as serviceability. The Italian design manual CNR-DT200

(NRC, 2004) does not include specific provisions for prestressed FRP composites. The Canadian Highway Bridge Design Code (CHBDC, 2006) specifications include brief guidelines of prestressed FRP tendons for bridge superstructure (e.g., maximum permissible stresses in FRP tendons at jacking and transfer for concrete members).

Although some codes and design manuals have recently been published to aid use of prestressed FRP composites for concrete structures (ACI, 2004; CHBDC, 2006; ISIS, 2007), there still exists a dearth of understanding of the detailed structural behavior of FRP-prestressed concrete members. In addition, published design manuals have not included up-to-date innovative prestressed FRP applications such as post-tensioned FRP laminates and near-surface-mounted (NSM) FRP bars. Some review papers have reported on the application of prestressed FRPs (Soudki, 1998; Lees, 2001; El-Hacha *et al.*, 2001); however, they have focused more on the conceptual behavior of FRP-prestressed members and material developments.

Despite these problems, prestressed FRP composites are increasingly used for various structural applications, for instance:

- concrete beams prestressed with internally bonded and unbonded FRP tendons (Sen *et al.*, 1993; Abdelrahman and Rizkalla, 1995; Soudki *et al.*, 1997; Saafi and Toutanji, 1998; Grace and Singh, 2003)
- externally unbonded prestressed FRP tendons (Grace *et al.*, 2004; Ghallab and Beeby 2005)
- external strengthening of existing concrete members using prestressed FRP laminates or bars (Triantafillou *et al.*, 1992; Wight *et al.*, 2001; El-Hacha *et al.*, 2004; Kim *et al.*, 2005, 2006, 2007b; Nordin and Taljsten, 2006).

Commercially available FRP products are summarized in Table 2.1. Contrary to conventional seven-wire steel strands for prestressed concrete applications, various shapes of FRP tendons are available, such as strand (CFCC), indented (Leadline), deformed (Technora), round (Parafil), and rectangular (Arapree) FRPs.

Field applications of prestressed FRP composites have been increasingly accepted by the civil engineering community around the world. Prestressed FRP composites have been used to construct new bridges and to repair existing bridges (Basler and Clenin, 2004; Grace *et al.*, 2004; Kim *et al.*, 2006; ISIS, 2009). FRP composites provide a unique feature in that fiber optic sensors can be incorporated with FRPs to monitor the real time performance of bridges (Rizkalla and Tadros, 1994). Some pictures of field applications are shown in Fig. 2.1. Further examples are available elsewhere (ACI, 2007; ISIS, 2009). The following sections provide a comprehensive overview of various

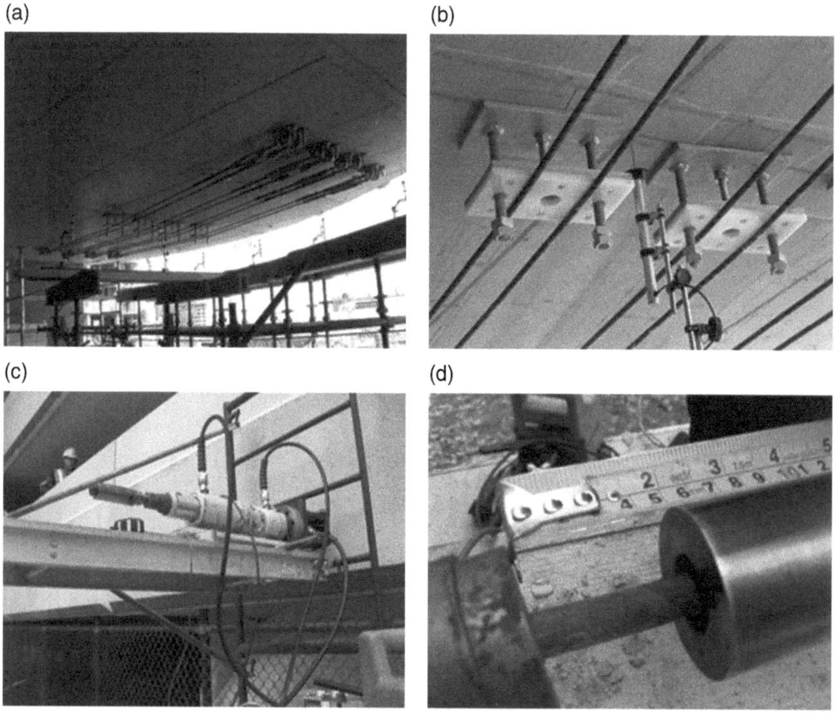

2.1 Site applications of CFRP tendons. (*Source*: Courtesy of Hughes Brothers, Inc. and Vector Construction.) (a) Post-tensioning system; (b) deviators; (c) jacking operation; (d) elongation.

applications of prestressed FRP composites from a practical perspective, such as prestressing operation and flexural responses of concrete members with prestressed FRP composites.

2.4 Internally bonded FRP tendons

The use of internally bonded FRP tendons is the most common method for FRP-prestressed concrete beams, as shown in Fig. 2.2a and Table 2.2. FRP-prestressed beams are designed in such a way that concrete sections are not cracked in service load (Grace and Singh, 2003), which is an analogous design approach to steel-prestressed beams. In some cases, unstressed FRP tendons are added to an FRP-prestressed beam to provide additional load-carrying capacity and ductility of the beam (Saafi and Toutanji, 1998; Grace and Singh, 2003). Conventional section analysis (i.e., strain compatibility) may be used for the application of internally bonded FRP tendons.

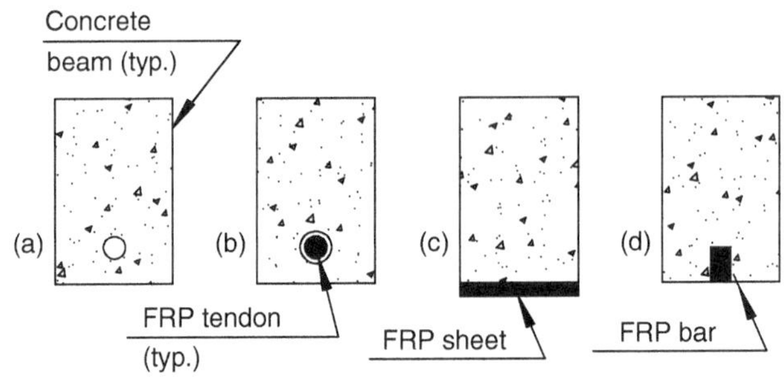

2.2 Typical applications of prestressed FRP composites for concrete beams: (a) internally bonded FRP tendon; (b) internally unbonded FRP tendon; (c) externally post-tensioned FRP sheet; (d) NSM post-tensioned FRP bar.

2.4.1 Prestressing operation

A prestressing operation may be conducted using hydraulic jacks reacting against anchorage. Special attention should be paid to gripping methods for FRP tendons during a jacking operation, to avoid premature failure of the tendons near the FRP-anchorage junction where stress concentrations occur. The reason is attributed to the relatively weak strength of FRP tendons in the radial direction (i.e., longitudinal strength of an FRP tendon is approximately ten times greater than radial strength). Various types of anchor systems have been developed to effectively prestress FRP tendons (Fig. 2.3), for example, clamp type (Malvar and Bish, 1995), plug-and-cone type (Burgoyne, 1988; Nanni *et al.*, 1996; Sayed-Ahmed and Shrive, 1998), and bond type (Zhang and Benmokrane, 2002) anchors. Further details of anchorage are available elsewhere (ACI, 2004). Typical prestressing levels are 50–65% of the ultimate strength for AFRP and CFRP tendons, respectively (ACI, 2004). The reason is owing to the stress-rupture of FRP tendons, which may result in abrupt failure of FRP-prestressed concrete members subjected to long-term load. A stress-rupture phenomenon is illustrated in Fig. 2.4a. The long-term behavior of FRP composites depends on the level of prestress, as shown in Fig. 2.4b. Synergies from various factors may accelerate stress-rupture, including alkaline environment and temperature (Budelmann and Rostasy, 1993). An increase in the initial prestress level of FRP tendons increases transfer length in the concrete (Nanni and Tanigaki, 1992). Numerical comparisons of various predictive models on the transfer length of FRP tendons are to be discussed in Section 2.9.1.. Allowable stresses in concrete and FRP tendons at transfer of the prestress and under service conditions are summarized in ACI-440.4R-04 (ACI, 2004).

Table 2.2 Flexural behavior of selected prestressed FRP composites for concrete structures

Reference	Test specimen		Prestressed FRP		Reinforcement		Material property[a]				Flexural behavior[b]				
	ID	Size (ft)*	Fiber	Type	ρ_F (%)†	Type[c]	f_c' (MPa)	f_{pu} (MPa)	ε_{pu} (%)	E_p (GPa)	Pressure level (%)‡	P_{cr} (kN)	P_u (kN)	P_{cr}/P_u (%)	Failure mode
Nanni and Tanigaki (1992)	A2	13.1	AFRP	Tendon	0.33	IB	37	1640	2.2	76	23	6§	12§	50.0	Slip
	B1	13.1	AFRP	Tendon	0.36	IB	38	1460	2.1	68	24	6§	13§	46.2	Crushing
	B2	13.1	AFRP	Tendon	0.71	IB	39	1460	2.1	68	24	9§	23§	39.1	Crushing
	C1	13.1	AFRP	Tendon	0.71	IB	34	1390	2.2	63	53	12§	27§	44.4	Slip
	D1	13.1	AFRP	Tendon	0.36	IB	39	1460	2.1	68	48	9§	15§	60.0	Crushing
McKay and Erki (1993)	B1	6.9	AFRP	Tendon	0.13	IB	41	2500	3.7	53	45	17§	35§	48.6	Rupture
	B2	6.9	AFRP	Tendon	0.13	IB	40	2500	3.7	53	46	17§	33§	51.5	Rupture
	B3	6.9	AFRP	Tendon	0.13	IB	40	2500	3.7	53	46	18§	40§	45.0	Rupture
Taerwe and Matthys (1995)	1	10.5	AFRP	Tendon	0.60	IB	69	3000	3.0	91	55	7	19	37.2	Crushing
	2	10.5	AFRP	Tendon	0.60	IB	69	3000	3.0	91	55	8	N/A	N/A	N/A
	3	10.5	AFRP	Tendon	0.60	IB	65	3000	3.0	91	55	9	18	50.0	Crushing
	4	10.5	AFRP	Tendon	0.60	IB	99	3000	3.0	91	55	9	21	42.6	Rupture
	7	10.5	AFRP	Tendon	0.60	IB	69	3000	3.0	91	55	8	23	34.6	Crushing
	8	10.5	AFRP	Tendon	0.60	IB	69	3000	3.0	91	55	8	28	28.6	Crushing
Gowripalan *et al.* (1996)	Arpr40	10.5	AFRP	Tendon	0.44	IB	35	1300	2.3	56	31	20§	50§	40.0	Crushing
	Arpr80	10.5	AFRP	Tendon	0.44	IB	85	1300	2.3	56	35	25§	50§	50.0	Rupture
Niitani *et al.* (1997)	BA-1M	39.4	AFRP	Tendon	1.3	IB	58	1250	2.0	64	45	55	131	42.3	Crushing
	BA-6Y	39.4	AFRP	Tendon	1.3	IB	58	1250	2.0	64	45	54	129	41.9	Crushing
Saafi and Toutanji (1998)	B1	8.2	AFRP	Tendon	0.32	IB	45	1330	2.1	50	50	36	51	71.1	Rupture
	B2	8.2	AFRP	Tendon	0.32	IUB	45	1330	2.1	50	50	31	44	70.7	Rupture
	B3	8.2	AFRP	Tendon	0.32	IB&IUB	45	1330	2.1	50	50	39	43	91.7	Rupture
	B4	8.2	AFRP	Tendon	0.48	IUB	45	1330	2.1	50	50	32	49	65.5	Rupture
Toutanji and Saafi (1999)	C1	16.4	AFRP	Tendon	0.48	IB	45	1330	2.1	50	50	85	195	43.7	Rupture
	C2	16.4	AFRP	Tendon	0.68	IUB	45	1330	2.1	50	50	85	185	46.1	Rupture
	C3	16.4	AFRP	Tendon	0.48	IB&IUB	45	1330	2.1	50	50	86	155	55.6	Rupture
	C4	16.4	AFRP	Tendon	0.48	IUB	45	1330	2.1	50	50	85	143	59.7	Rupture

(*Continued*)

Table 2.2 Continued

Reference	Test specimen		Prestressed FRP		Reinforcement		Material property[a]				Flexural behavior[b]				
	ID	Size (ft)*	Fiber	Type	ρ_F (%)†	Type[c]	f'_c (MPa)	f_{pu} (MPa)	ε_{pu} (%)	E_p (GPa)	Pressure level (%)‡	P_{cr} (kN)	P_u (kN)	P_{cr}/P_u (%)	Failure mode
Lees and Burgoyne (1999)	TB(1)	9.2	AFRP	Tendon	0.13	IB	59	1800	3.3	54	68	N/A	8	N/A	Rupture
	TB(2)	9.2	AFRP	Tendon	0.13	IB	59	1800	3.3	54	68	N/A	8	N/A	Rupture
	FB	9.2	AFRP	Tendon	0.17	IB	69	1430	2.1	69	59	N/A	8	N/A	Rupture
	TUB	9.2	AFRP	Tendon	0.17	IUB	60	1800	3.3	54	63	N/A	6	N/A	Crushing
	FUB	9.2	AFRP	Tendon	0.17	IUB	60	1430	2.1	69	63	N/A	6	N/A	Crushing
	TIB1	9.2	AFRP	Tendon	0.13	IPB	56	1800	3.3	54	65	N/A	7	N/A	Crushing
	TIB2	9.2	AFRP	Tendon	0.13	IPB	56	1800	3.3	54	65	N/A	8	N/A	Crushing +Rupture
	FIB1	9.2	AFRP	Tendon	0.17	IPB	56	1430	2.1	69	62	N/A	7	N/A	Crushing
	FIB2	9.2	AFRP	Tendon	0.17	IPB	56	1430	2.1	69	65	N/A	7	N/A	Rupture
	TAB	9.2	AFRP	Tendon	0.13	IB	58	1800	3.3	54	65	N/A	6	N/A	Crushing
	FAB	9.2	AFRP	Tendon	0.17	IB	58	1430	2.1	69	64	N/A	8	N/A	Crushing +Rupture
Stoll *et al.* (2000)	Beam2	40.0	CFRP	Tendon	0.24	IB	71	2600	1.6	147	77	182	569	32.1	Rupture
Dolan and Swanson (2002)	C1-H	32.8	CFRP	Tendon	0.12	IB	43	1820	1.2	146	60	N/A	166§	N/A	Rupture
	C2-H	32.8	CFRP	Tendon	0.12	IB	43	1820	1.2	146	60	N/A	158§	N/A	Rupture
Elrefai *et al.* (2007)	OL-DD	11.5	CFRP	Tendon	0.37	EUB	35	2160	1.5	144	40	30	110	27.2	Crushing
Nordin and Taljsten (2006)	BPS1	13.1	CFRP	Rod	0.17	NSM	68	2800	1.8	160	10	20	121	16.6	Rupture
	BPS2	13.1	CFRP	Rod	0.17	NSM	68	2800	1.8	160	10	21	121	17.3	Rupture
	BPS3	13.1	CFRP	Rod	0.17	NSM	66	2800	1.8	160	12	23	120	19.0	Rupture
	BPS4	13.1	CFRP	Rod	0.17	NSM	64	2800	1.8	160	12	23	123	18.5	Rupture
	BPS5	13.1	CFRP	Rod	0.17	NSM	67	2800	1.8	160	20	26	122	21.2	Rupture
	BPS6	13.1	CFRP	Rod	0.17	NSM	67	2800	1.8	160	20	N/A	148	N/A	Rupture
	BPM1	13.1	CFRP	Rod	0.17	NSM	63	2000	0.8	250	16	25	128	19.5	Rupture
	BPM2	13.1	CFRP	Rod	0.17	NSM	62	2000	0.8	250	16	25	132	18.9	Rupture

	BPM3	13.1	CFRP	Rod	0.17	NSM	65	2000	0.8	250	27	32	131	24.2	Rupture
	BPM4	13.1	CFRP	Rod	0.17	NSM	66	2000	0.8	250	19	23	123	18.5	Rupture
Wu *et al.* (2007)	3	6.9	CFRP	Tendon	0.29	NSM	46	2100	2.2	95	15	23	137	16.6	Debond
	4	6.9	CFRP	Tendon	0.29	NSM	46	2100	2.2	95	30	30	147	20.4	Debond
	5	6.9	CFRP	Tendon	0.29	NSM	46	2100	2.2	95	30	32	148	21.4	Crushing
Wight *et al.* (2001)	C	16.4	CFRP	Sheet	0.17	EB	35	1480	1.2	125	14	75	435	17.2	Rupture
	D	16.4	CFRP	Sheet	0.17	EB	35	1480	1.2	125	14	75	410	18.3	Rupture
El-Hacha *et al.* (2004)	RT1	14.8	CFRP	Sheet	0.06	EB	46	3400	1.4	230	50	48	130	37.1	Rupture
	RT2	14.8	CFRP	Sheet	0.06	EB	46	3400	1.4	230	50	49	151	32.5	Rupture
	RT3	14.8	CFRP	Sheet	0.06	EB	46	3400	1.4	230	50	32	102	31.6	Rupture
	LT1	14.8	CFRP	Sheet	0.06	EB	49	3400	1.4	230	50	46	123	37.3	Rupture
	LT2	14.8	CFRP	Sheet	0.06	EB	49	3400	1.4	230	50	47	162	29.2	Rupture
Kim *et al.* (2007)	B-3	11.8	CFRP	Sheet	0.06	EB	37	3800	1.7	227	48	54	99	54.5	Rupture
Kim *et al.* (2008)	J-1	6.0	CFRP	Sheet	0.50	EB	38	3800	1.7	227	16	60	170	35.4	Anchor
	J-2	6.0	CFRP	Sheet	0.50	EB	38	3800	1.7	227	16	N/A	112	N/A	Debond
	J-5	6.0	CFRP	Sheet	0.50	EB	38	3800	1.7	227	16	N/A	154	N/A	Anchor
	J-7	6.0	CFRP	Sheet	0.50	EB	38	3800	1.7	227	16	N/A	96	N/A	Slip
	J-9	6.0	CFRP	Sheet	0.50	EB	38	3800	1.7	227	16	N/A	84	N/A	Slip

* Length of test specimen.

† $\rho_F = A_{frp}/A_{concrete}$.

‡ Percentage of prestress level with respect to ultimate property.

§ Moment capacity (kNm).

[a] f'_c = concrete strength, f_{pu} = ultimate FRP strength, E_p = FRP tensile modulus, ε_{pu} = FRP rupture strain.

[b] *Pcr* = cracking loads, *Pu* = ultimate loads.

[c] IB = internally bonded, IUB = internally unbonded, EB = externally bonded, EUB = externally unbonded, NSM = near-surface mounted.

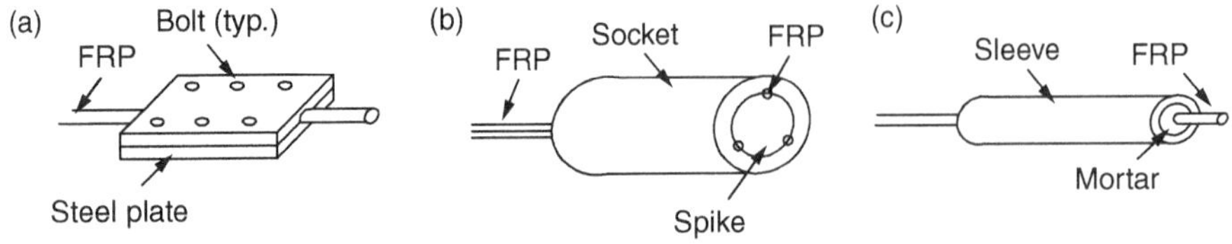

2.3 Typical anchorages for prestressing FRP tendons: (a) clamp type; (b) plug-and-cone type; (c) bond type.

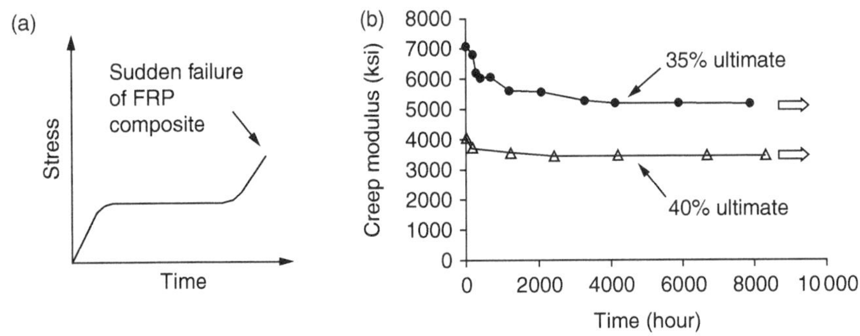

2.4 Typical time-dependent behavior of FRP composites: (a) schematic view of stress rupture; (b) reduced creep modulus of GFRP (experiment based on Abdel-Magid *et al.*, 2003).

2.4.2 Flexural behavior

The behavior of an internally bonded FRP-prestressed concrete beam depends on a reinforcement ratio that is categorized into three groups (Burke and Dolan, 2001): over-reinforced ($\rho > \rho_b$; where ρ is the reinforcement ratio and ρ_b is the balanced reinforcement ratio), under-reinforced ($\rho < \rho_b$), and significantly under-reinforced ($\rho < 0.5\rho_b$) sections. The balanced reinforcement ratio of ρ_b is defined as the reinforcement ratio at which concrete crushes and FRP ruptures simultaneously. As shown in Table 2.2, reinforcement ratios of prestressed FRP members are well below those of conventional reinforced concrete members having 1–1.5%.

FRP-prestressed beams show almost identical behavior to steel-prestressed beams prior to cracking (Sen *et al.*, 1993; Abdelrahman and Rizkalla, 1995) because the response of prestressed concrete beams until cracking is mainly governed by concrete sections, rather than reinforcement. To prevent sudden flexural failure immediately after the cracking of a beam, minimum reinforcement should be placed (ACI, 2004). Post-cracking deflections and crack widths of FRP-prestressed beams are greater than those of steel-prestressed beams, as shown in Fig. 2.5, because of the low modulus of FRP

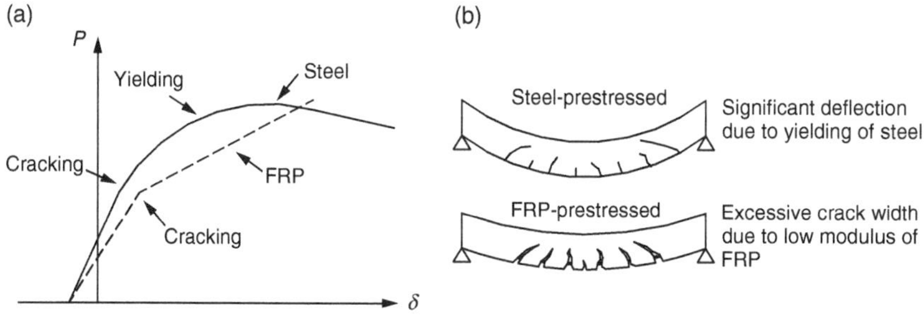

2.5 Typical flexural behavior of FRP-prestressed and steel-prestressed concrete beams: (a) load and deflection response; (b) warning of imminent failure. Solid line represents steel and dotted line represents FRP.

tendons (i.e., 62–150 GPa for FRPs, while 200 GPa for typical steel strands) (Sen *et al.*, 1993; Dolan *et al.*, 2000; Dolan and Swanson, 2002; ACI, 2004). Post-cracking deflection is also influenced by a reinforcement ratio (Grace and Singh, 2003). An increase of stresses in FRP tendons is observed at the locations of flexural cracks along the beam span (Nanni and Tanigaki, 1992) due to a lack of stress-sharing between the concrete and FRP tendon at cracked locations. An increase of stress levels in bonded FRP tendons is significant when compared to that of unbonded FRP tendons (Sections 2.5 and 2.6) (Pisani, 1998). This is related to the strain compatibility of the section. FRP-prestressed beams demonstrate higher stiffness than steel-prestressed beams after yielding of the steel, as shown in Fig. 2.5a.

For under-reinforced beams, rupture of FRP tendons controls the failure of the beams, rather than concrete crushing. A progressive rupture failure may occur in a concrete beam prestressed with vertically-distributed FRP tendons (Dolan and Swanson, 2002). This failure characteristic of FRP tendons is different from that of steel strands, which exhibit a yield plateau. The FRP tendon located furthest from the extreme compression fiber of a concrete section, thus, governs the strain capacity of the beam, so that a sudden decrease in load-carrying capacity may take place when the governing tendon ruptures. In lieu of showing a yield load (Fig. 2.5a), FRP-prestressed beams exhibit significant crack width (e.g., about 5 mm at ultimate) as an indicator of imminent failure, as shown in Fig. 2.5b. FRP-prestressed beams may absorb less energy than steel-prestressed beams due to the absence of yield characteristics (Zou, 2003). For over-reinforced beams, concrete crushing is the primary contribution to the failure of the beams (Nanni and Tanigaki, 1992). Such a compression-controlled failure mode may be recommended for the design of FRP-prestressed concrete beams, considering safety issues (Lees, 2001; Grace and Singh, 2003). A ratio between cracking and ultimate loads (i.e., P_{cr}/P_u) was influenced by the level of prestress, as shown in Table 2.2; in other words, the P_{cr}/P_u ratio increased when a prestress level increased.

Sustained load levels significantly influence the deflection of an FRP-prestressed beam. For example, a difference between long-term (288 days) and short-term deflections was up to 540% under a sustained load equivalent to the cracking load (Zou, 2003). The low modulus of FRP tendons causes lower long-term prestress losses than those of steel strands; however, such a characteristic may cause a significant increase in camber of the prestressed member (Youakim and Karbhari, 2007). The creep response of FRP-prestressed members was greater than that of steel-prestressed cases (Soudki *et al.*, 1997). The coefficients for predicting time-dependent behavior of steel-prestressed concrete beams may thus not be applicable to FRP-prestressed beams. Correspondingly, ACI 440.4R-04 (ACI, 2004) suggests long-term coefficients to estimate the deflection of concrete members with FRP tendons.

2.5 Internally unbonded FRP tendons

FRP tendons may be unbonded from concrete by using plastic ducts or sheaths, as shown in Fig. 2.2b. This type of prestressing provides global deformation compatibility, rather than section compatibility (Rao and Mathew, 1996; Naaman *et al.*, 2002), taking into account the slippage between concrete and FRP tendons. For designing such a prestressing application, a strain reduction factor may be used (Naaman and Alkhairi, 1991). Part of the concept of the strain reduction factor is that strains in a tendon are calculated using a conventional sectional-analysis (i.e., fully-bonded tendons), and then the calculated strains are multiplied by a reduction factor to account for the reduced strain effect of the unbonded FRP tendons. The reduction factor varies depending upon the level of cracking of a beam and load configurations (ACI, 2004). When a prestressed member is loaded in flexure, an increase of stresses in unbonded FRP tendons is not significant compared to that of fully-bonded FRP tendons. For design purposes, the ultimate stress in an unbonded FRP tendon is conveniently expressed as (ACI, 2004)

$$f_p = f_{pe} + \Delta f_p \qquad [2.1]$$

where f_p is the ultimate stress in the FRP tendon, f_{pe} is the effective prestress prior to applying live load, and Δf_p is the stress increase due to live load.

2.5.1 Prestressing operation

Post-tensioning is performed to prestress a concrete beam with internally unbonded FRP tendons. Sheaths for prestressing are placed inside the

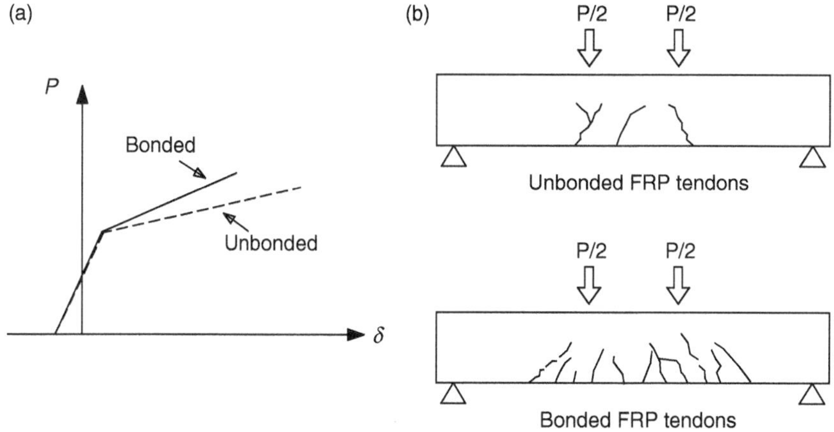

2.6 Comparison of flexural behavior between internally bonded and internally unbounded FRP tendons for prestressed concrete beams: (a) load–deflection responses; (b) cracking patterns at ultimate load.

beam and the concrete is cast. After adequate hardening of the concrete (i.e., usually 7–10 days after casting), FRP tendons located inside a sheath are post-tensioned using hydraulic jacks and permanently anchored at the ends of the beam. Contrary to steel strand applications, the sheaths may not be grouted because FRP tendons are non-corrosive. Allowable prestressing requirements are available in ACI440.4R-04 (ACI, 2004).

2.5.2 Flexural behavior

A concrete beam with internally unbonded FRP tendons exhibits greater deflections, fewer flexural cracks, and lower ultimate loads than a beam with fully-bonded FRP tendons, as shown in Fig. 2.6 (Saafi and Toutanji, 1998; Lees and Burgoyne, 1999; Toutanji and Saafi, 1999). Possible reasons for such a relatively low load-carrying capacity are attributed to the slippage of unbonded FRP tendons and strain localization at cracked locations. The localized cracks of a concrete beam with unbonded FRP tendons (Fig. 2.6b) are due to a lack of horizontal shear stress interaction between the concrete and unbonded FRP tendons along the beam span. The mode of failure was not affected by concrete strength, whereas it was substantially influenced by the method of prestressing (e.g., unbonded or bonded), as shown in Table 2.2.

2.6 Externally unbonded FRP tendons

Prestressed FRP tendons may be located outside a concrete beam to provide tensile resistance. Externally unbonded FRP tendons may particularly

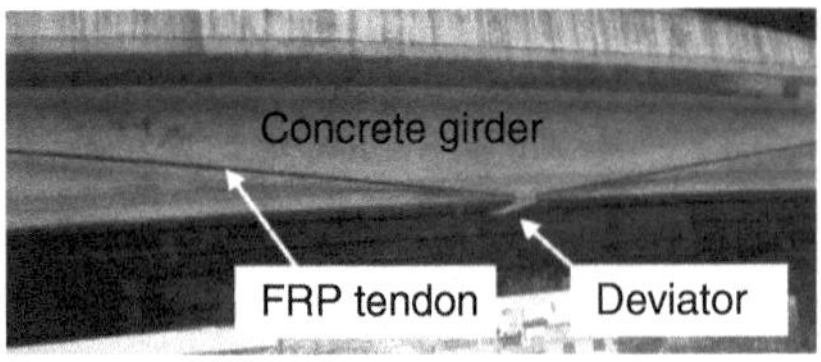

2.7 Externally unbonded FRP tendon for strengthening a bridge girder. (*Source*: Courtesy of Kim.)

be suitable for strengthening existing concrete beams to increase insufficient load-carrying capacity. Figure 2.7 shows a bridge girder strengthened with an unbonded FRP tendon. Benefits of such an externally unbonded FRP tendon application include an insignificant increase in dead load and rapid inspections, followed by an easy replacement of damaged (if any) FRP tendons (ACI, 2004). For analytical prediction of the behavior of a beam with externally unbonded FRP tendons, global displacement compatibility should be accounted for, rather than section compatibility. For practical design and site applications of externally unbonded FRP tendons, attention should be paid to a prestress level and the number of deviators (to be discussed).

2.6.1 Prestressing operation

An anchor system is required at the ends of a target beam to post-tension FRP tendons, as shown in Fig. 2.8. Post-tensioning is conducted after adequate aging of the concrete, thus a possibility of the failure of bearing concrete at the anchorage region may be lower. During post-tensioning, care must be exercised to provide safety to site workers because the stressed FRP tendons can rupture prior to reaching a target prestress level. It is thus suggested that safety brackets be placed along the beam span during prestressing FRP tendons. In the case of post-tensioning multiple FRP tendons, a gradual increase in prestressing force may be recommended to preserve the lateral stability of the concrete beam. Transfer of prestress to the concrete should be gradual. Unbonded FRP tendons experience constant stresses in the longitudinal direction except at the locations of deviators and anchorage where stress concentrations occur.

2.6.2 Flexural behavior

The behavior of a concrete beam prestressed with externally unbonded FRP tendons is different from that with internally bonded or unbonded FRP tendons, because of a change in the effective depth of FRP tendons induced by bending of the beam (Ghallab and Beeby, 2005) as schematically shown

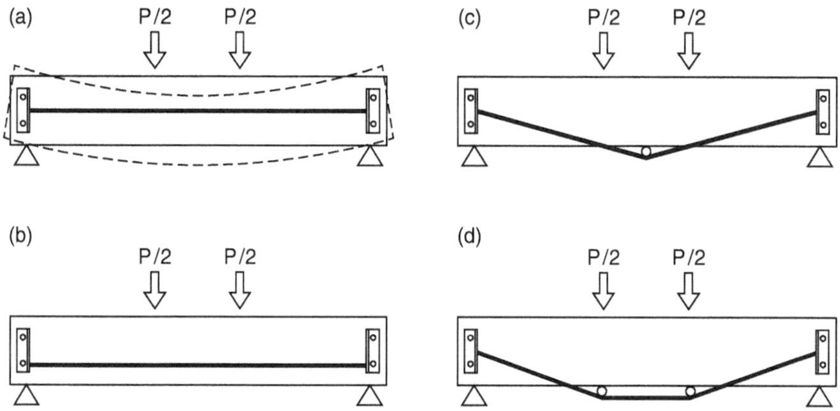

2.8 Various strengthening methods using externally unbounded FRP tendons: (a) without eccentricity; (b) with eccentricity; (c) with one deviator; (d) with two deviators.

in Fig. 2.8a. The change of beam-eccentricity may provide less structural efficiency because it reduces a moment-arm between compression and tension forces. The contribution of FRP tendons to the flexural strength of a beam decreases accordingly. A depth reduction factor in the strength prediction of such a beam is proposed to take into account variable tendon eccentricity (Aravinthan and Mutsuyoshi, 1997). Variation of the effective depth of a beam strengthened with externally unbonded FRP tendons may be controlled by deviators that provide a constant tendon eccentricity especially within a high moment zone (Fig. 2.8c and 2.8d). Stress concentrations of FRP tendons at deviators should be taken into consideration for such strengthening applications. A cushioning material may be placed at the interface between FRP tendons and deviators (Grace, 1999). The slippage of FRP tendons at deviation points (Rao and Mathew, 1996) may induce the abrasion of FRP tendons, which can accelerate stress concentrations. An increase in the number of deviators (Fig. 2.8c and 2.8d) increases the load-carrying capacity of a concrete beam prestressed with externally unbonded FRP tendons (Rao and Mathew, 1996; Ghallab and Beeby, 2005) as well as stresses in the deviated tendons (Ghallab and Beeby, 2005). Typical cracking patterns of a beam with externally unbonded FRP tendons are analogous to those of a beam with internally unbonded FRP tendons (i.e., a few localized cracks, Fig. 2.6b). When the beam is loaded in flexure an increase in stress levels of externally unbonded FRP tendons is insignificant in comparison to that of bonded FRP tendons (Pisani, 1998). This is related to a strain compatibility effect in the section. Equation [2.1] may, thus, be used for externally unbonded FRP applications. Prestressed beams using externally unbonded FRP tendons, in general, fail due to concrete crushing, rather than FRP rupture under monotonic flexural load (Grace and Abdel-Sayed, 2000; Ghallab and Beeby, 2005). Although the long-term performance of a concrete beam with externally unbonded FRP tendons is an important issue because of the stress concentrations and damage propagation of the tendons, limited research has been reported. Experimental investigations are, therefore, recommended to examine local damage propagation under various load configurations, including fatigue load.

2.7 Externally bonded post-tensioned FRP laminate

In order to strengthen or repair an existing concrete member in flexure, externally bonded post-tensioned FRP laminates may be used (Meier, 1995; Wight *et al.*, 2001; El-Hacha *et al.*, 2004; Kim *et al.*, 2008a,b,c,d). CFRP laminates are recommended for this application because of their favorable time-dependent behavior and high strength and stiffness compared to AFRP or GFRP composites (Meier, 1995). For such an innovative strengthening method, FRP laminates are tensioned (prior to bonding on the surface of the concrete) and then permanently bonded to the tensile side of the target structure using an epoxy adhesive, as shown in Fig. 2.2c. Notable benefits of the concrete members strengthened with prestressed FRP laminates are an active load-carrying mechanism (i.e., prestressed FRP laminates can sustain not only live load but also dead load), restored prestress losses in a prestressed concrete member, improved serviceability, and stress redistribution of existing concrete structures (Triantafillou *et al.*, 1992; Wight *et al.*, 2001; Kim *et al.*, 2007a). The most critical concerns of prestressed FRP-sheet application are prestressing methods and premature delamination failure. An adequate anchor system may prevent the premature delamination failure of the prestressed FRP laminates (Wight *et al.*, 2001; Kim *et al.*, 2005, 2008b).

2.7.1 Prestressing operation

A typical prestress level of 15%–50% of the ultimate tensile capacity has been applied to FRP laminates (Wight *et al.*, 2001; El-Hacha *et al.*, 2004; Kim *et al.*, 2007a). Recent research recommends that a level of prestress in CFRP laminates be 20% of the ultimate design strength, taking into account an adequate strength increase and ductility of strengthened members (Kim *et al.*, 2010). Three different methods are available for prestressing FRP laminates (Wight *et al.*, 2001). A schematic summary is given in Fig. 2.9. The first method (Fig. 2.9a) includes the artificial camber of a concrete beam (Saadatmanesh and Ehsani, 1991). The target beam is first loaded upward using a hydraulic jack and then FRP laminates are bonded to the bottom

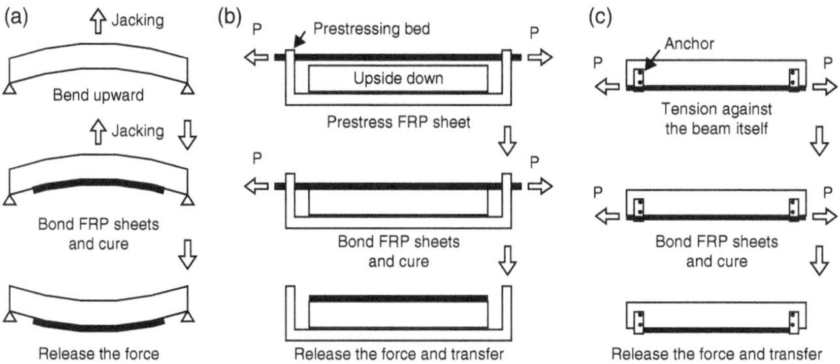

2.9 Typical methods to prestress FRP sheets: (a) artificial camber; (b) external apparatus; (c) direct tensioning.

of the beam. After curing of the FRP laminates, the jack is removed so that a prestress is naturally applied to the sheets by gravity. This method, however, provides a very low level of prestress to the FRP laminates (e.g., less than 3% of the ultimate strength of FRP (Saadatmanesh and Ehsani, 1991)). The second method (Fig. 2.9b) uses an external prestressing apparatus (Triantafillou *et al.*, 1992; Meier, 1995). FRP laminates are prestressed in an independent prestressing frame and bonded to the soffit of a target beam. The external frame is then removed after curing of the strengthening system. This method is suitable for laboratory investigations; however, it may not be practical for site application because of the need for a prestressing frame. The third method (Fig. 2.9c) is called the direct-tensioning method (Wight *et al.*, 2001; El-Hacha *et al.*, 2004; Kim *et al.*, 2005, 2007, 2008a,d). An anchor system is mounted in a beam and FRP laminates are directly tensioned against the beam itself (Kim *et al.*, 2009). Steel anchors are commonly used to sustain a high level of prestressing force. The most notable advantage of this prestressing method is its applicability on site (i.e., no prestressing frames are necessary for a prestressing operation, Fig. 2.9b and 2.9c). A field application has been recently conducted using the direct-tensioning method to repair an impact-damaged prestressed concrete girder bridge (Kim *et al.*, 2006). Steel anchorage used for prestressing FRP laminates may be left permanently (Wight *et al.*, 2001; El-Hacha *et al.*, 2004; Kim *et al.*, 2007, 2008a) or may be removed after curing of the post-tensioned FRP laminates (Kim *et al.*, 2008b).

2.7.2 Flexural behavior

Externally bonded prestressed FRP laminates may significantly enhance the serviceability of existing concrete beams. A strengthened beam with

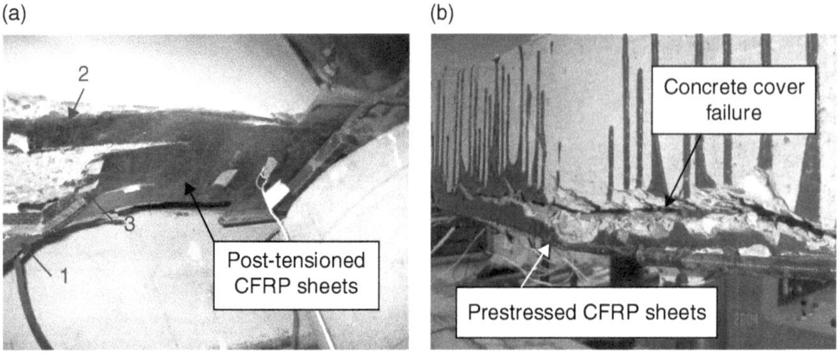

2.10 Typical failure of beams with externally bonded prestressed FRP sheets: (a) progressive rupture of the sheets (numbers indicate the failure sequence; Courtesy of Shi); (b) shear-off failure. (*Source*: Courtesy of Kim.)

prestressed FRP laminates exhibits a remarkably higher cracking load with improved stiffness when compared to a beam strengthened with unprestressed FRP laminates (Triantafillou *et al.*, 1992; Wight *et al.*, 2001; El-Hacha *et al.*, 2004; Kim *et al.*, 2008a). The beams strengthened with prestressed FRP laminates show many cracks of small width, whereas unstrengthened control beams show a few large cracks (Wight *et al.*, 2001; El-Hacha *et al.*, 2004). This observation implies that prestressed FRP laminates can effectively distribute applied stresses along the strengthened beam (i.e., reduced stress localization). Prestressed FRP laminates may close existing cracks (Kim *et al.*, 2008a), which is a benefit in terms of durability (i.e., reduced possibility of moisture ingress into the concrete beam). The level of stress in existing internal reinforcement subjected to external load is reduced after strengthening because of a stress redistribution mechanism that may result in an increased yield load capacity of a strengthened beam (Wight *et al.*, 2001; Kim *et al.*, 2005). Failure of a strengthening system is attributed to either rupture of post-tensioned FRP laminates (Wight *et al.*, 2001; El-Hacha *et al.*, 2004; Kim *et al.*, 2007a) or shear-off of FRP laminates (Triantafillou *et al.*, 1992; Kim *et al.*, 2008b), as shown in Fig. 2.10. Progressive rupture of prestressed FRP laminates may be observed (Kim *et al.*, 2007a) because of uneven prestressing forces across the FRP laminates (Fig. 2.10a). These uneven prestressing forces are mainly attributed to three reasons, namely, misaligned FRP laminates at anchorage when the laminates are post-tensioned, deformation of the anchorage under a high level of prestress, and uneven jacking forces during a prestressing operation. Care should thus be exercised, when a prestress is applied to FRP laminates, to avoid uneven stress distribution across the laminates, which may result in localized failure of the laminates (Fig. 2.10a). The shear-off failure of FRP laminates may be associated with a number of shear-off cracks at the level of concrete cover (Fig. 2.10b) because prestressed FRP laminates tend to return to their unstressed state when the strengthened beam is loaded in flexure.

2.8 Near-surface-mounted post-tensioned FRP bars

The NSM strengthening method is a relatively new technology for strengthening concrete beams. This method includes cutting narrow grooves along the tension side of a target beam, and inserting FRP bars or strips into the slits, then bonding them using an adhesive (*fib,* 2001), as shown in Fig. 2.2d. To improve strengthening effects, NSM FRP composites may be prestressed. Advantages of such prestressed NSM FRP application include enhanced force transfer from FRP to concrete, improved durability, and increased impact resistance (Nordin and Taljsten, 2006). Limited investigations of prestressed NSM FRPs have been reported previously (Nordin and Taljsten, 2006; Taljsten and Nordin, 2007; Wu *et al.*, 2007).

2.8.1 Prestressing operation

A prestressing operation for NSM FRP bars is typically conducted using an external apparatus (Nordin and Taljsten, 2006; Wu *et al.*, 2007). Contrary to the application of prestressed FRP laminates mentioned in a previous section, no additional anchorage may be required to prevent premature bond failure of the FRP composites. The reason is that prestressed NSM reinforcement is located inside a strengthened beam (Fig. 2.2d) and thus better bond is achieved as compared to externally bonded FRP laminates (Fig. 2.2c). Practical prestressing methods for NSM FRP bars should be developed for site application.

2.8.2 Flexural behavior

The flexural behavior of a concrete beam strengthened with prestressed NSM FRP bars is analogous to that of a beam with prestressed FRP laminates. Although post-cracking stiffness of beams with prestressed and non-prestressed NSM bars is similar, the cracking load of the beam with prestressed NSM bars is higher than that of the non-prestressed case due to a prestressing effect (Nordin and Taljsten, 2006). Yield capacity of beams strengthened with prestressed NSM FRP bars is significantly improved when compared to that of an unstrengthened control beam. Ductility of the strengthened beams, however, decreases when a prestressing level increases (Nordin and Taljsten, 2006). A similar observation was made in the case of beams strengthened with prestressed CFRP laminates (Kim *et al.*, 2010). Yield load of a strengthened beam is influenced by the level of prestress in NSM bars and corresponding tensile modulus. The failure of beams strengthened with prestressed NSM FRP bars is, in general, governed by fiber failure in NSM bars, rather than premature debonding or slip failure of the bars (Nordin and Taljsten, 2006). More testing is necessary for this type of strengthening technique.

2.9 Bond characteristics and deformability

Bond and deformability are crucial parameters to consider for FRP application. Inadequate bond significantly influences the performance of an FRP system. Sufficient deformability is necessary to ensure the safety of FRP structures. This section discusses the significance of these two subjects, including their practical considerations.

2.9.1 Bond characteristics

The surface of FRP tendons varies depending upon manufacturers. Surface conditions may affect the bond quality of FRP tendons embedded in concrete. FRP tendons (Table 2.1) are usually treated to have a rough surface – for instance, sand-coated (Arapree), fiber-wrapped (CFCC), and indented (Leadline) tendons. Bond resistance of FRP tendons is substantiated by friction and mechanical interlock (Zhang and Benmokrane, 2002). Bond characteristics of FRP tendons for concrete members may thus be different from those of steel strands. ACI 440.4R-04 (ACI, 2004) reports empirical equations to estimate adequate development and transfer lengths for prestressed FRP tendons. Figure 2.11 compares various existing models to predict the bond characteristics of FRP tendons. It should be noted that the models with solids dots in Fig. 2.11 were originally developed for seven-wire steel tendons. The material properties assumed for the numerical comparison were an ultimate strength of CFRP (f_{pu}) of 2450 MPa, effective and initial prestressing levels of 0.6 f_{pu} and 0.8 f_{pu}, respectively, and a concrete strength of 40 MPa. The predictive models based on steel strands were relatively conservative when compared to those based on FRP tendons. This is possibly due to the rough surface of FRP tendons; in other words, seven-wire steel strands do not have rough surface. Bond failure of FRP tendons causes a significant loss in load-carrying capacity of the beam prestressed with internally bonded FRP tendons. The evidence of such a beam failure mode is observed by excessive longitudinal cracks at the level of prestressed FRP tendons (Zou, 2003). Although such a bond failure mode is not an issue for unbonded FRP tendon application, bond of FRP tendons at anchorage should still be accounted for to avoid local failure.

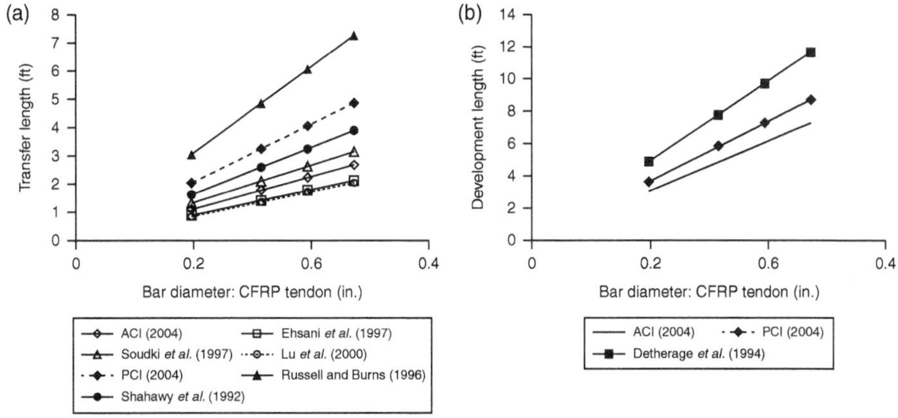

2.11 Comparison of predictive models for bond of FRP tendons: (a) transfer length; (b) development length.

For strengthening concrete members, bond is a critical consideration for the application of post-tensioned FRP laminates, taking into account their premature delamination failure. An adequate anchor system precludes the premature bond failure of prestressed FRP laminates bonded to the soffit of a concrete member (Wight *et al.*, 2001; El-Hacha *et al.*, 2004; Kim *et al.*, 2008a,d). Typical bond failure of prestressed FRP laminates due to a lack of adequate anchorage is available in Triantafillou *et al.* (1992). A bond critical zone near the cut-off points of prestressed FRP laminates may reasonably be assumed to be the shear stress concentration region where excessive horizontal shear deformation of the matrix resin (used to bond FRP laminates to the concrete) occurs (Kim *et al.*, 2008c). Prediction of a shear concentration region is available in literature (Taljsten 1997; Smith and Teng, 2001). Prestressed NSM FRP bars provide better bond resistance than externally bonded prestressed FRP laminates. This is due to the fact that more areas of FRP composites are interacted with concrete (i.e., prestressed NSM bars are located inside the concrete beam). Flexural testing indicates that no bond failure of prestressed NSM FRP bars has been observed until failure of test beams takes place (Nordin and Taljsten, 2006).

2.9.2 Deformability

Conventional ductility concepts (i.e., a ratio between the yield and ultimate response of a concrete beam under flexural load) are not applicable to a beam with FRPs, because FRP composites do not exhibit yield characteristics. Deformability is an alternative to quantify the flexural behavior of a concrete beam with FRP composites. For a beam prestressed with FRP composites, deformability may be defined as the ability to resist applied load after cracking of a beam section without a significant loss of load-carrying capacity. For a design purpose, deformability indices are proposed by CHBDC (2006) and ACI 440.4R-04 (ACI, 2004) as shown in Equations [2.2] and [2.3], respectively.

$$\mathrm{DI} = \frac{M_{\mathrm{ult}}\varphi_{\mathrm{ult}}}{M_{0.001}^{c}\varphi_{0.001}^{c}} \quad [2.2]$$

$$\mathrm{DI} = \omega\frac{\varepsilon_{pu}}{\varepsilon_{ps}} \quad \text{with} \quad \omega = \frac{(1-k)}{\left(1-\dfrac{\rho f_{\mathrm{pu}}}{0.85\beta_1 f_c'}\right)} \quad [2.3]$$

where: DI is the deformability index; M_{ult} is the ultimate moment capacity of the section; φ_{ult} is the curvature at M_{ult}; $M_{0.001}^{c}$ is the moment when the maximum concrete compressive strain reaches 0.001; $\varphi_{0.001}^{c}$ is the curvature at $M_{0.001}^{c}$; k is the ratio of neutral axis depth to FRP tendon depth; β_1 is the stress block factor; $\varepsilon_{\mathrm{ps}}$ and $\varepsilon_{\mathrm{pu}}$ are the strains at service and ultimate in the prestressed FRP composites, respectively; f_{pu} is the ultimate strength of the FRP tendon; f_c' is the specified compressive strength of the concrete; and ρ is the reinforcement ratio. ISIS Canada (ISIS, 2007) accepts the ACI approach shown in Equation [2.3].

CHBDC (2006) recommends that a deformability index be greater than 4.0 for rectangular sections and 6.0 for T-sections. ACI 440.4R-04 (ACI, 2004) does not provide specific limits of deformability index; however, it states that an acceptable deformability index shall be obtained unless the stress levels of FRP tendons exceed the allowable stress limits at jacking. A numerical comparison of deformability indices among various prestressed FRP applications is made using Equation [2.3], as shown in Table 2.3. It should be noted that the CHBDC approach (Equation [2.2]) was not included in Table 2.3 because most test data did not report detailed flexural responses that were necessary to calculate deformability indices using Equation [2.2]. The service strain ($\varepsilon_{\mathrm{ps}}$) in Equation [2.3] was assumed to be the prestressing strain of FRP composites, provided that the difference between service and prestressing strains might be insignificant (ACI, 2004). As shown in Table 2.3, deformability indices significantly vary depending upon prestress levels and particularly high deformability indices were observed for NSM FRP applications because of the contribution of the internal steel reinforcement in those beams. More research is necessary to suggest acceptable deformability indices for concrete members with various FRP systems.

Table 2.3 Variation of deformability index based on the ACI 440.4R-04 approach

Reference	Specimen	FRP	ε_{ps} (10^{-6})	ε_{pu} (10^{-6})	B (mm)	d (mm)	A_{frp} (mm^2)	f'_c (MPa)	ρ (10^{-3})	β_1	k	ω	DI
McKay and Erki (1993)*	B1	A	21 435	41 549	150	260	57.6	40	1.48	0.76	0.13	0.99	1.93
	B2	A	21 794	39 660	150	260	57.6	39	1.48	0.76	0.14	0.97	1.77
	B3	A	21 586	49 103	150	260	57.6	40	1.48	0.76	0.10	1.06	2.41
Saafi and Toutanji (1998)*	B1	A	13 340	21 000	150	260	120	45	3.08	0.72	0.28	0.84	1.33
Nordin and Taljsten (2006)†	BPS1	C	1708	17 500	200	295	100	68	1.70	0.65	0.16	0.96	9.86
	BPS2	C	1816	17 500	200	295	100	68	1.70	0.65	0.16	0.96	9.26
	BPS3	C	2003	17 500	200	295	100	66	1.70	0.65	0.16	0.96	8.42
	BPS4	C	2101	17 500	200	295	100	64	1.70	0.65	0.16	0.97	8.05
	BPS5	C	3273	17 500	200	295	100	66	1.70	0.65	0.17	0.95	5.08
	BPS6	C	3166	17 500	200	295	100	66	1.70	0.65	0.17	0.95	5.25
	BPM1	C	1180	8000	200	295	100	65	1.70	0.65	0.31	0.77	5.22
	BPM2	C	1184	8000	200	295	100	62	1.70	0.65	0.31	0.77	5.21
	BPM3	C	1761	8000	200	295	100	65	1.70	0.65	0.32	0.75	3.39
	BPM4	C	1368	8000	200	295	100	66	1.70	0.65	0.31	0.76	4.44

Notes: A = AFRP; C = CFRP; ε_{ps} = FRP strain in service; ε_{pu} = ultimate FRP strain; B = width of beam; d = effective depth of beam; A_{frp} = cross-sectional area of FRP; f'_c = concrete strength; ρ = reinforcement ratio; β_1 = stress block factor for concrete; k = ratio of neutral axis depth to FRP depth; ω = constant shown in Equation [2.3]; DI = deformability index.

* Internally bonded.

† NSM bars.

2.10 Conclusions and future trends

This chapter has presented a comprehensive overview of the application of prestressed FRP composites for concrete structures in flexure with emphasis on practical perspectives. Commercially available FRP products were reviewed with their engineering properties. Existing provisions of codes and design manuals were evaluated, including their applications and limitations. Detailed flexural responses of concrete members with prestressed FRP composites were examined. The following conclusions can be drawn:

- FRP materials for prestressed concrete application (either bonded or unbonded) should carefully be selected for their usage. AFRP tendons were susceptible to ultraviolet rays; however, they had good resistance to fatigue. GFRP demonstrated significant creep deformation. CFRP exhibited favorable temperature resistance and excellent long-term behavior.
- Various techniques were used to prestress FRP composites, namely, artificial camber, external apparatus, and direct-tensioning methods. A high level of prestressing force (e.g., 50–60% ultimate) was applied for internally bonded cases, whereas a relatively low level (e.g., 20–40% ultimate) was used for externally bonded and unbonded applications. Prestressing levels significantly influenced the flexure of concrete members with prestressed FRP composites, such as deformability, cracking, deflection, and failure modes.
- Excessive crack widths warned of the imminent failure of FRP-prestressed concrete members, rather than significant deflections that were typically observed in steel-prestressed concrete members. Long-term prestress losses of FRP-prestressed members were lower than those of steel-prestressed members because of the low modulus of FRP composites.
- Concrete members with unbonded FRP composites demonstrated larger deflections and fewer, but wide, crack widths when compared to the members with bonded FRP composites. These observations were attributed to the stress-sharing mechanism between the FRP and the concrete.
- Bond characteristics of prestressed FRP composites affected the flexure of concrete members. FRP tendons required shorter transfer and development lengths than seven-wire steel strands. This was due to the specially-treated surface of FRP tendons, such as sand-coated, fiber-wrapped, or indented. Insufficient bond triggered premature failure of prestressed FRP composites from strengthened concrete members. Adequate anchor systems precluded such an undesirable failure mode.

- Attention should be paid to the deformability of FRP-prestressed members because FRP composites do not exhibit yield characteristics. The level of initial prestress was the most contributing factor to deformability. Additional internal reinforcement substantially improved the deformability of concrete members with prestressed FRP composites.

Although the application of prestressed FRP composites for concrete structures has been well established over a decade, there still exists a paucity of understanding of the behavior of such structures. The following recommendations are made to further advance the state-of-the-art of prestressed concrete structures:

- Detailed stress transfer mechanisms from prestressed FRP tendons to concrete should be examined. Progressive damage propagation induced by stress concentrations along the FRP-concrete interface needs to be studied. Meso-scale finite element modeling may be a recommended approach.
- More experimental investigations are necessary to promote the cutting-edge technologies (e.g., NSM FRP and prestressed FRP laminates) to the industry sector. A practical anchor system should be developed for prestressed NSM FRP methods for site application. Predictive models for such innovative methods need to be developed for design professionals.
- Durability of externally unbonded and bonded FRP composites should be examined, including a combination of environmental effects and stress concentrations at deviators and anchorage. Long-term characteristics of FRP, including bond degradation, also need a due consideration to warrant the field application of prestressed FRP composites.
- Code and design manuals should include the state-of-the-art application of prestressed FRP composites. An upgrade of existing provisions may be necessary; for example, the ACI-440.4R-04 document should provide specific limits for the deformability index and should include provisions for long-term performance.
- More field application incorporated with advanced sensing technologies is recommended to monitor the actual performance of prestressed FRP composites.

2.11 Acknowledgment

The authors gratefully acknowledge financial support provided by the Natural Sciences and Engineering Research Council of Canada (NSERC) and Intelligent Sensing for Innovative Structures (ISIS Canada).

2.12 References

Abdel-Magid, B., Lopez-Anido, R., Smith, G. and Trofka, S. (2003). Flexure creep properties of E-glass reinforced polymers, *Compos. Struct.*, **62**, 247–253.

Abdelrahman, A.A. and Rizkalla, S.H. (1995). Serviceability of concrete beams prestressed by carbon fibre plastic rods, *Proc. 2nd Int. RILEM symp. (FRPRCS-2), Non-metallic (FRP) Reinforcement for Concrete Structures*, Ghent, Belgium, 403–412.

ACI. (2004). Prestressing concrete structures with FRP tendons (ACI-440.4R-04), American Concrete Institute Committee 440, Farmington hills, MI.

ACI. (2007). Report on fiber reinforced polymer (FRP) reinforcement for concrete structures (ACI 440R-07), American Concrete Institute Committee 440, Farmington hills, MI.

Aravinthan, T. and Mutsuyoshi, H. (1997). Prediction of the ultimate flexural strength of externally prestressed PC beams, *Trans. Jpn. Concrete Inst.*, **19**, 225–230.

Bakis, C.E., Bank, L.C., Brown, V.L., Cosenza, E., Davalos, J.F., Lesko, J.J., Machida, A., Rizkalla, S.H. and Triantafillou, T.C. (2002). Fiber-reinforced polymer composites for construction state-of-the-art review, *J. Compos. Constr.*, **6**(2), 73–87.

Basler, M. and Clenin, R. (2004). Desrochers M. Bridge strengthening with prestressed CFRP plate systems, *4th Int. Conf. Advanced Composite Materials in Bridges and Structures*, Calgary, AB, Canada (CD-Rom).

Budelmann, H. and Rostasy, F.S. (1993). Creep rupture of FRP elements for prestressed concrete-phenomenon, results, and forecast models, *Fiber Reinforced Plastic Reinforcement for Concrete Structures (ACI-SP 138)*, ACI, 87–100.

Burgoyne, C.J. (1988). Engineering applications of Parafil rope. *Symposium on Engineering Applications of Parafil Rope*, Imperial College, London, UK, 39–47.

Burke, C.R. and Dolan, C.W. (2001). Flexural design of prestressed concrete beams using FRP tendons, *PCI J.*, **46**(2), 76–87.

CHBDC. (2006). CSA S6–06: Canadian Highway Bridge Design Code. CSA International, Toronto, Ontario, Canada.

Deatherage, J.H., Burdette, E.G. and Chew, C.K. (1994). Development length and lateral spacing requirements of prestressing strand for prestressed concrete bridge girders, *PCI J.*, **39**(1), 70–83.

Dolan, C.W. (1989). Prestressed Concrete using KEVLAR reinforced tendons. PhD dissertation, Cornell University, Ithaca, NY.

Dolan, C.W. (1999). FRP prestressing in the USA, *Conc. Int.*, **21**(10), 21–24.

Dolan, C.W., Hamilton, H.R., Bakis, C.E. and Nanni, A. (2000) Design recommendations for concrete structures prestressed with FRP tendons, final report. Department of Civil and Architectural Engineering Report DTFH61–96-C-0019, University of Wyoming, Laramine, Wyo.

Dolan, C.W. and Swanson, D. (2002). Development of flexural capacity of a FRP prestressed beam with vertically distributed tendons, *Compos. Part B*, **33**, 1–6.

Ehsani, M.R., Saadatmanesh, H. and Nelson, C.T. (1997). Transfer and flexural bond performance of aramid and carbon FRP tendons, *PCI J.*, **42**(1), 76–86.

El-Hacha, R., Wight, R.G. and Green, M.F. (2001). Prestressed fibre-reinforced polymer laminates for strengthening structures, *Prog. Struct. Eng. Mat.*, **3**, 111–121.

El-Hacha, R., Wight, R.G. and Green, M.F. (2003). Innovative system for prestressing fiber-reinforced polymer sheets, *Struct. J., ACI*, **100**(3), 305–313.

El-Hacha, R., Wight, R.G. and Green, M.F. (2004). Prestressed carbon fiber reinforced polymer sheets for strengthening concrete beams at room and low temperatures, *J. Compos. Constr.*, **8**(1), 3–13.

Elrefai, A., West, J. and Soudki, K. (2007). Strengthening of RC beams with external post-tensioned CFRP tendons, *Case Histories and Use of FRP for Prestressing Applications (SP-245)*, 123–142.

fib. (2001). Externally Bonded FRP Reinforcement for RC Structures (*fib* Task Group 9.3), *International Federation for Structural Concrete*, Lausanna, Switzerland.

Ghallab, A. and Beeby, A.W. (2005). Factors affecting the external prestressing stress in externally strengthened prestressed concrete beams, *Cem. Conc. Compos.*, **27**, 945–957.

Gowripalan, N., Zou, X.W. and Gilbert, R.I. (1996). Flexural behaviour of prestressed beams using AFRP tendons and high strength concrete, *2nd Int. Conf. Advanced Composite Materials in Bridges and Structures (ACMBS-II)*, 325–333.

Grace, N.F. (1999). Continuous CFRP prestressed concrete bridges, *Conc. Int.*, **21**(10), 42–47.

Grace, N.F. and Abdel-Sayed, G. (2000). Behavior of carbon fiber-reinforced prestressed concrete skew bridges, *Struct. J., ACI*, **97**(1), 26–35.

Grace, N.F., Roller, J.J., Navarre, F.C., Nacey, R.B. and Bonus, W. (2004). Load testing a CFRP-reinforced bridge, *Conc. Int.*, **26**(7), 51–57.

Grace, N.F. and Singh, S.B. (2003). Design approach for carbon fiber-reinforced polymer prestressed concrete bridge beams, *Struct. J., ACI*, **100**(3), 365–376.

ISIS Canada. (2007). ISIS Canada Design Manual No. 5-prestressing concrete structures with FRPs, Winnipeg, MB, Canada.

ISIS Canada. (2009). http://www.isiscanada.com (accessed 9 February 2009).

Kim, Y.J., Bizindavyi, L., Wight, R.G. and Green, M.F. (2005). Anchoring techniques for strengthening reinforced concrete beams with prestressed CFRP sheets, *3rd Int. Conf. Construction Materials (ConMat05)*, Vancouver, BC, Canada (CD-Rom).

Kim, Y.J., Green, M.F., Fallis, G.J., Wight, R.G. and Eden, R. (2006). Damaged bridge girder strengthening: field application of prestressed fiber-reinforced polymer sheets, *Conc. Int.*, **28**(11), 47–52.

Kim, Y.J. and Heffernan, P.J. (2008). Fatigue behavior of structures strengthened with fiber reinforced polymers: state-of-the-art, *J. Compos. Constr.*, **12**(3), 246–256.

Kim, Y.J., Shi, C., Bizindavyi, L. and Green, M.F. (2007a). Applying prestressed CFRP sheets to restore prestress losses in prestressed concrete beams, *Case Histories and Use of FRP for Prestressing Applications (SP-245)*, ACI, 105–122.

Kim, Y.J., Green, M.F. and Wight, R.G. (2007b). Flexural behaviour of reinforced or prestressed concrete beams strengthened with prestressed CFRP sheets: application of a fracture mechanics approach, *Can. J. Civ. Eng.*, **34**(5), 664–677.

Kim, Y.J., Shi, C. and Green, M.F. (2008a). Ductility and cracking behavior of prestressed concrete beams strengthened with prestressed CFRP sheets, *J. Compo. Constr.*, **12**(3), 274–283.

Kim, Y.J., Wight, R.G. and Green, M.F. (2008b). Flexural strengthening of RC beams with prestressed CFRP sheets: development of non-metallic anchor systems, *J. Compos. Constr.*, **12**(1), 35–43.

Kim, Y.J., Wight, R.G. and Green, M.F. (2008c). Flexural strengthening of RC beams with prestressed CFRP sheets: using non-metallic anchor systems, *J. Compos. Constr.*, **12**(1), 44–52.

Kim, Y.J., Longworth,.J, Wight, R.G. and Green, M.F. (2008d). Flexure of two-way slabs strengthened with prestressed or non-prestressed CFRP sheets, *J. Compos. Constr.*, **12**(4), 366–374.

Kim, Y.J., Green, M.F. and Wight, R.G. (2009). Bond and short-term prestress losses of prestressed composites for strengthening PC beams with integrated anchorage, *J. Rein. Plas. Compos.*, **29**(9), 1277–1294.

Kim, Y.J., Green, M.F. and Wight, R.G. (2010). Effect of prestress levels in prestressed CFRP sheets for strengthening prestressed concrete beams: numerical parametric study, *PCI J.*, **55**(2), 96–108.

Lees, J.M. and Burgoyne, C.J. (1999). Experimental study of influence of bond on flexural behavior of concrete beams pretensioned with aramid fiber reinforced plastics, *Struct. J., ACI* **96**(3), 377–385.

Lees, J.M. (2001). Fibre-reinforced polymers in reinforced and prestressed concrete applications: moving forward, *Prog. Struct. Eng. Mat.*, **3**, 122–131.

Lu, Z., Boothby, T,E,, Bakis, C. and Nanni, A. (2000). Transfer and development lengths of FRP prestressing tendons, *PCI J.*, **45**(2), 84–95.

Malvar, L.J. and Bish, J. (1995). Grip effects in tensile testing of FRP bars, *Proc. 2nd Int. RILEM symposium (FRPRCS-2), Non-metallic (FRP) Reinforcement for Concrete Structures*, Ghent, Belgium, 108–115.

MBrace. (1998). *MBrace TM Composite Strengthening System Design Guide*, Master Builders, Inc., Cleveland, OH.

McKay, K.S. and Erki, M.A. (1993). Flexural behavior of concrete beams pretensioned with aramid fibre reinforced plastic tendons, *Can. J. Civ. Eng.*, **20**, 688–695.

Meier, U. (1995). Strengthening of structures using carbon fibre/epoxy composites, *Constr. Build. Mat.*, **9**(6), 341–351.

Naaman, A.E. and Alkhairi, F.M. (1991). Stress at ultimate in unbonded prestressing tendons-Part II: proposed methodology, *Struct. J., ACI*, **89**(6), 683–692.

Naaman, A.E., Burns, N., French, C., Gamble, W.L. and Mattock, A.H. (2002). Stresses in unbonded prestressing tendons at ultimate: recommendation, *Struct. J., ACI*, **99**(4), 518–529.

Nanni, A. and Tanigaki, M. (1992). Pretensioned prestressed concrete members with bonded fiber reinforced plastic tendons: development and flexural bond lengths (static), *Struct. J.*, ACI, **89**(4), 433–441.

Nanni, A., Bakis, C., Neil, E.F.O. and Dixon, T.O. (1996). Short-term sustained loading of FRP tendon-anchor systems, *Constr. Build. Mat.*, **10**(4), 255–266.

Niitani, K., Tezuka, M. and Tamura, T. (1997). Flexural behaviour of prestressed concrete beams using AFRP pre-tensioning tendons, *3rd Int. Sym. Non-metallic (FRP) Reinforcement for Concrete Structures (FRPRCS-2)*, **2**, 663–670.

Nordin, H. and Taljsten, B. (2006). Concrete beams strengthened with prestressed near surface mounted CFRP, *J. Compos. Constr.*, **10**(1), 60–68.

NRC. (2004). Guide for the design and construction of externally bonded FRP systems for strengthening existing structures, National Research Council, Rome, Italy.

PCI. (2004). *PCI Design Handbook*, 9th Edn. Precast Prestressed Concrete Institute, Chicago, IL.

Pisani, M.A. (1998). A numerical survey on the behaviour of beams pre-stressed with FRP cables, *Constr. Build. Mat.*, **12**, 221–232.

Rao, P.S. and Mathew, G. (1996). Behavior of externally prestressed concrete beams with multiple deviators, *Struct. J., ACI*, **93**(4), 1–10.

Rizkalla, S.H. and Tadros, G. (1994). A smart highway bridge in Canada, *Conc. Int.*,**16**(6), 42–44.

Russell, B. and Burns, N.H. (1996). Measured transfer lengths of 0.5 and 0.6 in. strands in pretensioned concrete, *PCI J.*, **41**(5), 44–65.

Saadatmanesh, H. and Ehsani, M.R. (1991). RC beams strengthened with GFRP plates. I: experimental study, *J. Struct. Eng.*, **117**(11), 3417–3433.

Saafi, M. and Toutanji, H. (1998). Flexural capacity of prestressed concrete beams reinforced with aramid fiber reinforced polymer (AFRP) rectangular tendons, *Constr. Build. Mat.*, **12**, 245–249.

Sayed-Ahmed, E. and Shrive, N.G. (1998). A new steel anchorage system for post-tensioning applications using carbon fibre reinforced plastic tendons, *Can. J. Civ. Eng.*, **25**, 113–127.

Sen, R., Mariscal, D. and Shahawy, M. (1993). Durability of fiberglass pretensioned beams, *Struct. J., ACI*, **90**(5), 525–533.

Shahawy, M.A., Issa, M. and Batchelor, B. (1992). Strand transfer lengths in full scale AASHTO prestressed concrete girders, *PCI J.*, **37**(3), 84–96.

Smith, S.T. and Teng, J.G. (2001). Interfacial Stresses in Plate Beams, *Eng. Struct.*, **23**(7), 857–571.

Soudki, K.A. (1998). FRP reinforcement for prestressed concrete structures, *Prog. Struct. Eng. Mat.*, **1**(2), 135–142.

Soudki, K.A., Green, M.F. and Clapp, F. (1997). Long-term behavior of concrete beams prestressed with carbon fibre tendons, *Non-metallic (FRP) Reinforcement for Concrete Structures*, **2**, 647–654.

Stoll, F., Saliba,.J.E. and Casper, L.E. (2000). Experimental study of CFRP-prestressed high-strength concrete bridge beams, *Compos. Struct.*, **49**, 191–200.

Taerwe, L. and Matthys, S. (1995). Structural behaviour of concrete slabs prestressed with AFRP bars, *1st Int. Sym. Non-metallic (FRP) reinforcement for Concrete Structures*, 421–429.

Taljsten, B. (1997). Strengthening of beams by plate bonding, *J. Mat. Civ. Eng.*, **9**(4), 206–212.

Taljsten, B. and Nordin, H. (2007). Concrete beams strengthened with external prestressing using external tendons and near-surface-mounted reinforcement (NSMR), *Case Histories and Use of FRP for Prestressing Applications (SP-245)*, 143–164.

Teng, J.G., Chen, J.F., Smith, T. and Lam, L. (2003). Behavior and strength of FRP-strengthened RC structures: a state-of-the-art review, *Struct. Build.*, **156**(1), 51–62.

Toutanji, H. and Saafi, M. (1999). Performance of concrete beams prestressed with aramid fiber-reinforced polymer tendons, *Compos. Struct.*, **44**, 63–70.

Triantafillou, T.C., Deskovic, N. and Deuring, M. (1992). Strengthening of concrete structures with prestressed fiber reinforced plastic sheets, *Struct. J., ACI*, **89**(3), 235–244.

Wight, R.G., Green, M.F. and Erki, M.A. (2001). Prestressed FRP sheets for post-strengthening reinforced concrete beams, *J. Compos. Constr.*, **5**(4), 214–220.

Wu, Z., Iwashita, K. and Sun, X. (2007). Structural performance of RC beams strengthened with prestressed near-surface-mounted CFRP tendons, *Case Histories and Use of FRP for Prestressing Applications (SP-245)*, 165–178.

Youakim, S.A. and Karbhari, V.M. (2007). An approach to determine long-term behavior of concrete members prestressed with FRP tendons, *Constr. Build. Mat.*, **21**, 1052–1060.

Zhang, B., Benmokrane, B. and Chennouf, A. (2000). Prediction of tensile capacity of bond anchorages for FRP tendons, *J. Compos. Constr.*, **4**(2), 39–47.

Zhang, B. and Benmokrane, B. (2002). Pullout bond properties of fiber-reinforced polymer tendons to grout, *J. Mat. Civ. Eng.*, **14**(5), 399–408.

Zou, P.X.W. (2003). Flexural behavior and deformability of fiber reinforced polymer prestressed concrete beams, *J. Compos. Constr.*, **7**(4), 275–284.

3
Analyzing bond characteristics between composites and quasi-brittle substrates in the repair of bridges and other concrete structures

C. CARLONI, University of Hartford, USA

DOI: 10.1533/9780857097019.1.61

Abstract: During the last two decades, externally bonded fiber-reinforced polymer (FRP) composites have been widely used for strengthening, repairing, and rehabilitating reinforced concrete (RC) structural members. The bond characteristics contribute to the effectiveness of the stress transfer achieved between the FRP composite and the concrete substrate. Debonding of the FRP composite reinforcement is the most critical concern in this type of application. Under monotonic and fatigue-loading conditions, FRP–concrete shear debonding has been idealized as a Mode-II fracture problem along the bi-material interface. A cohesive material law is used to describe the interfacial stress transfer at the macroscopic level. The area under the entire curve represents the fracture energy, and is related to the load-carrying capacity of the interface. In this chapter, previous experimental results published by the author are discussed to show how the fracture energy can be considered a true fracture parameter. The results are instrumental in discussing the strain limits provided in international codes and guidelines. Future research needed and newly developed composites are introduced at the end of the chapter.

Key words: fiber-reinforced polymer (FRP), concrete, fracture energy, debonding, Mode-II, strain limit, interfacial crack propagation.

3.1 Introduction

The increasing use of fiber-reinforced polymer (FRP) composites for strengthening and repairing existing reinforced concrete (RC) structures has promoted a remarkable research interest over the last two decades (Bank, 2006). When applied to the tension side of a beam, the FRP composite laminate/sheet increases the internal moment capacity of the strengthened section. Similarly, when attached to the web of a beam in the high-shear region, the FRP composite contributes to the nominal shear capacity of the section. Particular attention was paid to the study of the debonding mechanism in beams retrofitted with FRP laminates/sheets (Buyukozturk and Hearing, 1998; Teng *et al.*, 2003). In fact, debonding is one of the most important causes of inefficiency, and certainly one of the most dangerous because it is a brittle phenomenon (Arduini *et al.*, 1997), which might occur with no visible warning at a load level significantly lower than the flexural or shear capacity of the strengthened member corresponding to the rupture of the FRP reinforcement.

Debonding in beams takes place in regions of high-stress concentration (Wu *et al.*, 1997; Leung, 2001). With respect to flexural FRP reinforcement, two different types of debonding are considered. (1) End plate debonding (EPD), which occurs at the curtailment of the FRP composite and typically involves the concrete cover, although it can also propagate at the FRP–concrete interface. EPD is also termed *cover delamination* (ACI 440.2R-08, 2008); EPD could be prevented by providing anchorages at the ends of the FRP reinforcement, for example in the form of FRP stirrups (Chen and Teng, 2003). (2) Intermediate crack-induced debonding (ICD), which initiates at flexural or flexural/shear cracks near the region of maximum moment and propagates in the direction of decreasing moment; ICD typically occurs in a thin mortar-rich layer of concrete (Carloni and Subramaniam, 2010), where the epoxy impregnates the substrate, and does not propagate into the concrete cover.

This chapter will focus on ICD, although some results and observations can be applied to EPD. Moreover, some of the concepts herein presented are applicable to the debonding of FRP laminates/sheets applied to the web of a beam as external stirrups. For flexural reinforcement, guidelines and codes adopted in different countries limit the value of the effective strain in the FRP composite in order to impede the onset of ICD. However, in this chapter the cohesive nature of the debonding mechanism will be discussed, and it will be observed how some assumptions, which are commonly accepted in the design of the FRP reinforcement, require additional investigation and discussion.

In this chapter, there will be a brief discussion of newly-developed fiber-reinforced cementitious matrix (FRCM) composites. FRCMs employ a cement-based matrix instead of epoxy. FRCM composites have some potential advantages when compared to FRPs: (1) resistance to high temperatures and UV radiation; (2) the inorganic matrix is water-based, which facilitates its application and allows for reuse of tools; (3) permeability compatibility with the concrete substrate; and (4) unvarying workability time with temperature between 4°C and 40°C.

3.2 Experimental investigation of debonding

Full-scale beam tests were, and still are, the optimal tests (Sebastian, 2001; Rosenboom and Rizkalla, 2008) to investigate the characteristic of the FRP–concrete bond. However, the high cost of the set-up and the complexity of the flexural behavior of a RC strengthened beam, which includes load sharing between internal steel reinforcement and external FRP laminates/sheets, in addition to the quasi-brittle behavior of concrete, and the subsequent formation of a not easily predicted crack pattern, have induced several research groups to develop an *ad hoc* set-up to investigate the FRP–concrete debonding. Ritchie *et al.* (1991), Sharif *et al.* (1994), and Arduini *et al.* (1997) were among the first experimental studies to focus on the debonding phenomenon in beams, whereas Chajes *et al.* (1996), Taljsten (1997), and Bizindavyi and Neal (1999) were among the first researchers to use a single-lap shear test set-up to study the stress transfer between FRP and concrete. A literature review on FRP–concrete debonding can be found in Hearing (2000), Buyukozturk *et al.* (2004), Gunes (2004), Han (2009), Grace *et al.* (2012), and Kang *et al.* (2012). Chen and Teng (2001) and Yao *et al.* (2005) described in their work the most common direct-shear set-ups proposed in the literature, which are still used although it should be pointed out that some of those set-ups have gained more popularity than others in the last decade, and some others have been proposed or adjusted from the original ones (Mohamed Ali *et al.*, 2008; Gartner *et al.*, 2011). Researchers are still debating the differences among the set-ups in order to identify which one(s) are capable of capturing the debonding phenomenon in beams. It is recognized that among the different set-ups, the single-lap and the double-lap shear test set-ups capture the debonding phenomenon better.

3.2.1 Single-lap direct-shear tests

In order to foster the discussion on the *quasi-brittle* nature of the debonding mechanism, comment on the strain limits provided in guidelines and codes, and eventually highlight the need for future research, in this section part of the previous experimental work conducted by the author in collaboration with Prof. Kolluru V. Subramaniam (Indian Institute of Technology, Hyderabad, India) will be briefly revisited (Subramaniam *et al.*, 2007). Single-lap direct-shear tests were used to evaluate the FRP–concrete debonding using the classical pull–push configuration. The tensile load was applied to the FRP sheet, while the concrete block was restrained against movement. This set-up is also referred to as the *near-end supported single-shear test* (Yao *et al.*, 2005).

Figure 3.1 shows the specimen dimensions and the loading arrangement (Carloni and Subramaniam, 2012b). The nominal dimensions of the

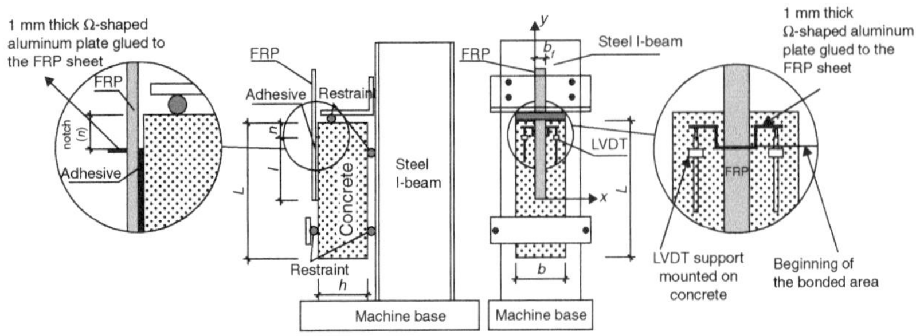

3.1 Specimen dimensions and loading arrangement.

prismatic concrete block were as follows: length L = 330 mm, width b = 125 mm, height h = 125 mm. The coarse aggregates consisted of gravel with a maximum size of 10 mm. River sand was used as fine aggregate. The 28-day compressive strength of concrete, determined as per ASTM C 39 (2011), was 39 MPa. The nominal thickness t_f of the carbon fibers contained in the FRP composite sheet was equal to 0.167 mm. The tensile strength and the Young's modulus E_f of the composite sheet based on the fiber thickness were equal to 3.83 and 230 GPa, respectively. The FRP sheet was bonded in the center on one side of the concrete block using the wet-layup procedure. The bonded length ℓ of the sheet was kept fixed at 152 mm for all specimens. A length (n) of the FRP equal to 35 mm was left unbonded to provide an initial notch. Different nominal widths, b_f, of the FRP sheet equal to 12, 19, 25, 38, and 46 mm were tested. Two linear variable differential transducers (LVDTs) were mounted on the concrete surface close to the edge of the bonded area. The LVDTs reacted off of a thin aluminum Ω-shaped plate, which was glued to the FRP surface at the beginning of the bonded area as shown in Fig. 3.1. The average of the two LVDT readings was named *global slip*. Tests were conducted in displacement control by increasing the global slip at a constant rate equal to 0.00065 mm/s, up to failure. The modality of failure of all direct-shear test specimens was associated with progressive debonding of the FRP sheet from the concrete substrate.

The strain components on the surface of the FRP and surrounding concrete during the monotonic quasi-static tests were determined from the displacement field, which was measured using a full-field optical technique known as digital image correlation (DIC). Details about DIC can be found in Sutton *et al.* (1983, 2009). The strain analysis reported in the next section refers to the Cartesian system shown in Fig. 3.1.

Figure 3.2 shows a typical load vs global slip response obtained from a direct-shear test. The load response of Fig. 3.2 corresponds to test W_6 (b_f = 25 mm) reported in Subramaniam *et al.* (2007). The load responses

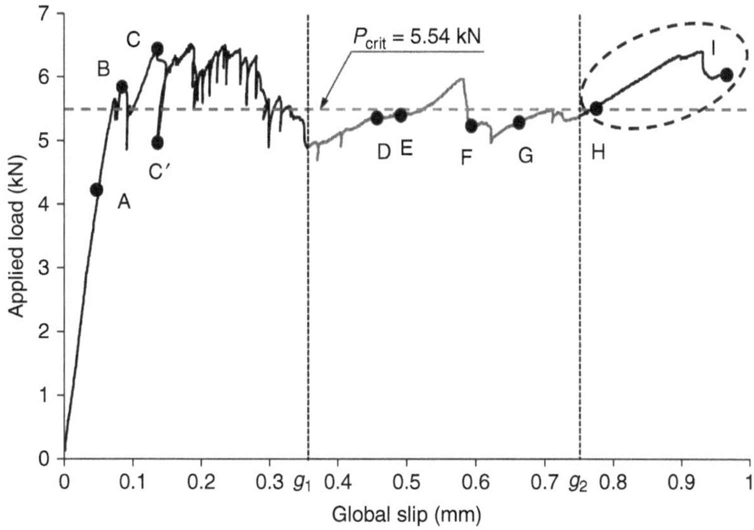

3.2 Applied load vs global slip response for test W_6 (Subramaniam *et al.*, 2007).

obtained from specimens with different FRP widths were nominally similar in shape. Additional tests were published by the same authors (Subramaniam *et al.*, 2011) using the same test set-up and concrete blocks with a different width (b = 52 mm) and height (h = 100 mm). The load responses of those specimens were also similar in shape. The load response was initially linear up to point A. A nonlinear behavior was observed in the region AC. The strain analysis, presented in the next section, showed that the interfacial crack initiated in the nonlinear part AC of the load response (Ali-Ahmad *et al.*, 2006; Subramaniam *et al.*, 2007, 2011. The difference between crack initiation and crack propagation in a strengthened beam is outlined in Sebastian (2001). A sudden drop (CC') in the load marked the onset of the interfacial crack propagation. Load drops before point C (for example at point B) indicated that the debonding process initiated. As the crack propagated (after C), the load remained nominally constant and its value P_{crit} was determined as the mean value of the load corresponding to the range of the global slip (g_1 = 0.35 mm, g_2 = 0.75 mm). The global slip range (g_1, g_2) slightly varied for each test and was identified on the basis of the strain analysis results that will be presented in the next section. The load response between points H and I will be justified by the strain analysis that follows. The load P_{crit} is often termed *load-carrying capacity* or *bond strength*. Bond strength is probably used improperly in this context, because strength is usually associated with a stress component rather than a load, which depends on the size of the specimen.

To provide a rational basis for comparison, the load-carrying capacity was expressed in terms of the ultimate nominal stress (σ_{crit}) in the FRP at debonding:

$$\sigma_{\text{crit}} = \frac{P_{\text{crit}}}{b_f t_f} \qquad [3.1]$$

The variation of σ_{crit} with the ratio b_f / b is reported in Subramaniam *et al.* (2007). σ_{crit} increased linearly with the ratio b_f / b, suggesting the existence of a width effect. This relationship appeared to be independent of the width of the concrete block (Subramaniam *et al.*, 2011). In Subramaniam *et al.* (2007), it was found that the specimens with $b_f = 12$ mm exhibited significant scatter and provided results that did not agree with the trend. The reason for the nonconformity of the $b_f = 12$ mm specimens became clear with the results of the strain analysis.

3.2.2 Strain analysis in direct-shear tests

Figure 3.3 shows the variation of the axial strain ε_{yy} on the surface of the FRP along the bonded length for point E of the load response of Fig. 3.2. The axial strain distribution along the FRP obtained from all specimens tested was nominally similar. The experimental strain values are represented by '×'s. The axial strain values were determined along the center line of the FRP sheet by averaging the strain across a 10-mm-wide strip for each value of y. Averaging the strain across a 10 mm width allows for taking into account the variation of the strain due to the presence of a non-homogeneous substrate and local material variations in the FRP sheet (Ali-Ahmad *et al.*, 2006). A 10-mm strip was chosen based on the size of the aggregate. It can be noticed that at $y = 116$ mm the axial strain has a relative maximum. The axial strain along the FRP at point E was approximated using the procedure outlined in Ali-Ahmad *et al.* (2006). The experimental nonlinear strain distribution along the bonded length of Fig. 3.3 was approximated using the following expression (solid line in Fig. 3.3):

$$\varepsilon_{yy} = \varepsilon_0 + \frac{\alpha}{1 + e^{-\frac{y - y_0}{\beta}}} \qquad [3.2]$$

where α, β, y_0, and ε_0 were determined using nonlinear regression analysis of the strains obtained from DIC. The choice of the function [3.2] is not unique (Dai *et al.*, 2005; 2006; Zhou *et al.*, 2010; Liu and Wu, 2012) and its

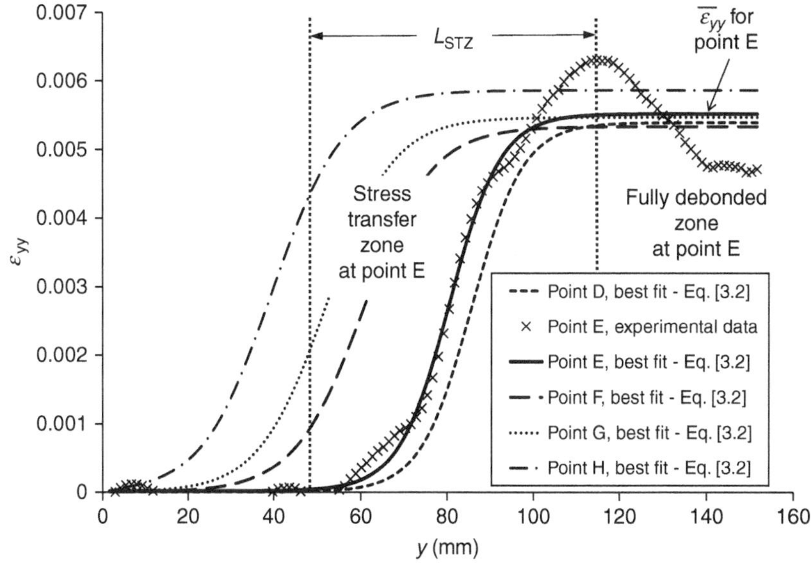

3.3 Axial strain profiles corresponding to points D, E, F, G, and H of the load response of Fig. 3.2.

influence on the results presented in the next section was not investigated in the study. The observed strain distribution along the FRP was essentially equal to zero close to the unloaded end. There was a rapid increase in strain upon approaching the loaded end. The strain leveled off at a value $\bar{\varepsilon}_{yy}$ approximately equal to 5520 $\mu\varepsilon$. The observed strain distribution was divided into three main regions: (1) the stress-free zone (SFZ); (2) the stress-transfer zone (STZ); and (3) the fully-debonded zone (FDZ). In the FDZ, the strains were essentially constant and were found to remain unchanged as the global slip increased, which is consistent with the observation that the load remained nominally constant (P_{crit}) after the debonding process propagated. Analysis of the strain data revealed that the STZ was fully established when the load response attained P_{crit}. In other words, the 'S' shape of the strain profile was fully established for points of the load response after C. For each test, the global slip range (g_1, g_2) within which the STZ was fully established was determined and used to compute the load-carrying capacity P_{crit}. The fitting curves, obtained using Equation [3.2], for points D, F, G, and H, are shown in Fig. 3.3 for comparison. It can be observed that a simple translation of the STZ further along the length of the FRP sheet occurred as the global slip increased while its shape remained constant. The translation of the STZ indicated self-similar crack growth.

The STZ at point H reached the end of the bonded length. Any further increment of the global slip corresponded to a different fracture path that

Table 3.1 Fracture parameters of the tests published in Subramaniam *et al.* (2007)

Test	b_f (mm)	P_{crit} (kN)	L_{STZ} (mm)	$\bar{\varepsilon}_{yy}$ (με)	G_F (MPa × mm)
W_1	46	12.90	80	7210	0.874
W_2	46	12.05	76	6180	0.634
W_3	46	13.20	75	5890	0.563
W_4	38	10.09	81	6430	0.692
W_5	38	10.02	73	6240	0.652
W_6	25	5.54	80	5560	0.546
W_7	25	5.44	76	5630	0.530
W_8	25	5.36	69	6490	0.705
W_9	19	4.27	75	6400	0.686
W_10	19	4.05	78	5890	0.579

involved a larger amount of concrete as the crack kinked in the direction of the concrete block. The portion of the load response H-I in Fig. 3.2 is related to this different behavior of the FRP–concrete interface (Carrara *et al.*, 2011). It is important to notice that the best-fit curve provided a maximum strain $\bar{\varepsilon}_{yy}$ that varied slightly from point to point along the load response. These variations were consistent with the variation of the load. The mean values of the maximum strain $\bar{\varepsilon}_{yy}$, calculated from ten points of the load response in the range (g_1, g_2), are shown in Fig. 3.4 and reported in Table 3.1 for all the tests published in Subramaniam *et al.* (2007). The mean value (5560 με) of test W_6 is shown with a black-filled marker in Fig. 3.4.

The progressive debonding of the FRP composite sheet from concrete is associated with a STZ of a fixed length L_{STZ}, which translates as the crack advances (Fig. 3.3). L_{STZ} is also termed the *effective bond length* (Chen and Teng, 2001) or the *development length* (ACI 440.2R-08, 2008). Values of L_{STZ} are reported in Table 3.1 and summarized in Fig. 3.5. Each value plotted in Fig. 3.5 represents the mean value for each test and was calculated from ten points of the load response within (g_1, g_2). Some of the proposed formulas for L_{STZ} (Nuebauer and Rostasy, 1999; Chen and Teng, 2001; Monti *et al.*, 2003; Karbhari *et al.*, 2006) are plotted in Fig. 3.5 for comparison. The experimental values of L_{STZ} are not affected by the width of the FRP sheet.

Figure 3.6 shows the variation of the axial strain ε_{yy} across the width of the specimen W_6 (Subramaniam *et al.*, 2007) at point E (Fig. 3.2) for different values of y. Figure 3.6 reveals the presence of a *central region* of the FRP where the axial strain is nominally constant across the width. b_s is the measurement of the central region of the FRP. The width b_d of the concrete strip involved in the stress transfer is larger than the FRP width. Only for those specimens with $b_f > 12$ mm the central region was

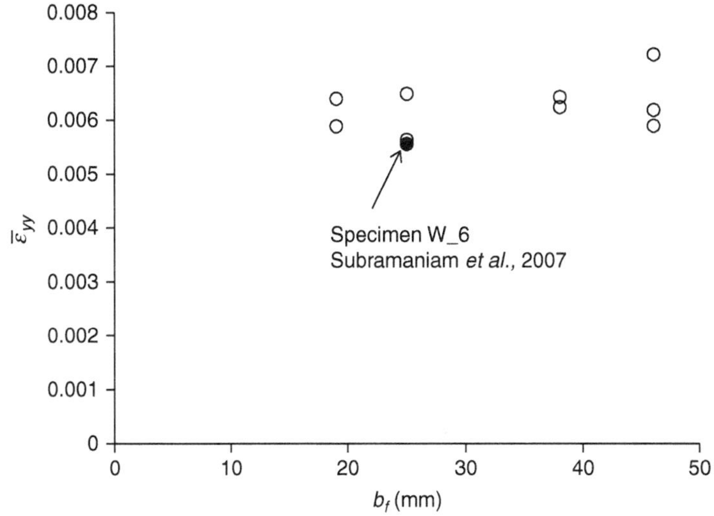

3.4 Mean values of the maximum strain ε_{yy} vs FRP width b_f for all tests reported in Subramaniam *et al.* (2007).

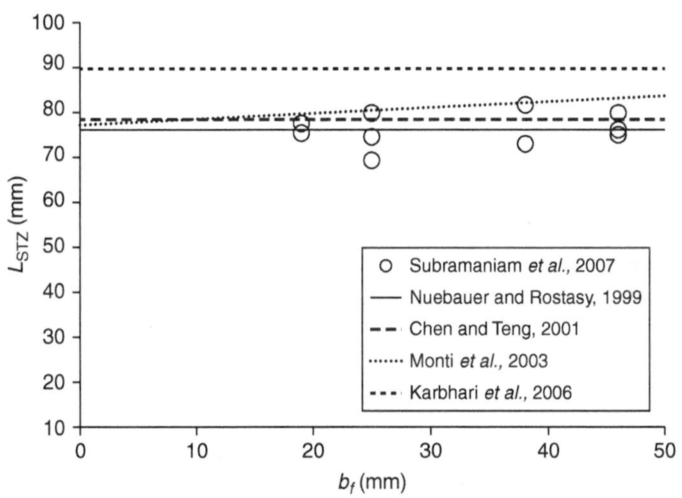

3.5 Mean values of the length of the STZ vs FRP width b_f for all tests reported in Subramaniam *et al.* (2007).

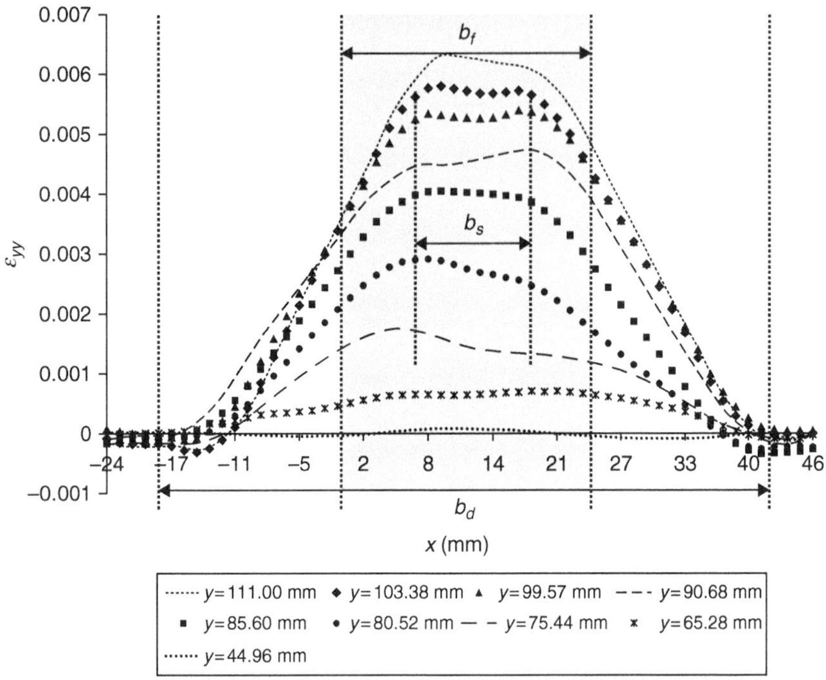

3.6 Variation of the axial strain ε_{yy} at point E (Fig. 3.2) as a function of *x* for different values of *y*.

observed; therefore, it was concluded that specimens with $b_f = 12$ mm did not follow the trend of σ_{crit} vs b_f / b due to the absence of the central region. A plot of the strain ε_{yy} for a specimen with $b_f = 12$ mm can be found in Subramaniam *et al.* (2007). The width of the central region of the FRP increased as the FRP width increased, while the quantity $(b_d - b_s)$ remained constant. Hence, the increase in the nominal stress at debonding with an increase in the FRP width was explained by the increase of the central region. Available results in the literature (Chen and Teng, 2001; Yao *et al.*, 2005) did not confirm the conclusions reported in Subramaniam *et al.* (2007). Chen and Teng (2001) reported a decrease of the normalized load-carrying capacity with an increase of the FRP-to-concrete width ratio. A coefficient β_p that takes into account the effect of the FRP-to-concrete width ratio was proposed by the authors. It should be noticed, however, that the specimens considered by the authors (Chen and Teng, 2001) to validate the model had a minimum FRP width equal to 25.4 mm. Size effect was not considered in their study, and some of the data, obtained from the literature and used to calibrate the model, were from bond tests between steel plates and concrete.

Subramaniam *et al.* (2007) showed that the plot of the shear strain ε_{xy} across the width of the specimen for different values of y also revealed the presence of edge regions in the FRP sheet that were characterized by high-shear strain gradients. The central region in which the axial strain was constant coincided with the central region of the strip where the shear strain was zero.

3.3 Fracture mechanics approach to the analysis of debonding

Fracture propagation during ICD occurs in concrete; therefore, it is reasonable to assume that the *quasi-brittle behavior* of concrete governs the debonding process. It seems appropriate to model the debonding mechanism within the framework of fracture mechanics applied to quasi-brittle materials (Bazant and Planas, 1997). In linear elastic fracture mechanics (LEFM) it is well known that a singularity in the stress and displacement fields occurs at the crack tip (Anderson, 2004; Bazant and Planas, 1997). In LEFM the nonlinearity of the material is neglected, although in practice only a limit value of the stress can be reached and in a zone near the crack tip a stress re-distribution occurs, which is related to the nonlinearity of the material. The nonlinearity near the crack tip can be due to hardening or softening of the material. In order to overcome the inapplicability of LEFM for quasi-brittle materials such as concrete, the concept of the *fracture process zone* (FPZ) is introduced. The FPZ represents a bridging zone between the cracked and uncracked regions, where progressive softening occurs. It is important to point out that in quasi-brittle materials the FPZ is most likely coincident with, or close in size to, the region of material nonlinearity. In other materials, such as steel, the softening part of the nonlinear zone is negligible and the nonlinear hardening zone is predominant. The characteristics and size of the FPZ depend on the material. In concrete, the FPZ is related to progressive damage that is associated with several complex phenomena (microcracking, void formation, etc.). As the fracture process progresses in concrete, coalescence of microcracks in the FPZ gives continuity to the already existing crack and consequently the crack propagates.

The concept of cohesive zone or cohesive crack is associated with the work of Hillerborg *et al.* (1976). The cohesive crack model is a simple model for the FPZ in concrete, and can be taken as a reference to compare other models (Elices *et al.*, 2002). For a Mode-I (Anderson, 2004) condition, it is assumed that within the FPZ a crack tip should not be defined. Conversely, the stress σ transferred through the cohesive crack is a function of the crack opening w. The function $\sigma = f(w)$ is characteristic of the material and is often called the *softening function*. When $w = 0$, $\sigma = f_t$, which is the tensile strength of concrete.

3.3.1 Cohesive material law

Several researchers attempted to study the ICD phenomenon and the debonding in direct-shear tests as a Mode-II fracture problem (Anderson, 2004) where the interface region is idealized to be of zero thickness with well-defined material properties (Ali-Ahmad *et al.*, 2006; Wu and Niu, 2007; Mazzotti *et al.*, 2008). In this idealization, the interfacial crack, associated with initiation and propagation of debonding, is subject to a Mode-II loading condition. The quasi-brittle behavior of the interfacial crack, in the spirit of the cohesive crack model, is described by introducing a cohesive material law, which relates the interfacial shear stress (τ_{zy}) to the relative slip (s) between FRP and concrete. It is important to highlight that the cohesive material law τ_{zy}–s represents the constitutive law of a fictitious material that links the FRP strip to the concrete substrate (interface). The shear stress should not be associated with the shear stress that occurs in concrete at a particular distance from the concrete surface. Researchers (Abdel Baky *et al.*, 2012) attempted to describe the variation of the shear stress through the depth of the concrete substrate, and observed that the normal stress σ_{yy} in concrete influences the value of the maximum shear stress (τ_{max}). Although it is reasonable to assume that the stress field in concrete near the interface has an important role in the stress transfer, and therefore in the debonding mechanism, the cohesive material law herein introduced aims to describe the interfacial debonding at the macro-scale; hence its parameters should not refer to the actual stress state in concrete at the microscopic level.

Fracture along the interface does not occur on an ideal plane parallel to the FRP strip but follows a tortuous path, which is in part controlled by the distribution of the aggregates and in part by the mixed-mode nature of the fracture process at the microscopic level. In fact, the crack continuously kinks to follow the path that requires the least amount of energy, and is related to the fracture properties of the two materials (Hutchinson and Suo, 1992; Gunes, 2004; Gunes *et al.*, 2009). At the macroscopic level, the microscopic mixed-mode fracture can be considered a Mode-II fracture at the FRP–concrete interface.

Several analytical/numerical procedures to estimate the cohesive interface behavior from the load response of direct-shear tests were developed. For example, Ali-Ahmad *et al.* (2006) established an experimental procedure to directly determine the Mode-II interfacial fracture law using DIC measurements. The cohesive material law for the interface, when implemented in a numerical analysis procedure, allowed for predicting the load response of concrete beams strengthened with externally bonded FRP sheets (Wu and Yin, 2003; Ali-Ahmad *et al.*, 2007; Wu and Niu, 2007). Among others, Ferracuti *et al.* (2007), Mazzotti *et al.* (2008), and Carrara *et al.* (2011) used a procedure similar to Ali-Ahmad *et al.* (2006). In those studies, the authors used strain gage readings along the FRP surface to obtain the interfacial law. Pellegrino *et al.* (2008) and Pellegrino and Modena (2009) used a double-lap shear test and a small-beam test to investigate the effect of the axial stiffness of the composite on the cohesive material law and indicated the need of more research to study this aspect.

From the measured strain ε_{yy} along the bonded length, the interface shear stress was calculated as (Taljsten, 1996, 1997):

$$\tau_{zy} = E_f t_f \frac{d\varepsilon_{yy}}{dy} \tag{3.3}$$

The following assumptions were made: (1) the FRP sheet was homogenous and linear elastic; (2) the thickness and the width of the FRP sheet were constant along the bonded length; (3) the interface was subject only to shear loading; (4) the interface between the FRP and the concrete was assumed to be of infinitesimal thickness; and (5) the concrete substrate was rigid. The relative slip, $s(y)$, between FRP and concrete at a given location on the FRP was obtained by integrating the axial strain in the FRP up to that point. Lu *et al.* (2005) commented on the possible ways to obtain the $\tau_{zy}-s$ curves, observing that the violent local variations of strain measured by strain gages entailed substantial differences in the fracture parameters. The procedure followed by Subramaniam *et al.* (2007, 2011) used the strain contours obtained from DIC, which allowed to identify the fluctuations of the strain profile due to the local variations of the FRP and the substrate (Ali-Ahmad *et al.*, 2006). The cohesive material law curves corresponding to points D, E, F, and G of the load response of Fig. 3.2 are shown in Fig. 3.7. An important aspect of this procedure should be pointed out. The cohesive material law is a local constitutive law and should be obtained from strain measurements at a specific point along the bonded length for the entire load response and eventually compared with the constitute laws obtained from other points along the bonded area. Conversely, the procedure adopted in Ali-Ahmad *et al.* (2006) and in Subramaniam *et al.* (2007) used the strain values along the bonded length. Inherently, points along the bonded length have different interfacial properties, as suggested by the load response depicted in Fig. 3.2.

Several expressions of the cohesive material law are available in the literature. For example, Ferracuti *et al.* (2007) proposed the following relationship:

$$\tau_{zy}(s) = \bar{\tau}\frac{s}{\bar{s}}\frac{n}{(n+1)+\left(\frac{s}{\bar{s}}\right)^n} \tag{3.4}$$

where $\bar{\tau}$ is the maximum shear stress and $\bar{s}$ is the corresponding slip. Other researchers indicated these parameters as τ_{max} and s_0, respectively. The parameter n (>2) mainly governs the softening branch of the softening curve.

3.3.2 Interfacial fracture energy

The interfacial fracture energy G_F is the energy required to create and fully break the elementary unit area of the cohesive crack. G_F corresponds to the area under the entire $\tau_{zy}-s$ curve (Bazant and Planas, 1997):

$$G_F = \int_0^{s_f} \tau_{zy}(s)\,ds \qquad [3.5]$$

where s_f is the slip corresponding to complete separation of the interface. The mean values of G_F for all tests published in Subramaniam *et al.* (2007) are shown in Fig. 3.8 and reported in Table 3.1. The values plotted in Fig. 3.8 were obtained for each test by processing the same ten images corresponding to ten points of the load response used to calculate the values of Figs 3.4 and 3.5. The experimental data in Fig. 3.8 are compared with some of the available formulas in the literature for G_F. Some of the formulas used as comparison in Fig. 3.7 include a width effect coefficient that depends on b_f / b.

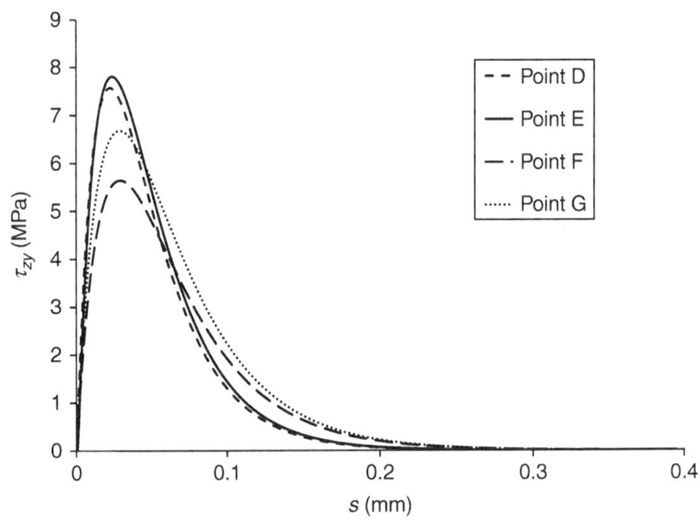

3.7 $\tau_{zy}-s$ curves for points D, E, F, and G in Fig. 3.2.

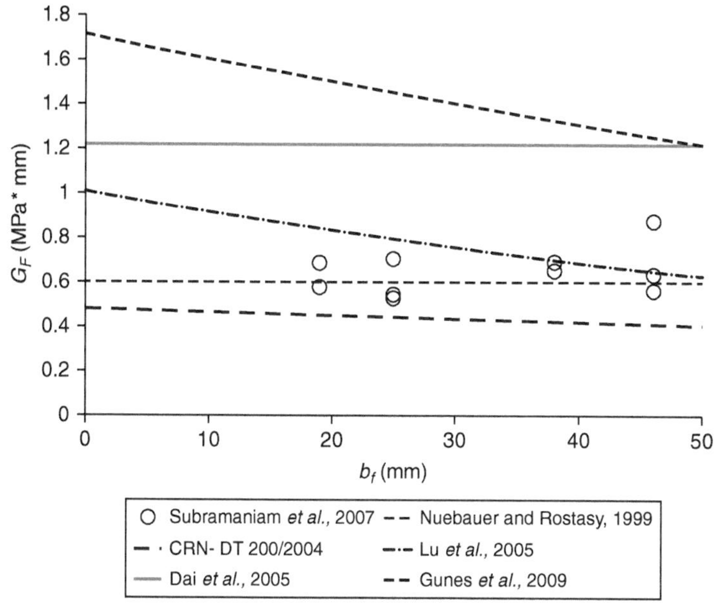

3.8 Mean values of the fracture energy G_F vs FRP width b_f (Subramaniam *et al.*, 2007).

The relationship between the interfacial fracture energy and the fracture energy of concrete (Mode-I) is still an open discussion among researchers (Achintha and Burgoyne, 2008, 2011; Carrara *et al.*, 2011). Although the fracture process in ICD occurs in concrete, it propagates in a mortar-rich thin layer (Carloni and Subramaniam, 2010) in which the mechanical and fracture properties are not easily defined. Undoubtedly, the two fracture energies are related, although a convincing relationship has not yet been found. Rabinovitch (2004) successfully used the Mode-I fracture energy of concrete to study the EPD using the fracture mechanics concept of energy release rate. In the context of EPD, the energy required to create and fully break the elementary unit area of cohesive crack should be closely related to the fracture energy of concrete as the debonding typically occurs in the concrete cover and peeling stresses are not negligible.

Taljsten (1996) obtained a relationship between the interfacial fracture energy and the load-carrying capacity in direct-shear tests by considering the energy release during the advancement of the interfacial crack a by an amount da (Fig. 3.9). The *energy release rate* G per unit width b_f of the composite is obtained as:

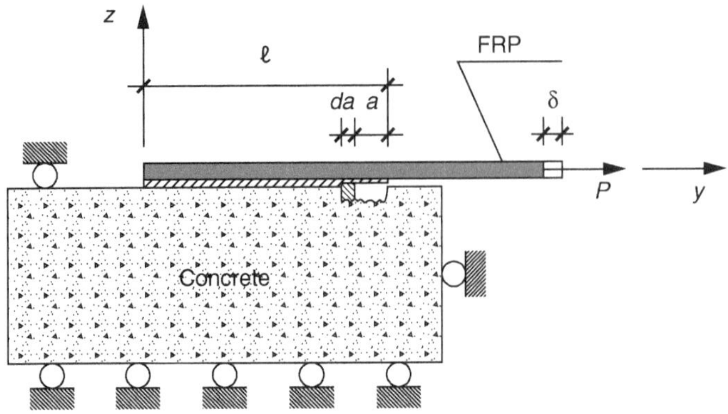

3.9 Crack propagation in direct-shear tests.

$$G = \frac{1}{b_f}\left[\frac{d}{da}(F - U_e)\right] \qquad [3.6]$$

where F is the work done by the external load P and U_e is the elastic energy. When debonding propagates, $G = G_f$.

If δ is the displacement of the point of application of the applied force P (Fig. 3.9) and C is the compliance of the system, then:

$$U_e = \frac{1}{2}P^2C \qquad [3.7]$$

and when the interfacial crack propagates:

$$G_F = \frac{1}{b_f}\left[P\frac{d\delta}{da} - \frac{dU_e}{da}\right] = \frac{1}{b_f}\frac{P^2}{2}\frac{\partial C}{\partial a} \qquad [3.8]$$

Therefore:

$$P = \sqrt{\frac{2G_F b_f}{\partial C / \partial a}} \qquad [3.9]$$

If the substrate is considered rigid and the adhesive layer is idealized as a zero-thickness layer:

$$\frac{\partial C}{\partial a} = \frac{1}{E_f b_f t_f} \qquad [3.10]$$

E_f and t_f are the elastic modulus and the thickness of the composite, respectively. From Equations [3.9] and [3.10], the interface fracture energy G_F is related to the load-carrying capacity (Hearing, 2000; Yuan *et al.*, 2001; Wu *et al.*, 2002; Liu and Wu, 2012; Wu *et al.*, 2012):

$$P_u = b_f\sqrt{2G_F E_f t_f} \quad [3.11]$$

Equation [3.11] can be obtained through an energy balance approach (Focacci *et al.*, 2000; Hearing, 2000; Liu and Wu, 2012) and is based on the assumption that a pure Mode-II interfacial crack propagation occurs across the entire width of the composite. The theoretical load-carrying capacity under pure Mode-II was indicated as P_u in Equation [3.11] to distinguish it from the experimental value P_{crit}. Wu *et al.* (2002) used an equilibrium approach to show how the load-carrying capacity depends only on the fracture energy and not on the shape of the $\tau_{zy}-s$ curve. An analytical approach can be found in Wu *et al.* (2012) in which a cohesive material law $\tau_{zy}(s)$ was assumed. The authors were able to predict the snapback phenomenon observed by other researchers (Ali-Ahmad *et al.*, 2007). The load-carrying capacity calculated from Equation [3.11] should not be expected to match the experimental results ($P_u \neq P_{\text{crit}}$). In fact, the width effect shown in Fig. 3.6 indicated that only in the central region of the strip the axial strain is constant (and the shear strain is zero), which entailed that the Mode-II loading condition should be assumed only in the central region of the FRP sheet.

It is interesting to notice that different experimental set-ups are available in literature (Yao *et al.*, 2005) for direct-shear tests that might lead to significantly different results and complicate the analysis of the data. It was shown that the cohesive material model developed from direct-shear tests is directly applicable to beam tests (Chen and Teng, 2003). However, Gunes *et al.* (2009) observed that the Mode-II fracture energy G_F under flexural loading should be higher than that measured from direct-shear tests, due to the curvature effect.

3.3.3 Discussion of interfacial fracture energy as a true fracture parameter

Contradictory results (Chen and Teng, 2001; Subramaniam *et al.*, 2007; Mazzotti *et al.*, 2008; Carloni and Subramaniam, 2012b) complicate the interpretation of the interfacial fracture energy G_F as a true fracture parameter. Han (2009) reported that G_F increases with decreasing ratio b_f / b. Similarly, Czaderski *et al.* (2010) reported that G_F increases with decreasing width of the FRP. The interfacial cohesive law was obtained by measuring the strain along the FRP strip close to the edge. The authors also questioned whether G_F is a material property by making a parallelism with the Mode-I and Mode-II fracture energies of concrete (Carpinteri *et al.*, 1993; Gunes, 2004). Chen and Teng (2001) reviewed some of the fracture mechanics approaches, which provided a value of G_F that depended on the width of the FRP and concrete. Conversely, Dai *et al.* (2006) affirmed that G_F is a material property and has a certain value for a particular combination of FRP and concrete types. Mazzotti *et al.* (2008) obtained the fracture energy from the load-carrying capacity of the interface of near-end supported single-shear tests. They observed that the interfacial fracture energy did not scale with the width of the FRP sheet due to the small size of the aggregate. Ching and Buyukozturk (2006) conducted direct-shear tests to study the influence of the effect of moisture on G_F. It was concluded that a different debonding mechanism occurred if the specimens were conditioned at relative humidity (RH) 100%, with a progressive decrease of the fracture toughness with increasing exposure. Subramaniam *et al.* (2007, 2011) indicated that the interfacial fracture parameters do not scale with the width if determined from measurements in the central region of the FRP sheet.

The discussion on the nature of the interfacial fracture energy as a true fracture parameter in part arises from the misuse of Equation [3.11]. From Equation [3.11], researchers used the experimental load-carrying capacity P_{crit} to determine the fracture energy by assuming that $P_u = P_{\text{crit}}$:

$$G_F = \frac{P_{\text{crit}}^2}{2b_f^2 E_f t_f} \qquad [3.12]$$

This approach led to an erroneous evaluation of G_F, because the experimental value of the load-carrying capacity includes the width effect and in particular the mixed-mode behavior of the edges (Subramaniam *et al.*, 2007). The fracture energy cannot be directly related to the experimental load-carrying capacity of the interface. The values of G_F determined via Equation [3.12] scale with the width of the FRP sheet. Hence, the experimental values of the fracture energy determined through Equation [3.12] should not be considered as *true* values. The experimental values in Fig. 3.8 indicated that the fracture energy is nominally invariant with the width of the FRP sheet. Therefore, G_F is a true fracture parameter, provided that it is obtained from strain measurements in the central region of the FRP sheet where a Mode-II fracture occurs.

3.3.4 Mode-I interfacial loading and alternative fracture approaches

In addition to the discussion regarding the nature of the fracture energy, concerns remain on the applicability of the direct-shear test results to describe

ICD, mainly because of the presence of a Mode-I component (peeling stresses). The Mode-I should not be confused with the one observed before at the microscopic level, but in the spirit of the macroscopic approach of the fictitious interface. The Mode-I opening is described by the relationship between the normal stress (peeling) σ_{zz} and the opening of the crack w (Martinelli *et al.*, 2011; Carrara and Ferretti, 2013). The Mode-I interfacial fracture energy, corresponding to the area of the $\sigma_{zz} - w$ curve, is considerably lower than the Mode-II fracture energy (Taljsten, 1996; Gunes, 2004), thus even a small component of the load perpendicular to the FRP sheet could potentially reduce the load-carrying capacity of the interface. A Mode-I component is always present in the direct-shear test measurements, due to the relationship between shear and moment. A limited number of experimental works reported the study of Mode-I and mixed-mode debonding (Wan *et al.*, 2004; Davalos *et al.*, 2006; Alam *et al.*, 2012). Some authors (Yao *et al.*, 2005) recognized that the effect of a small loading angle (offset) was insignificant for relatively long bonded lengths. The presence of a Mode-I condition in beams can be explained by considering the opening of a flexural/shear crack as illustrated in Fig. 3.10 (Garden and Hollaway, 1998). As the crack opens, the two faces of the crack will undergo a relative vertical displacement that will cause a mixed-mode condition for the FRP–concrete interface. Rabinovitch (2008, 2012) used a fracture mechanics approach that considered the Mode-I and Mode-II cohesive material laws and their coupling. A set of nonlinear differential equations was derived by considering a multi-layer description of the strengthened beam. A different length of the

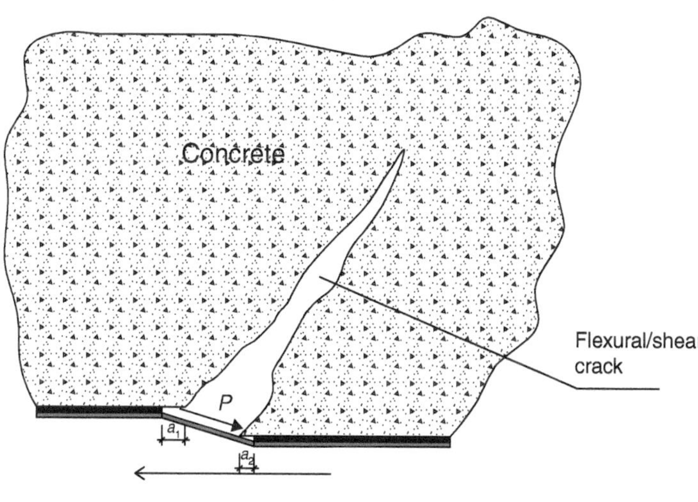

3.10 Mixed-mode debonding propagation in beams.

STZ for Mode-I and Mode-II can be observed in these studies. Mazzucco *et al.* (2012) used a similar approach to capture the coupling of the shear and peeling stresses, but introduced a contact-damage model for the adhesion between layers. Gunes *et al.* (2009) reported that if the strengthened beam was sufficiently strong in shear, the flexural/shear crack mouth displacement would be limited and consequently the mixed-mode nature of debonding fracture would quickly merge into a Mode-II condition. It is interesting to notice that the results published by Alam *et al.* (2012) showed that the effective bond length increases if the Mode-I component is significant.

Alternative approaches within the framework of fracture mechanics are available in the literature. Achintha and Burgoyne (2008, 2011), for example, studied the debonding phenomenon as a Mode-I problem, by considering that the debonding often occurs in the concrete just above the interface. The Mode-I fracture energy of concrete was used in their approach. The authors observed that none of the existing studies available in the literature provided a reliable estimate of the interfacial fracture energy G_F. Gunes *et al.* (2009) proposed a global energy balance model to predict FRP debonding failure. The amount of energy dissipated in the system during debonding was determined by calculating the change in the potential energy of the system. The component of the energy dissipation due to the debonding process was calculated by means of the interfacial fracture energy. An expression of G_F was provided that included the geometric factor k_b to take into account the influence of the width of the FRP sheet.

3.4 Numerical analysis of the fiber-reinforced polymer (FRP)–concrete interface

A one-dimensional model, developed by Ali-Ahmad *et al.* (2007), was used in Carloni and Subramaniam (2012b) to carry out a numerical analysis of the FRP–concrete interfacial crack initiation and propagation. The main goal of this simplistic numerical model was to confirm that the interfacial stress transfer is not a one-dimensional problem. Moreover, the results of the numerical analysis showed that even though the load-carrying capacity is only related to the fracture energy G_F, as expressed in Equation [3.11], the shape of the τ_{zy} $-s$ curve affects the interfacial response and in particular could entail for a load corresponding to crack initiation that is significantly lower than the load-carrying capacity. The FRP sheet was described by using one-dimensional axial elements. The bond between FRP and substrate was modeled using spring elements with a simplified bilinear response, which took into account the interfacial cohesive material law (Subramaniam *et al.*, 2007, 2011). A linear ascending shear stress vs relative slip response was assumed up to the peak shear stress, τ_{max}, and relative slip

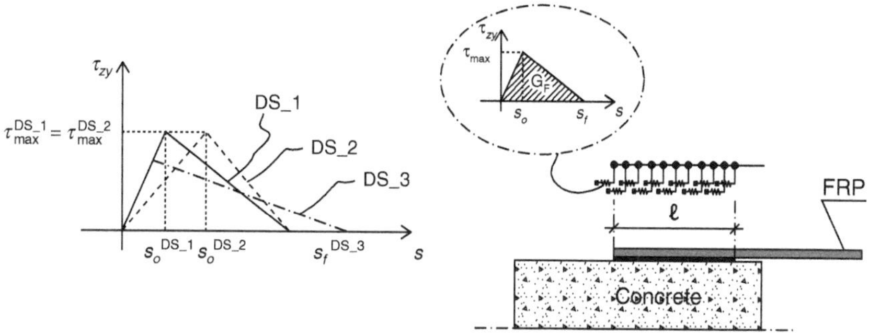

3.11 Finite element model.

equal to s_o; following this, a linear softening was considered up to a value of the slip s_f, beyond which complete separation of the FRP from concrete occurred. The value of s_o was determined by considering the peak shear stress and the initial slope of the cohesive law (Carloni and Subramaniam, 2012a). The slip s_f was obtained by enforcing the value of G_F. The value of G_F was assumed to be equal to the mean fracture energy of test W_6 in Subramaniam *et al.* (2007). The numerical model is shown in Fig. 3.11. The approximation of the τ_{zy}–s curve with a bilinear response is discussed in Lu *et al.* (2005).

In the numerical representation, the concrete substrate was assumed to be rigid, based on the observed experimental results (Carloni and Subramaniam, 2010). The material response of the carbon FRP (CFRP) was considered to be linear elastic up to rupture. The engineering properties of the composite were based on the thickness of the fibers contained in the composite sheet and not on the gross composite area. The properties of the axial elements were determined based on the nominal width of the CFRP composite sheet. Details about the numerical model can be found in Ali-Ahmad *et al.* (2007).

Three different numerical simulations were performed to study the influence of the fracture parameters on the load response of the FRP–concrete interface (Fig. 3.11). The fracture parameters used in the simulations are summarized in Table 3.2. The parameters $(\tau_{max}, G_F, \tau_{max}/s_o)$ used in DS_1 corresponded to the mean values computed from the τ_{zy} –s curves associated with the aforementioned ten points of the load response of test W_6 (Subramaniam *et al.*, 2007). In DS_2, the slope of the ascending response was decreased with respect to DS_1 without varying τ_{max} and G_F. Finally, in DS_3 τ_{max} was reduced while s_o and s_f were computed in order to maintain the slope of the ascending response and the fracture energy equal to DS_1.

In Fig. 3.12, the load responses of the numerical simulations are compared with the load response of Fig. 3.2. Since the model is one-dimensional, it is

Table 3.2 Fracture parameters of the numerical simulations

Simulation	τ_{max} (MPa)	s_o (mm)	s_f (mm)	G_F (MPa mm)
DS_1	6.65	0.026	0.164	0.546
DS_2	6.65	0.060	0.164	0.546
DS_3	4.00	0.016	0.273	0.546

implicitly assumed that the cohesive material law determined from the center of the FRP sheet applies to the entire width of the FRP. Therefore, the load-carrying capacity P_u^{num} of DS_1, DS_2, and DS_3 is the same but should be lower than the experimental value (P_{crit}), because the model neglects the width effect. The value of P_u^{num} computed from Equation [3.11] utilizing the mean fracture energy of test W_6 (Subramaniam *et al.*, 2007), provided in Tables 3.1 and 3.2, is:

$$P_u^{num} = b_f \sqrt{2G_F E_f t_f} = 5.20 \text{ kN} \qquad [3.13]$$

P_u^{num} matches the numerical results but is lower than the experimental load-carrying capacity as observed above. The maximum strain, corresponding to debonding, can be obtained from Equation [3.13]:

$$\bar{\varepsilon}_{yy}^{num} = \frac{P_u^{num}}{b_f t_f E_f} = 5330 \ \mu\varepsilon \qquad [3.14]$$

The strain value calculated in Equation [3.14] is in good agreement with the mean value of the maximum strain $\bar{\varepsilon}_{yy}$ of test W_6, which is equal to 5560 με (Fig. 3.4 and Table 3.1). However, it should be observed that the experimental value of the strain is slightly higher than that obtained in Equation [3.14], because the applied load is affected by the presence of the edge regions and by the restraining effect of the adjacent concrete. Furthermore, the value of G_F used in Equation [3.13] is calculated as the mean value from ten images corresponding to ten different points of the load response.

A comparison between the numerical simulations shows that even though the load-carrying capacity is only related to the area under the curve of the cohesive material law, the fracture parameters and therefore the shape of the curve influence the linear and nonlinear part of the load response. If the slope of the ascending branch decreases (DS_2), then the slope of the linear portion of the load response decreases. A decrease of the slope corresponds to a decrease of the stiffness of the interface and determines an upward shift of the end of the linear portion of the load response, which in turn delays

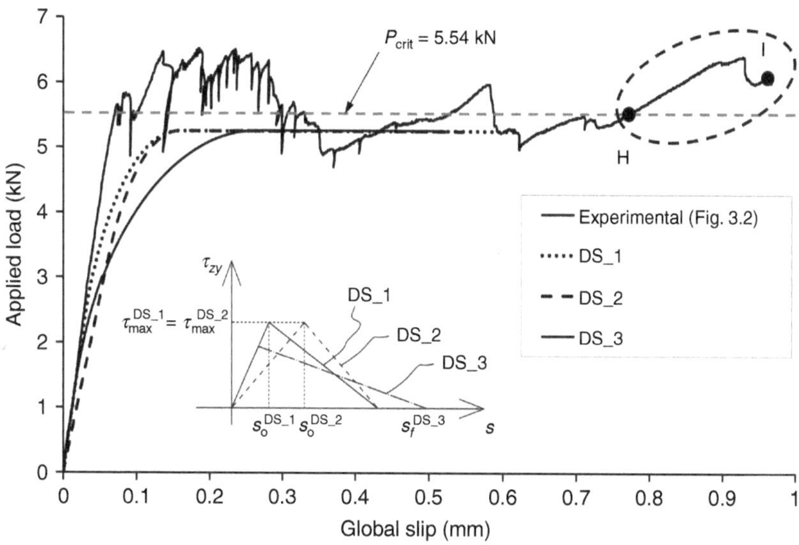

3.12 Comparison between the experimental load response of Fig. 3.2 and the numerical analysis.

the formation of microcracking in concrete. The strain value corresponding to the formation of the STZ in concrete for DS_2 is higher than DS_1. A decrease of the maximum shear stress without varying the fracture energy entails for an increase of the L_{STZ}, which produces a downward shift of the end of the linear portion of the load response. The strain value corresponding to the formation of the STZ in concrete for DS_3 is lower than DS_1.

3.5 Design aspects related to debonding

In Section 10.1.1 of the ACI 440.2R-08 (2008) document, the flexural failure modes of an RC strengthened beam are described. In order to prevent ICD, in the same section the effective strain ε_{fd} in the FRP reinforcement is limited to the strain level at which debonding may occur. The formula used in the ACI 440.2R-08 document is a modification of the original formula proposed by Chen and Teng (2001):

$$\bar{\varepsilon}_{yy}^{\text{Chen and Teng}} = 0.427\beta_L\beta_p\sqrt{\frac{\sqrt{f_c'}}{E_f t_f}} \quad [3.15]$$

where $\beta_L = 1$ if the bonded length of FRP is greater than the effective bond length. β_p takes into account the effect of the width ratio b_f / b. f_c' is the specified compressive strength of concrete.

A similar approach is proposed in the Italian guidelines (CRN-DT 200/2004), in which a limit for the strain in the FRP is provided in order to prevent debonding:

$$\varepsilon_{\text{fdd}} = \frac{k_{\text{cr}}}{\gamma_{f,d}\sqrt{\gamma_c}}\sqrt{\frac{2\Gamma_{\text{FK}}}{E_f t_f}} \tag{3.16}$$

where k_{cr} can be taken equal to 3.0, Γ_{FK} is the fracture energy G_F, γ_c is the *partial safety factor* for concrete that can be obtained from the Italian design code for RC structures, and $\gamma_{f,d}$ is the partial safety factor for FRP materials (CRN-DT 200/2004). A formula to compute the fracture energy is provided in the Italian document (CRN-DT 200/2004) and was used in Fig. 3.8:

$$\Gamma_{\text{FK}} = 0.03k_b\sqrt{f_{\text{ck}}f_{\text{ctm}}} \tag{3.17}$$

In Equation [3.17] the fracture energy is related to the mechanical properties of concrete but it is also recognized to be dependent on the ratio $b_f\,/\,b$ through the coefficient k_b. f_{ck} is the characteristic compressive strength of concrete and f_{ctm} is the mean value of the tensile strength of concrete. In this chapter, since the tensile strength was not measured in the experimental work herein presented, it was assumed (Gunes *et al.*, 2009):

$$f_{\text{ctm}} = 0.53\sqrt{f_c'} \tag{3.18}$$

The strain at debonding for test W_6 in Subramaniam *et al.* (2007) determined from Equations [3.15] and [3.16] is equal to 6600 and 6700 με, respectively. The term $k_{cr}\,/\left(\gamma_{f,d}\sqrt{\gamma_c}\right)$ was not applied in Equation [3.16] because it contains safety factors. The width factors β_p and k_b were equal to 1.22 and 1.30, respectively. The experimental mean value of $\bar{\varepsilon}_{yy}$, equal to 5560 $\mu\varepsilon$, is lower than the values of the strain obtained from Equations [3.15] and [3.16]. This can be justified in part by taking into account that most of the formulas provided in the past were calibrated by using the approach of Equation [3.12].

The strain at debonding could be calculated from Equation [3.11]:

$$\bar{\bar{\varepsilon}}_{yy} = \frac{P_u}{b_f t_f E_f} = \sqrt{\frac{2G_F}{t_f E_f}} = \bar{\varepsilon}_{yy}^{\text{num}} = 5330\ \mu\varepsilon \tag{3.19}$$

The value of the strain obtained from Equation [3.11] is indicated with a double-bar sign. The fracture energy used in Equation [3.19] is the mean value from test W_6 (Subramaniam *et al.*, 2007). The value obtained in Equation [3.19] equals the one in Equation [3.14] because the load-carrying capacity depends on the fracture energy and not on the shape of the cohesive material law.

It is important to discuss the applicability of a strain limit. Wu and Niu (2007) noticed that the strain compatibility-based approach cannot be applied at those cross-sections where debonding occurs. In fact, the cross-sectional analysis entails a perfect bond between FRP and concrete. Gunes *et al.* (2009) applied the global energy approach and observed that the ACI strain limit does not take into account the interaction between the shear and flexural external reinforcements. The numerical analysis previously presented indicated that crack initiation could occur at a load level in the FRP much lower than the one corresponding to the strain in Equations [3.15] or [3.16]. Hence, the applicability of the cross-sectional analysis is questionable even before the debonding failure is reached. In order to partially overcome this issue, a maximum shear stress limit could be added to the strain limits of Equations [3.15] and [3.16]. Several expressions are available in the literature to evaluate the maximum shear stress τ_{max} in the cohesive material law (Han, 2009; Ueno *et al.*, 2013). The inclusion in guidelines and codes of formulas to calculate and limit the maximum shear stress at the interface could help the designer evaluate at what load level microcracking of concrete occurs and eventually decide whether an additional reduction factor should be applied to the strain limit in order to prevent brittle failure of the interface.

The issue related to the strain limit can be considered from another point of view. In this chapter, only the debonding mechanism initiating from a single crack was considered. The presence of a multi-crack pattern corresponds to the possibility of crack-induced debonding at different cross-sections (Pan *et al.*, 2010). If one considers that the FRP sheet between two flexural cracks is subject to two tensile forces in opposite directions, two STZs form between two cracks that could potentially overlap if the sum of the lengths of the STZs is smaller than the crack spacing.

3.6 Future trends

There are still unresolved issues that need to be addressed in the near future to improve international guidelines and codes. In this last section, some of those issues will be briefly presented. Finally, a newly-developed type of fiber-reinforced composites will be introduced.

3.6.1 Unresolved issues

Durability studies of the FRP strengthening technique are still limited, and additional data are necessary. Among the others, the effect of environmental conditions on the interfacial fracture properties was studied in Green *et al.* (2000), Ching and Buyukozturk (2006), and Subramaniam *et al.* (2008). It was recognized that freeze–thaw cycles determined a reduction of the load-carrying capacity. Conversely, a recent study by Dash *et al.* (2013) showed that the combined effect of moisture, temperature, and sustained loading could be beneficial for the ultimate load-carrying capacity of the interface.

The fatigue behavior of the FRP interface has still not been fully investigated (Kim and Heffernan, 2008; Carloni *et al.*, 2012; Carloni and Subramaniam, 2013). The limited studies available in the literature are contradictory; however, it appears that fatigue loading even at low ranges could be detrimental for the stress transfer between FRP and concrete (Aidoo *et al.*, 2004). The influence of the frequency (Diab *et al.*, 2009) and the explicit dependence of the fatigue life on the load range and its mean value are not fully understood. Carloni and Subramaniam (2013) observed that fatigue-induced debonding could occur at the epoxy–fiber interface rather than in a thin layer of concrete. The shift of the debonding interface entails reduction of the effective bond length.

The relationship between the results of direct-shear tests and the actual performance of FRP strips bonded to RC beams is still an open issue. The effect of the curvature of the beam and the relative vertical displacement at flexural/shear cracks on the interface behavior is not easily captured by direct-shear tests. Sebastian (2001) observed that a strain distribution is present in the FRP strip prior to initiation of debonding, which is not taken into account in direct-shear tests. He also observed that measuring the strain on the FRP surface could be misleading in beams because of the presence of curvature. Rosenboom and Rizkalla (2008) pointed out that the strain limit provided in codes might not be sufficient to prevent debonding and noticed that the width effect in beams and direct-shear tests could be different. The relationship between the debonding mechanism in direct-shear tests and beams is also complicated by the fact that several direct-shear test set-ups are available and a standardized test is not available.

3.6.2 Fiber-reinforced cementitious matrix (FRCM) composites

Newly-developed composites that represent a sustainable alternative to FRPs are the so-called FRCM composites. In FRCM composites, fibers are typically bundled, and modified cement-based mortar is used instead of epoxy as matrix. The limited available literature (D'Ambrisi and Focacci, 2009; Ombres, 2012) suggests that FRCM composites can be used effectively for strengthening of RC structures. However, in FRCM composites debonding is typically observed at the matrix–fiber interface rather than in the substrate or at the matrix–concrete interface, which is typically observed in FRP composites. A complete understanding of the interfacial stress-transfer mechanism in FRCM composites is not available yet, although researchers have applied the concept of cohesive crack to describe the matrix–fiber interfacial behavior (D'Ambrisi *et al.*, 2012; D'Antino *et al.*, 2013). The studies available in the literature indicate that the effective bond length of FRCMs is much longer than that typically observed for the FRP–concrete interface. Moreover, friction between fibers, and between fibers and matrix, during the debonding process complicate the phenomenon. If these observations are confirmed, the debonding would not be governed by the properties of the concrete but rather by the combination of fibers and mortar used in the application.

3.7 Acknowledgments

The author would like to thank Prof. Kolluru V. Subramaniam (Indian Institute of Technology, Hyderabad) for the fruitful collaboration over the last 10 years and for giving him the opportunity to perform the experimental work at The City College of New York where Prof. Subramaniam worked from 2000 until 2010. The writer gratefully acknowledges Prof. Angelo Di Tommaso (University of Bologna), for giving him the opportunity of working on this subject during his Master's thesis and for being a constant guide over the years. Finally, the author would like to thank Prof. Lesley H. Sneed (Missouri S&T) for the ongoing extensive research collaboration on FRCM composites.

3.8 References

Abdel Baky, H., Ebead, U.A. and Neale, K.W. (2012), 'Nonlinear micromechanics-based bond-slip model for FRP/concrete interfaces', *Engineering Structures*, **39**, 11–23.

Achintha, P.M.M. and Burgoyne, C.J. (2008), 'Fracture mechanics of plate debonding', *Journal of Composites for Construction*, **12**(4), 396–404.

Achintha, P.M.M. and Burgoyne, C.J. (2011), 'Fracture mechanics of plate debonding: validation against experiment', *Construction and Building Materials*, **25**(6), 2961–2971.

ACI 440.2R-08 (2008), *Guide for the Design and Construction of Externally Bonded FRP Systems for Strengthening Concrete Structures*, American Concrete Institute.

Aidoo, J., Harries, K.A. and Petrou, M. F. (2004), 'Fatigue behavior of carbon fiber reinforced polymer strengthened reinforced concrete bridge girders', *Journal of Composites for Construction*, **8**(6), 501–509.

Alam, M.S., Kanakubo, T. and Yasojima, A. (2012), 'Shear-peeling bond strength between continuous fiber sheet and concrete', *ACI Structural Journal*, **109**(1), 75–82.

Ali-Ahmad, M., Subramaniam, K.V. and Ghosn, M. (2006), 'Experimental investigation and fracture analysis of debonding between concrete and FRP', *Journal of Engineering Mechanics*, **132**(9), 914–923.

Ali-Ahmad, M., Subramaniam, K.V. and Ghosn, M. (2007), 'Analysis of scaling and instability in FRP-concrete shear debonding for beam-strengthening applications', *Journal of Engineering Mechanics*, **133**(1), 58–67.

Anderson, T.L. (2004), *Fracture Mechanics: Fundamentals and Applications*, Boca Raton, Florida, CRC Press.

Arduini, M., Di Tommaso, A. and Nanni, A. (1997), 'Brittle failure in FRP plate and sheet bonded beams', *ACI Structural Journal*, **94**(5), 363–370.

ASTM C39 / C39M (2011), *Standard Test Method for Compressive Strength of Cylindrical Concrete Specimens*, West Conshohocken, PA, ASTM International.

Bank, L.C. (2006) *Composite for Construction: Structural Design with FRP Materials*, Hoboken, New Jersey, J. Wiley and Sons.

Bazant, Z.P. and Planas, J. (1997), *Fracture and Size Effect in Concrete and Other Quasibrittle Materials*, Boca Raton, Florida, CRC Press.

Bizindavyi, L. and Neal, K.W. (1999), 'Transfer lengths and bond strengths for composites bonded to concrete', *Journal of Composites for Construction*, **3**(4), 153–159.

Buyukozturk, O. and Hearing, B. (1998), 'Failure behavior of precracked concrete beams retrofitted with FRP', *Journal of Composites for Construction*, **2**(3), 138–144.

Buyukozturk, O., Gunes, O. and Karaca, E. (2004), 'Progress on understanding debonding problems in reinforced concrete and steel members strengthened using FRP composites', *Construction and Building Materials*, **18**, 9–19.

Carloni, C. and Subramaniam, K.V. (2010), 'Direct determination of cohesive stress transfer during debonding of FRP from concrete', *Composite Structures*, **93**(1), 184–192.

Carloni C. and Subramaniam K.V. (2012a), 'FRP/masonry debonding: numerical and experimental study of mortar joints', *Journal of Composites for Construction*, **16**(5), 581–589.

Carloni C. and Subramaniam K.V. (2012b), 'Application of fracture mechanics to debonding of FRP from RC members', *ACI SP 286–10*.

Carloni, C., Subramaniam, K.V., Savoia, M. and Mazzotti, C. (2012), 'Experimental determination of FRP-concrete cohesive interface properties under fatigue loading', *Composite Structures*, **94**, 1288–1296.

Carloni, C. and Subramaniam, K.V. (2013), 'Sub-critical fatigue crack growth in FRP/concrete cohesive interface', *Composites Part B*, **51**, 35–43.

Carpinteri, A., Valente, S., Ferrara, G. and Melchiorri, G. (1993), 'Is mode-II fracture energy a real material property', *Computers and Structures*, **48**(3), 397–413.

Carrara, P., Ferreti, D., Freddi, F. and Rosati, G. (2011), 'Shear tests of carbon fiber plates bonded to concrete with control of snap-back', *Engineering Fracture Mechanics*, **79**, 2663–2678.

Carrara, P. and Ferretti, D. (2013), 'A finite-difference model with mixed interface laws for shear tests of FRP plates bonded to concrete', *Composites: Part B*, **54**, 329–342.

Chajes, M.J., Finch, W.W., Januszka, T.F and Thomson, T.A. (1996), 'Bond and force transfer of composite material plates bonded to concrete', *ACI Structural Journal*, **93**(2), 208–217.

Chen, J.F. and Teng, J.G. (2001), 'Anchorage strength models for FRP and steel plates bonded to concrete', *Journal of Structural Engineering*, **127**(7), 784–791.

Chen, J.F. and Teng J.G. (2003), 'Shear capacity of FRP strengthened RC beams: FRP debonding' *Construction and Building Materials*, **17**(1), 27–41.

Ching, A. and Buyukozturk, O. (2006), 'Peel and shear fracture characterization of debonding in FRP plated concrete affected by moisture', *Journal of Composites for Construction*, **10**(1), 35–47.

CRN-DT 200/2004 (2004), *Guide for the Design and Construction of Externally Bonded FRP Systems for Strengthening Existing Structures*, Rome, Italy, *National Research Council* (CRN), Advisory Committee on Technical Recommendations for Construction.

Czaderski, C., Soudki, K. and Motavalli, M. (2010), 'Front and side view image correlation measurements on FRP to concrete pull-off bond tests', *Journal of Composites for Construction*, **14**(4), 451–463.

D'Ambrisi, A. and Focacci, F. (2009), 'Flexural strengthening of RC beams with cement based composites', *Journal of Composites for Construction*, **15**(2), 707–720.

D'Ambrisi, A., Feo, L. and Focacci, F. (2012), 'Bond-slip relations for PBO-FRCM materials externally bonded to concrete', *Composites: Part B*, **43**(8), 2938–2949.

D'Antino, T., Sneed, L.H., Carloni, C. and Pellegrino, C. (2013), 'Bond behavior of the FRCM-concrete interface', In *Proceedings of 11th International Symposium on Fiber Reinforced Polymers For Reinforced Concrete Structures*, Guimarães, Portugal.

Dai, J.G., Ueda, T. and Sato, Y. (2005), 'Development of the nonlinear bond stress-slip model of fiber reinforced plastics sheet-concrete interfaces with a simple method', *Journal of Composites for Construction*, **9**(1), 52–62.

Dai, J.G., Ueda, T. and Sato, Y. (2006), 'Unified analytical approaches for determining shear bond characteristics of FRP-concrete interfaces through pullout tests', *Journal of Advanced Concrete Technology*, **4**, 133–145.

Dash, S., Jeong, Y., Lopez, M.M. and Bakis, C.E. (2013), 'Experimental characterization of moisture, temperature, and sustained loading on concrete-FRP bond performance', In *Proceedings of 11th International Symposium on Fiber Reinforced Polymers For Reinforced Concrete Structures*, Guimarães, Portugal.

Davalos, J.F., Kodkani, S.S. and Ray, I. (2006), 'Fracture mechanics method for Mode-I interface evaluation of FRP bonded to concrete substrates', *Journal of Materials in Civil Engineering*, **18**(5), 732–742.

Diab, H.M., Wu, Z. and Iwashita, K. (2009), 'Theoretical solution for fatigue debonding growth and fatigue life prediction of FRP-concrete interfaces', *Advances in Structural Engineering*, **12**(6), 781–792.

Elices, M., Guinea, G.V., Gómez, J. and Planas, J. (2002), 'The cohesive zone model: advantages, limitations and challenges', *Engineering Fracture Mechanics*, **64**, 137–163.

Ferracuti, B., Savoia, M. and Mazzotti, C. (2007), 'Interface law for FRP-concrete delamination', *Composite Structures*, **80**(4), 523–531.

Focacci, F., Nanni, A. and Bakis C.E. (2000), 'Local bond-slip relationship for FRP reinforcement in concrete', *Journal of Composites for Construction*, **4**(1), 24–31.

Garden, H.N. and Hollaway, L.C. (1998), 'An experimental study of the influence of plate end anchorage of carbon fibre composite plates used to strengthen reinforced concrete beams', *Composite Structures*, **42**, 175–188.

Gartner A., Dougals E.P., Donal, C.W. and Hamilton H.R. (2011), 'Small beam bond tests method for CFRP composites applied to concrete', *Journal of Composites for Construction*, **15**(1), 52–60.

Grace, C., Yang, Y. and Sneed, L.H. (2012), 'Fracture mechanics approach to predicting the behavior of reinforced concrete members with externally-bonded fiber reinforced polymer laminates', *ACI SP 286–7.*

Green, M.F., Bisby, L.A., Beaudoin, Y. and Labossiere, P. (2000), 'Effect of freeze-thaw cycles on the bond durability between fibre reinforced polymer plate reinforcement and concrete', *Canadian Journal of Civil Engineering*, **27**, 949–959.

Gunes, O. (2004), *A Fracture Based Approach to Understanding Debonding in FRP Bonded Structural Members*, PhD Thesis, Massachusetts Institute of Technology, Cambridge, MA.

Gunes, O., Buyukozturk, O. and Karaca, E. (2009), 'A fracture-based model for FRP debonding in strengthened beams', *Engineering Fracture Mechanics*, **76**, 1897–1909.

Han, M. (2009), *Fracture mechanics based-model of FRP sheets-concrete interface bond*, PhD Thesis, The University of Alabama in Huntsville, Huntsville, AL.

Hearing, B.F. (2000), *Delamination of Reinforced Concrete Retrofitted with Fiber Reinforced Plastics*, PhD Thesis, Massachusetts Institute of Technology, Cambridge, MA.

Hillerborg, A., Modéer, M. and Petersson, P.E. (1976), 'Analysis of crack formation and crack growth in concrete by means of fracture mechanics and finite elements', *Cement and Concrete Research*, **6**, 773–782.

Hutchinson, J.W. and Suo, Z. (1992), 'Mixed-mode cracking in layered materials', *Advances in Applied Mechanics*, **29**, 63–191.

Kang, T.H.-K., Howell, J., Kim, S. and Lee, D.J. (2012), 'A state-of-the-art review on debonding failures of FRP laminates externally adhered to concrete', *International Journal of Concrete Structures and Materials*, **6**(2), 123–134.

Karbhari, V.M., Niu, H. and Sikorsky, C. (2006), 'Review and comparison of fracture mechanics-based bond strength models for FRP-strengthened structures', *Journal of Reinforced Plastics and Composites*, **25**(17), 1757–1738 .

Kim, Y.J. and Heffernan, P.J. (2008), 'Fatigue behavior of externally strengthened concrete beams with fiber-reinforced polymers: state of the art', *Journal of Composites for Construction*, **12**(3), 246–256.

Leung, C.K.Y. (2001), 'Delamination failure in concrete beams retrofitted with bonded plate', *Journal of Materials in Civil Engineering,* **13**(2), 106–113.

Liu, K. and Wu, Y.F. (2012), 'Analytical identification of bond-slip relationship of EB-FRP joints', *Composite; Part B*, **43**, 1955–1963.

Lu, X.Z., Teng, J.G., Ye, L.P. and Jiang, J.J. (2005), 'Bond-slip models for FRP sheets/plates bonded to concrete', *Engineering Structures*, **27**, 920–937.

Martinelli, E., Czaderski, C. and Motavalli, M. (2011), 'Modeling in-plane and-of-plane displacement fields in pull-off tests FRP strips', *Engineering Structures*, **33**, 3715–3725.

Mazzotti, C., Savoia, M. and Ferracuti, B. (2008), 'An experimental study on delamination of FRP plates bonded to concrete', *Construction and Building Materials*, **22**, 1409–1421.

Mazzucco, G., Salomoni, V.A. and Majorana, C.E. (2012), 'Three-dimensional contact-damage coupled modeling of FRP reinforcements – simulation of the delamination and long term process', *Computer and Structures*, **110–111**, 15–31.

Mohamed Ali, M.S., Oehelers, D.J. and Griffith M. (2008), 'Shear transfer across cracks in FRP strengthened RC beams', *Journal of Composites for Construction*, **12**(4), 416–424.

Monti, G., Renzelli, M. and Luciani, P. (2003), 'Adhesion in uncracked and cracked concrete zones', in *Proceedings of the Sixth International Symposium on FRP Reinforcement for Concrete Structures*, **1**, 183–192 .

Neubauer, U. and Rostasy, F.S. (1999), 'Bond failure of concrete fiber reinforced polymer plates at inclined cracks – experiments and fracture mechanics model', in *Proceedings 4th International Symposium FRP Reinforcement for RC Structures*, ECS, Edinburgh, Scotland, 109–118 .

Ombres L. (2012), 'Debonding analysis of reinforced concrete beams strengthened with fibre reinforced cementitious mortar', *Engineering Fracture Mechanics*, **81**, 94–109.

Pan, J., Leung, C.K.Y. and Luo, M. (2010), 'Effect of multiple secondary cracks on FRP debonding from the substrate of reinforced concrete beams', *Construction and Building Materials*, **24**, 2507–2516.

Pellegrino, C., Tinazzi, D. and Modena, C. (2008), 'Experimental study on bond behavior between concrete and FRP reinforcement', *Journal of Composites for Construction*, **12**(2), 180–189.

Pellegrino, C. and Modena, C. (2009), 'Influence of axial rigidity on FRP-concrete bond behavior: an analytical study', *Advances in Structural Engineering*, **12**(5), 639–649.

Rabinovitch, O. (2004), 'Fracture mechanics failure criteria for RC beams strengthened with FRP strips – a simplified approach', *Composite Structures*, **64**, 479–492.

Rabinovitch, O. (2008), 'Debonding analysis of fiber-reinforced-polymer strengthened beams: cohesive zone modeling versus a linear elastic fracture mechanics approach', *Engineering Fracture Mechanics*, **75**, 2842–2859.

Rabinovitch, O. (2012), 'Dynamic debonding on concrete beams strengthened with composite materials', *International Journal of Solids and Structures*, **49**, 3641–3658.

Ritchie, P., Thomas, D., Lu, L. and Connelly, M. (1991), 'External reinforcement of concrete beams using fiber reinforced plastics', *ACI Structural Journal*, **88**(4), 490–500.

Rosenboom O. and Rizkalla S. (2008), 'Modeling of IC debonding of FRP-strengthened concrete flexural members', *Journal of Composites for Construction*, **12**(2), 168–179.

Sebastian, W. (2001), 'Significance of midspan debonding failure in FRP-plated concrete beams', *Journal of Structural Engineering*, **127**(1), 792–798.

Sharif, A., Al-Sulaimani, G.J., Basunbul, I.A., Baluch, M.H. and Ghaleb, B.N. (1994), 'Strengthening of initially loaded reinforced concrete beams using FRP plates', *ACI Structural Journal*, **91**(2), 160–168.

Subramaniam, K.V., Carloni, C. and Nobile, L. (2007), 'Width effect in the interface fracture during shear debonding of FRP sheets from concrete', *Engineering Fracture Mechanics*, **74**(4), 578–594.

Subramaniam, K.V., Ali-Ahmad, M. and Ghosn, M. (2008), 'Freeze–thaw degradation of FRP–concrete interface: impact on cohesive fracture response', *Engineering Fracture Mechanics*, **75**, 3924–3940.

Subramaniam, K.V., Carloni, C. and Nobile, L. (2011), 'An understanding of the width effect in FRP–concrete debonding', *Strain*, **47**, 127–137.

Sutton, M.A., Wolters, W.J., Peters, W.H., Ranson, W.F. and McNeill, S.R. (1983), 'Determination of displacements using an improved digital correlation method', *Image and Vision Computing*, **1**(3), 133–139.

Sutton M.A., Orteu J.J. and Shreier H.W. (2009), *Image Correlation for Shape, Motion and Deformation Measurements*, New York, Springer .

Taljsten, B. (1996), 'Strengthening of concrete prisms using the plate-bonding technique', *International Journal of Fracture*, **82**, 253–266.

Taljsten, B. (1997), 'Defining anchor lengths of steel and CFRP plates bonded to concrete', *International Journal of Adhesion and Adhesives*, **7**(4), 319–327.

Teng, J.G., Smith, S.T., Yao, J. and Chen, J.F. (2003), 'Intermediate crack-induced debonding in RC beams and slabs', *Construction and Building Materials*, **17**(6–7), 447–462.

Ueno, S., Toutanji, H. and Vuddandam, R. (2013), 'Prediction of FRP/concrete interface debonding using elasticity theory', In *Proceedings of 11th International Symposium on Fiber Reinforced Polymers For Reinforced Concrete Structures*, Guimarães, Portugal.

Wan, B., Sutton, M.A., Petrou, M.F., Harries, K.A. and Li, N. (2004), 'Investigation of bond between fiber reinforced polymer and concrete undergoing global mixed mode I/II loading', *Journal of Engineering Mechanics*, **130**(12), 1467–1475.

Wu, Z., Matsuzaki, T. and Tanabe, K. (1997), 'Interface crack propagation in FRP-strengthened concrete structures', *Non-Metallic (FRP) Reinforcement for Concrete Structures, Japan Concrete Institute*, **1**, 319–326.

Wu, Z. S., Yuan, H. and Niu, H. (2002), 'Stress transfer and fracture propagation in different kinds of adhesive joints', *Journal of Engineering Mechanics*, **128**(5), 562–573

Wu, Z. and Yin, J. (2003), 'Fracturing behaviors of FRP-strengthened concrete structures', *Engineering Fracture Mechanics*, **70**(10), 1339–1355.

Wu, Z. S. and Niu, H. (2007), 'Prediction of crack-induced debonding failure in R/C structures flexurally strengthened with externally bonded FRP composites', *Doboku Gakkai Ronbunshuu E*, **63**(4), 620–639.

Wu, Y.F., Xu, X.S., Sun, J.B. and Jiang, C. (2012), 'Analytical solution for the bond strength of externally bonded reinforcement', *Composite Structures*, **94**, 3232–3939.

Yao, J., Teng, J.G. and Chen, J.F. (2005), 'Experimental study on FRP-to-concrete bonded joints', *Composites: Part B*, **36**, 99–113.
Yuan, H., Wu, Z. and Yoshizawa, H. (2001), 'Theoretical solutions on interfacial stress transfer of externally bonded steel/composite laminates', *Structural Engineering/ Earthquake Engineering*, **18**(1), 27–39.
Zhou, Y.W., Wu, Y.F. and Yun, Y. (2010), 'Analytical modeling of the bond-slip relationship at FRP-concrete interfaces for adhesively-bonded joints', *Composites: Part B*, **41**, 423–433.

4

Identifying damage in honeycomb fiber-reinforced polymer (FRP) composite sandwich bridge decks

P. Z. QIAO and W. FAN, Washington State University, USA

DOI: 10.1533/9780857097019.1.94

Abstract: A strain energy-based damage identification method for plate-type structures is presented. The damage identification method employs a damage location factor matrix and a damage severity correction factor (DSCF) matrix for damage localization and quantification. An experimental modal test of an as-manufactured fiber-reinforced polymer sandwich deck panel is conducted in the laboratory to demonstrate the applicability and effectiveness of the proposed DSCF-based damage identification method. The composite deck panel at healthy and three damaged stages is tested using a surface-bonded polyvinylidene fluoride sensor array and impact hammer system. The present damage identification method can be used as a viable and effective technique for damage localization and quantification of plate-type structures.

Key words: damage identification, fiber-reinforced polymer, composites, bridge decks, plates, structural health monitoring, damage quantification, dynamics, sandwich structures.

4.1 Introduction

Most non-destructive damage identification methods can be categorized as either local or global damage identification techniques (Fan and Qiao, 2011). Local damage identification techniques, such as ultrasonic methods and X-ray methods, require that the vicinity of damage is known *a priori* and is readily accessible for testing, which cannot be guaranteed for most cases in civil or aerospace engineering. The vibration-based damage identification method as a global damage identification technique is developed to overcome these difficulties. The basic idea of vibration response-based damage detection is that the damage will induce changes in the physical properties (mass, damping, stiffness, etc.), which will in turn cause detectable changes in dynamic properties of the structure (natural frequencies, modal damping, mode shapes, etc.) (Doebling *et al.*, 1996). Therefore, it is intuitive that damage can be identified by analyzing the changes in vibration features of the structure.

Many damage identification methods have been proposed utilizing mode shapes (Ratcliffe, 1997; Hadjileontiadis *et al.*, 2005), frequencies (Cawley and Adams, 1979; Salawu, 1997), or modal parameters derived from mode shapes and frequencies, such as modal strain energy (Stubbs *et al.*, 1995; Stubbs and Kim, 1996), modal compliance (Choi *et al.*, 2005), modal flexibility, and uniform load surface (Zhang and Aktan, 1998; Wang and Qiao, 2007). Even though many methods have been developed, no single method is completely effective in all situations. A comparative study by Farrar and Jauregui (1996) based on experimental data shows that the strain energy-based damage index performed best among some popular existing damage identification methods.

The modal strain energy-based method is a widely used category of damage identification methods. This method uses the fractional modal strain energy change (MSEC) for damage detection. For beam-type or plate-type structures, the modal strain energy can be directly related to (and is usually derived from) the strain mode shape. Stubbs *et al.* (1995) and Stubbs and Kim (1996) developed a damage index (DI) method based on the modal strain energy. This method assumes that if the damage is primarily located at a single sub-region, then the fractional strain energy will remain relatively constant in sub-regions. For beam-type structures, the bending stiffness EI is assumed to be essentially constant over the length of the beam for both the undamaged and damaged modes. Shi and Law (1998) and Shi *et al.* (2000) presented a damage localization method for beam, truss, or frame type structures based on the MSEC. The MSEC at the element level is suggested as an indicator for damage localization. Law *et al.* (1998) applied this strain energy method to detect the damage location in a structure with incomplete and noisy measured modal data. The method consists of three stages: expansion of the measured mode shapes, localization of the damage domain using the elemental energy quotient difference, and damage quantification based on sensitivity of the modal frequency.

In this study, a modal strain energy-based damage severity correction factor (DSCF) method recently proposed for damage localization and quantification of plate-type structures (Fan, 2011; Fan and Qiao, 2012) is experimentally verified using a fiber-reinforced polymer (FRP) sandwich deck panel with a surface-boned piezoelectric sensor array. The method can be easily implemented in experimental testing and can be used to both locate and quantify damage. The proposed method has been thoroughly investigated based on a series of numerically simulated data. However, the damage identification method may fail sometimes due to the effects of noise, measurement errors, and sensor placement, even though this method has been verified with numerically simulated data. Therefore, the damage identification method should further be verified by experimental data on a real structure.

The objective of this chapter is to present an experimental verification of the DSCF-based damage identification method on an FRP sandwich deck panel with a polyvinylidene fluoride (PVDF) sensor array. The method is applied to both the experimental data from modal testing and numerical data from a detailed finite element model. The theoretical background of the proposed DSCF method is first introduced, followed by the experimental modal testing of an FRP sandwich panel using a PVDF sensor array. The DSCF-based damage identification method is then applied to the experimental data for damage localization and quantification. The numerical simulation of the FRP sandwich panel is then presented to verify the experimental test and the damage identification method. Finally, concluding remarks are provided.

4.2 The damage severity correction factor (DSCF) method for damage identification: theory

The equation of motion for free vibration of an undamped healthy structural dynamic system leads to the following eigenvalue problem:

$$(K - \lambda_i M)\phi_i = 0 \qquad [4.1]$$

where: K is the stiffness matrix; M is the mass matrix; λ_i is the eigenvalue of the ith mode, which is the square of the ith modal frequency; and ϕ_i is the vibration mode shape vector of the ith mode. Note that both K and M are symmetric matrices.

For damaged structure, assuming the mass loss is negligible, the eigenvalue problem of a damaged structure can be expressed as

$$\left[(K + \Delta K) - (\lambda_i + \Delta\lambda_i)M\right](\phi_i + \Delta\phi_i) = 0 \qquad [4.2]$$

Structural damage, such as cracks, delamination, debonding, and barely visible impact damage (BVID) in composite materials, usually causes a loss in stiffness but not a loss in mass. The change of structural stiffness matrix ΔK introduced by structural damage can be modeled by a proportional damage model, i.e., ΔK can be expressed as a function of the undamaged element stiffness by the equation

$$\Delta K = -\sum_{j=1}^{n} \alpha_j K_j \qquad [4.3]$$

where n is the total element number, K_j is the stiffness matrix of the jth element, and α_j is the damage severity of the jth element. The damage severity is defined as the fractional change in elemental stiffness ($\alpha_j = 0$ for a healthy element).

For a plate-type structure, the elemental modal potential energy can be expressed by the elemental modal strain energy. The potential energy of the jth element in the ith mode of the healthy plate can be expressed as

$$\phi_i^T K_j \phi_i = U_{ij} = \frac{1}{2} \int\int_{A_j} k_j \left[\left(\kappa_{xxi}^2 + \kappa_{yyi}^2 + 2\nu \kappa_{xxi} \kappa_{yyi} + 2(1-\nu) \kappa_{xyi}^2 \right) \right] \mathrm{d}x\mathrm{d}y = k_i \gamma_{ij} = k \gamma_{ij} \qquad [4.4]$$

where k_j = (EI)$_j$ is the bending stiffness of the jth element in the healthy plate, which is a constant k over the plate, and γ_{ij} is defined as

$$\gamma_{ij} = \int\int_{A_j} \left[\left(\kappa_{xxi}^2 + \kappa_{yyi}^2 + 2\nu \kappa_{xxi} \kappa_{yyi} + 2(1-\nu) \kappa_{xyi}^2 \right) \right] \mathrm{d}x\mathrm{d}y \qquad [4.5]$$

where the curvature mode shapes κ_{xx}, κ_{yy}, and κ_{xy} can be derived from the displacement mode shape data using the finite difference approximation, or they can be directly measured as the strain mode shape using piezoelectric sensors such as lead zirconate titanate (PZT) and PVDF. Deriving the curvature mode shape from the displacement mode shape data using the finite difference approximation may significantly magnify the effects of noise and measurement error. Therefore, using the directly measured strain mode shape data is advantageous (Fan and Qiao, 2011).

Combining Equations [4.3] and [4.4], we have

$$\phi_i^T \Delta K \phi_i = -\sum_{j=1}^{n} \alpha_j \phi_i^T K_j \phi_i = -k \sum_{j=1}^{n} \alpha_j \gamma_{ij} \qquad [4.6]$$

Similarly, the total potential energy of all elements in the ith mode of healthy plate is

$$\phi_i^T K \phi_i = \sum_{j=1}^{n} U_{ij} = k \sum_{j=1}^{n} \gamma_{ij} = k \gamma_i \qquad [4.7]$$

where γ_i is defined as

$$\gamma_{ij} = \sum_{j=1}^{n} \gamma_{ij} = \iint_A \left[\left(\kappa_{xxi}^2 + \kappa_{yyi}^2 + 2\nu\kappa_{xxi}\kappa_{yyi} + 2(1-\nu)\kappa_{xyi}^2\right)\right] \mathrm{d}x\mathrm{d}y \qquad [4.8]$$

The potential energy of the jth element in the ith mode of damaged plate can be expressed as

$$(\phi_i^T + \Delta\phi_i^T)(K_j + \Delta K_j)(\phi_i + \Delta\phi_i) = k_j^*\gamma_{ij}^* = (1-\alpha)k\gamma_{ij}^* \qquad [4.9]$$

where the superscript * denotes the damaged state.

Neglecting the high-order terms of $\Delta\varphi_i$ in Equation [4.9] and using Equations [4.3], [4.4], and [4.6], we have

$$\begin{aligned}(1-\alpha_j)k\gamma_{ij}^* &= \phi_i^T K_j\phi_i + 2\phi_i^T K_j \Delta\phi_i + \phi_i^T \Delta K_j \phi_i + 2\phi_i^T \Delta K_j \Delta\phi_i \\ &= (1-\alpha_j)k\gamma_{ij} + 2(1-\alpha_j)\phi_i^T K_j \Delta\phi_i\end{aligned} \qquad [4.10]$$

So we can derive

$$\phi_i^T K_j \Delta\phi_i = \frac{k(\gamma_{ij}^* - \gamma_{ij})}{2} \qquad [4.11]$$

It can be further derived from Equation [4.11] that

$$\phi_i^T K_j \Delta\phi_i = -k\sum_{j=1}^{n} \frac{\alpha_j(\gamma_{ij}^* - \gamma_{ij})}{2} \qquad [4.12]$$

$$\phi_i^T K \Delta\phi_i = \frac{k}{2}(\gamma_i^* - \gamma_i) \qquad [4.13]$$

Substituting Equations [4.6], [4.7], [4.12], and [4.13] into Equation [4.2] and simplifying the equation, we obtain

$$\sum_{j=1}^{n}\left(\frac{\gamma_{ij}^*/\gamma_{ij} + 1}{\gamma_i^*/\gamma_i + 1}\right)\frac{\gamma_{ij}}{\gamma_i}\alpha_j = -\frac{\Delta\lambda_i}{\lambda_i} \qquad [4.14]$$

where $\Delta\lambda_i/\lambda_i$ is the ith eigenvalue change ratio, which reflects the global effect of the damage on the ith mode. Here, γ_{ij}/γ_i is defined as a damage location factor (DLF), which reflects the sensitivity of the ith vibration mode to the damage at the jth element. The DLF is an important factor for damage localization and quantification. While a DSCF is defined as

$$\text{DSCF}_{ij} = \frac{\gamma_{ij}^*/\gamma_{ij} + 1}{\gamma_i^*/\gamma_i + 1} \tag{4.15}$$

DSCF represents the nonlinear effect of damage severity α_j at the jth element on the ith modal frequency change ratio. It is a function of damage location and severity and can be easily calculated from the modal strain energy in the healthy and damaged state. In the case of minimum damage, DSCF $\approx$ 1 and the eigenvalue change ratio is a linear function of damage severity.

It should be noted that if the curvature mode shape data from the healthy and damaged state are not normalized on the same basis (e.g., mass-normalized), the value of DSCF may differ depending on the normalization. In order to keep the DSCF consistent for unnormalized curvature mode shape, the healthy and damaged curvature mode shapes are normalized so that their total modal strain energies are the same. In this case, the DSCF can be approximated from the unnormalized curvature mode shape by

$$\text{DSCF}_{ij} = \frac{\left(\gamma_{ij}^*\gamma_i/\gamma_i^*\gamma_{ij}\right) + 1}{2} \tag{4.16}$$

The DSCF is a good damage location indicator, as well as an important factor in damage quantification.

4.3 DSCF-based damage identification method: key steps

The proposed DSCF-based damage identification method (Fan, 2011; Fan and Qiao, 2012) consists of three steps: mode selection, damage localization, and damage quantification.

4.3.1 Step 1: Selection of sensitive modes

The vibration modes for damage identification are selected based on the eigenvalue change ratio $\Delta\lambda_i/\lambda_i$. Before applying any damage identification algorithm to modal parameters obtained in modal testing, the sensitivity of the vibration modes to the existing damage should be evaluated to ensure the effectiveness and accuracy of damage localization and damage quantification. The eigenvalue change ratio $\Delta\lambda_i/\lambda_i$ is adopted as a mode sensitivity indicator.

4.3.2 Step 2: Localization of damage

Damage is then localized using the DSCF and DLF. A DI based on the DSCF is proposed as a good indicator for damage location. The DI β^*_{ij} at the jth element in the ith mode can be defined as:

$$\beta^*_{ij} = \begin{cases} \mathrm{DSCF}_{ij} & \text{if } \mathrm{DLF}_{ij} > \dfrac{5}{n}\% \\ 1 & \text{if } \mathrm{DLF}_{ij} \le \dfrac{5}{n}\% \end{cases} \qquad [4.17]$$

where n is the total element number. Since DLF_{ij} is a damage location factor that reflects the sensitivity of the ith vibration mode to the damage at the jth element, a vibration mode is insensitive to damage at an element with extremely low DLF in this mode. Therefore, this mode is inadequate to detect damage at this element, even if the damage exists. The DI with extremely low DLF_{ij} is directly set to 1 correspondingly to avoid the potential numerical error. In this study, the criterion for extremely low DLF_{ij} is set at 5% of the average of DLF_{ij} in the ith mode.

Then, assuming that the damage indices β^* at different elements is a normally distributed random variable, the DI β^*_{ij} can be normalized as

$$Z^*_{ij} = \frac{\beta^*_{ij} - \overline{\beta}^*_i}{\sigma^*_{\beta i}} \qquad [4.18]$$

where $\overline{\beta}_i$ and $\sigma_{\beta i}$ represent the mean and standard deviation of the damage indices in the ith mode, respectively. A damage detection criterion can be set as the normalized DI Z^*_{ij} larger than 2, to filter out small damage indices induced by measurement noise.

$$Z^*_{ij} = 0 \text{ if } \left|Z^*_{ij}\right| < 2 \qquad [4.19]$$

Furthermore, in order to eliminate false positive indications of damaged element induced by the measurement and numerical errors at those elements with low strain energy, a modified damage indices β_j at the jth element can be derived from DLF_{ij} and Z^*_{ij} as

$$\beta_j = \sum_{i=1}^{m} Z^*_{ij} \times DLF_{ij} \qquad [4.20]$$

Finally, a normalized DI Z_j can be generated from β_j to indicate the damaged elements,

$$Z_j = \left| \frac{\beta_j - \bar{\beta}}{\sigma_\beta} \right| \qquad [4.21]$$

and a similar damage detection criterion can be set as in Equation [4.19].

$$Z_j = 0 \text{ if } Z_j < 2 \qquad [4.22]$$

4.3.3 Step 3: Quantification of damage

Damage severity is finally quantified using the eigenvalue change ratio, DLF, and DSCF. Once the damaged element is detected in Step 2, the DSCF can be further used to calculate the severity of damage. If multiple modes are available, Equation [4.14] can be written in the matrix form as

$$RA = F \qquad [4.23]$$

$$R = \begin{bmatrix} DLF_{11} \times DSCF_{11} & DLF_{12} \times DSCF_{12} & \cdots & DLF_{1n} \times DSCF_{1n} \\ DLF_{21} \times DSCF_{21} & DLF_{22} \times DSCF_{22} & \cdots & DLF_{2n} \times DSCF_{2n} \\ \cdots & \cdots & \cdots & \cdots \\ DLF_{m1} \times DSCF_{m1} & DLF_{m2} \times DSCF_{m2} & \cdots & DLF_{mn} \times DSCF_{mn} \end{bmatrix}, A = \begin{Bmatrix} \alpha_1 \\ \alpha_2 \\ \cdots \\ \alpha_n \end{Bmatrix}, F = \begin{Bmatrix} -\dfrac{\Delta\lambda_1}{\lambda_1} \\ -\dfrac{\Delta\lambda_2}{\lambda_2} \\ \cdots \\ -\dfrac{\Delta\lambda_m}{\lambda_m} \end{Bmatrix}$$

where m is the total number of vibration modes, n is the total damaged element number, and DLF is the damage location factor γ_{ij}/γ_i.

For a unique solution, the number of the used vibration modes should be equal to the total number of unknown parameters (elemental damage severities). However, for the laboratory test or *in situ* application, the number of accessible vibration modes that are suitable for damage identification is very limited (typically the first few modes), and it is usually much less than the number of unknown parameters required. Furthermore, due to the existence of measurement errors, more measured modes than unknown parameters are required to construct a system of overdetermined equations to improve the accuracy of damage severities estimation. Therefore, to avoid non-uniqueness, appropriate assumptions on damage location or the damage severity distribution pattern have to be made to reduce the number of unknown parameters. For example, in damage identification with large sensor spacing, it is usually assumed that the damage only happens in a single element. For example, in the delamination identification case, it is reasonable to assume that the damage severity is the same for one delaminated area since delamination usually propagates between the same layers inside a composite plate.

When the number of measured modes is larger than the total number of unknown parameters, the least square fitting method is used to obtain the best solution.

$$A = (R^T R)^{-1} R^T F \qquad [4.24]$$

4.4 Experimental verification of the DSCF-based damage identification method

To verify the proposed DSCF-based damage identification method, a full-size as-manufactured FRP sandwich deck panel was experimentally tested in Smart Structural Laboratory at Washington State University.

4.4.1 Description of the fiber-reinforced polymer (FRP) sandwich deck panel

The FRP sandwich deck panel is a plate-type structure with a length of 1.981 m and a width of 1.495 m. The sandwich panel was connected to two underlying concrete supports using steel bolted connectors, which are considered to provide a simply supported boundary condition. The FRP sandwich panel consists of two face laminates and a sinusoidal honeycomb core, as shown in Fig. 4.1. The thickness of face laminates and honeycomb core are $t_f = 13.5$ mm and $t_c = 128$ mm, respectively. The constituent materials used for the composite sandwich panel (both face laminates and core) consist of E-glass fibers and polyester resin, and their mechanical properties are listed in Table 4.1.

The geometry of the sinusoidal core is shown in Fig. 4.2. The wave function of corrugated core wall can be defined as

Table 4.1 Mechanical properties of the constituent materials

Material	E (GPa)	G (GPa)	ν	Density ρ (g/cm^3)
E-glass fiber	72.4	28.8	0.255	2.55
Polyester resin	5.06	1.63	0.300	1.14

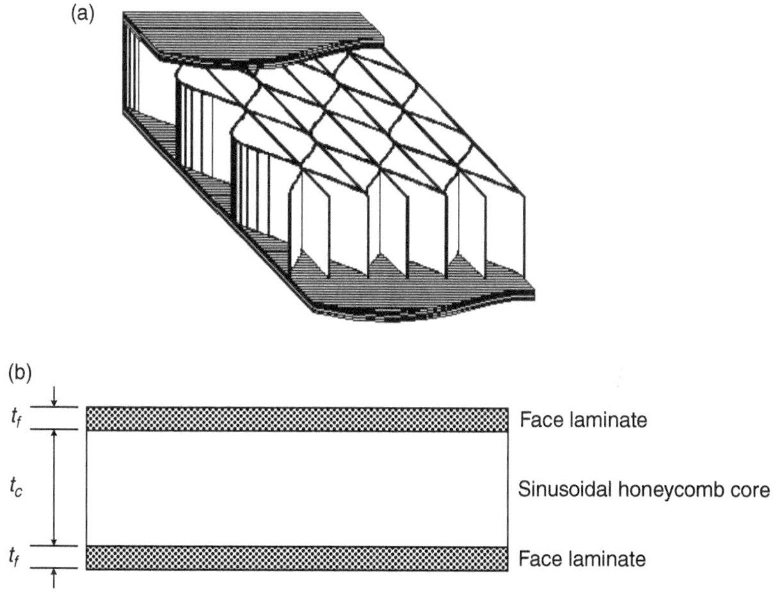

4.1 (a) Configuration of honeycomb FRP composite sandwich panel with sinusoidal core; (b) thickness notation of sandwich layers.

$$x = \frac{b}{2}\left(1 - \cos\frac{\pi y}{a}\right) \quad [4.25]$$

The dimensions of the sinusoidal core are a = 50.8 mm, b = 50.8 mm, and $t_1 = t_2$ = 3.34 mm (as shown in Fig. 4.2).

4.4.2 Experimental setup

The modal testing of the FRP sandwich plate was conducted using a polyvinylidene fluoride (PVDF) sensor network bonded to the top surface in the longitudinal direction. The PVDF sensors used in the experiment are

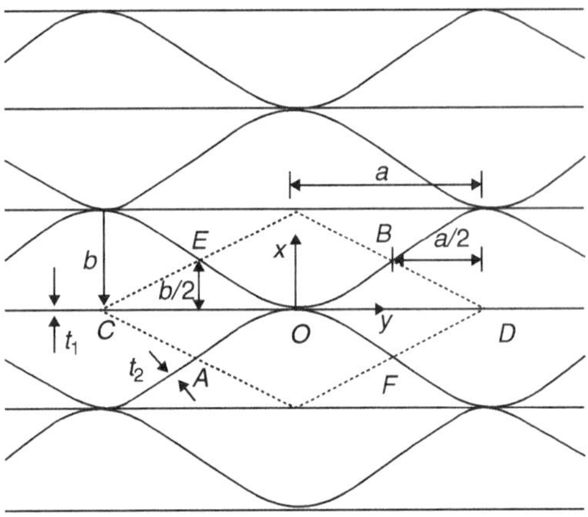

4.2 Geometry of sinusoidal core in honeycomb structures.

Model DT1–028K/L from the Measurement Specialties Company. A total of 285 nodes of PVDF sensors are uniformly distributed into 19 rows and 15 columns on the surface of the panel to build a sensor array. The plate is divided into 18 × 14 elements by the 19 × 15 sensors.

The panel is subjected to a dynamic pulse load applied at the central point using a modally tuned hammer (PCB 086C20). The response measurements are captured by the 19 × 15 sensor network to record the response of the structure subjected to the fixed point force excitation. A total of 19 × 15 node points are monitored corresponding to an actual spatial sampling distance of 101.6 mm. The coherent signals of the excitation and the responses are digitized and collected by the data acquisition system dSPACE CP1103 at a sampling frequency of 4000 Hz. Due to the fact that only 20 channels are available, only one column of nodes with 19 sensors is measured in each experiment. In this way, the 285 transfer functions between the hammer excitation and different PVDF sensors can still be measured for modal analysis in 15 experiments. The experimental setup using the distributed PVDF sensor array is shown in Fig. 4.3.

After the healthy FRP sandwich panel is tested, three stages of damage with increasing severity are artificially induced into the sandwich panel. In the first stage of induced damage (Damage Stage 1), a saw cut is induced between the top face sheet and the honeycomb core, to simulate debonding between the top face sheet and honeycomb core. A 203.2 mm × 203.2 mm area of face sheet-core debonding (cut) is shown in Fig. 4.4. In the second stage of increasing damage (Damage Stage 2), another saw cut is further induced between the bottom face sheet and the honeycomb core in the same area of the top face sheet-core debonding, to simulate debonding between the bottom face sheet and honeycomb core. Finally, in the third stage of the

4.3 Experimental setup of an FRP sandwich deck panel using the distributed PVDF sensor network.

largest damage in this study (Damage Stage 3), the two vertical sides of the core between the two face sheet-core debonding areas are cut to simulate the core crashing scenario, as shown in Fig. 4.4.

4.4.3 Modal testing results

The measurements for each sensor node are repeated 16 times, and a pre-trigger of 0.05 seconds at the level of 0.01 volt is set to synchronize the time-domain signal. The synchronized time history data from the excitation and response measurements are averaged to enhance the signal-to-noise ratio (SNR), as shown in Fig. 4.5. Then, the frequency-domain power spectral densities are generated from the excitation and response time-domain data using fast Fourier transform (FFT), as shown in Fig. 4.6. The frequency-response functions (FRFs) of these tested points are calculated from the excitation and response frequency-domain data. The FRF curve and its coherence curve, corresponding to time-domain data in Fig. 4.5 and frequency-domain data in Fig. 4.6, are illustrated in Fig. 4.7. Finally, these FRF curves are imported to the modal analysis program ME'Scope for curve fitting and modal extraction. The modal testing is conducted using a fixed point force excitation and a distributed PVDF sensor network, so the extracted mode shape should be the strain/curvature mode shape of the plate (Wang, 1998).

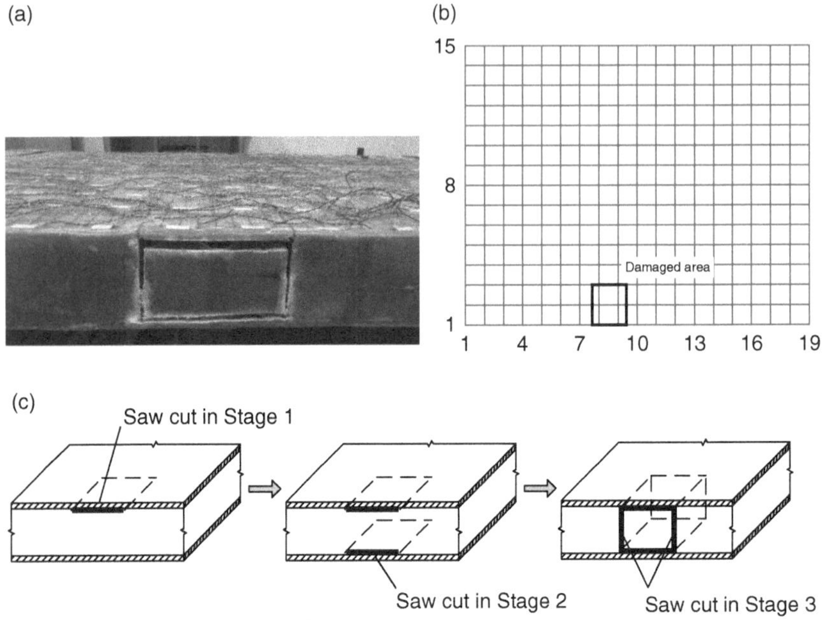

4.4 Artificially induced damage in the FRP sandwich deck panel. (a) Damage in Stage 3; (b) damaged area; and (c) schematic sketch of the three damage stages.

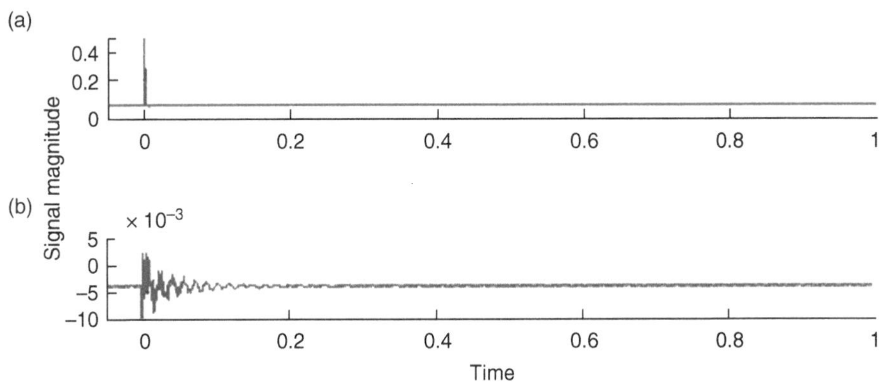

4.5 Time-domain data from (a) actuator and (b) sensor (both at node 143).

Since the plate is tested using a PVDF sensor network in the longitudinal direction (*y*-direction), only the curvature mode shapes in the longitudinal direction (κ_{yy}) are captured. Therefore, only the longitudinal bending modes, in which κ_{yy} is dominant, are suitable for damage identification. In the excitation frequency range of the modally tuned hammer, the first and third bending

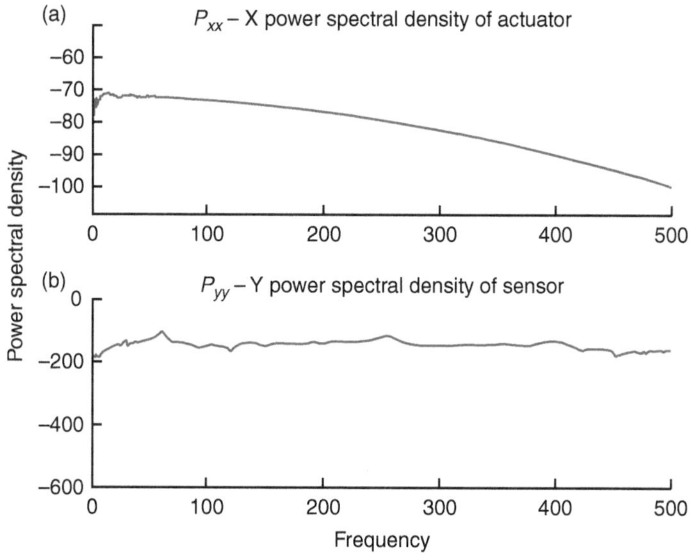

4.6 Frequency-domain data from (a) actuator and (b) sensor (both at node 143).

modes are captured. However, the second bending mode cannot be captured, because the fixed point force excitation is applied at the central point of the plate, which is at the nodal line of the second bending mode. Hence, the first and third bending modes in the longitudinal direction are investigated for damage identification. It should be noted that the modal frequency of the first bending mode lies closely around 60 Hz. Because the unshielded PVDF sensors are prone to electro-magnetic interference (EMI), their responses around 60 Hz could be noisy due to EMI from power system.

In the healthy stage and the three damage stages, the modal testing is conducted to extract the modal frequencies and curvature mode shapes of the FRP sandwich deck panel. The extracted curvature mode shapes and modal frequencies of the FRP sandwich deck panel for the first and third bending modes are shown in Plates I and II (in the color section between pages 172 and 173), respectively.

4.5 Implementing the DSCF-based damage identification method with the experimental data

The proposed three-step DSCF-based damage identification method is applied to the experimental data, and their results are presented and discussed in the following subsections.

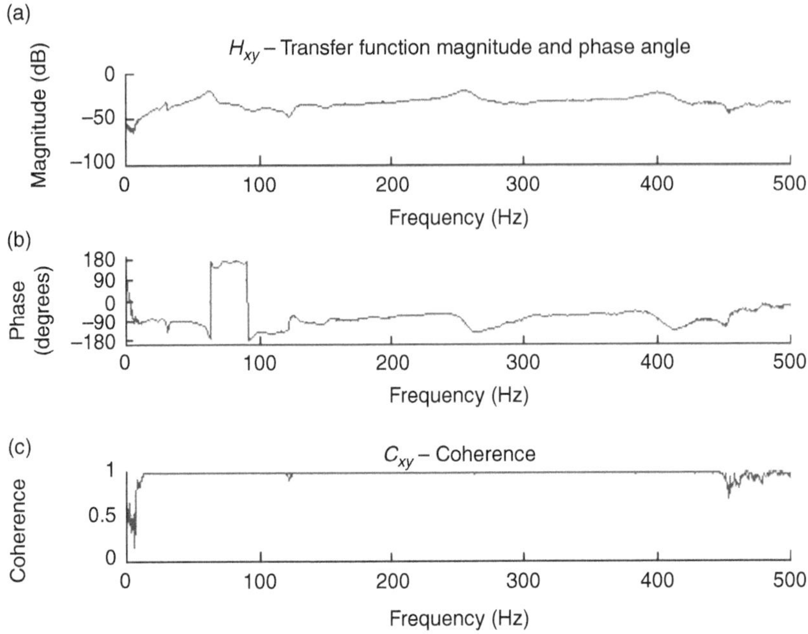

4.7 A frequency–response function ((a) and (b)) and its coherence functions (c) (both sensor and actuator are at node 143).

4.5.1 Step 1: Mode sensitivity

For the first longitudinal bending mode, the modal eigenvalue change ratios at three damage stages are 0.57%, 0.91%, and 3.59%, respectively. While for the third longitudinal bending mode, the modal eigenvalue change ratios at three damage stages are 0.71%, 1.55%, and 4.07%, respectively. It can be seen from the eigenvalue change ratio that both the modes are sensitive to the damage evolution (severity). Thus, both are used for damage identification.

4.5.2 Step 2: Damage localization

The modal strain energy of a plate can be derived from the curvature mode shapes of κ_{xx}, κ_{yy}, and κ_{xy}. However, one PVDF sensor can only be used to measure the strain response in one direction. Therefore, a sensor array positioned in one direction can only capture the strain/curvature mode shape in that direction at one time. Hence, for a longitudinal bending mode, the modal strain energy is approximated by its partial modal strain energy in the dominant direction. Assuming $\kappa_{xx} = \kappa_{xy} = 0$, the partial modal strain energy in the y-direction can be derived from Equation [4.5] as

$$\gamma_{ij}^{y} = \int\int_{A_j} \kappa_{yyi}^{2} \mathrm{d}x\mathrm{d}y \qquad [4.26]$$

The DLF and DSCF matrices for damage identification can be generated as follows. First, a bivariate cubic spline interpolation function is constructed for the curvature mode shape. The bivariate cubic spline is constructed as the tensor product of two univariate cubic splines. It can be expressed as the weighted sum of products of two cubic spline functions:

$$f(x,y) = \sum_i \sum_j a(i,j) g_i(x) h_j(y) \qquad [4.27]$$

where $g_i(x)$ and $h_j(y)$ are the cubic spline functions in x- and y-directions, respectively. Then, using the bivariate cubic spline function, γ_{ij} and γ_{ij}^{*} can be calculated from Equation [4.26]. Finally, the DLF and DSCF matrix can be calculated from γ_{ij} and γ_{ij}^{*}.

The DSCF-based damage indices using the modal strain energy from the first and third longitudinal bending modes are shown in Plate III (in the color section between pages 172 and 173). Although the damaged area does not exactly match the sensor grids, the DSCF damage localization method correctly approximates the location and area of the damage. It can be noticed that in Damage Stage 3 a false positive indication is given near the damaged area. There might be two reasons: (a) the effect of damage might have spread out to surrounding elements (Choi *et al.*, 2006); and (b) when cutting the two sides of the damaged area, the saw cut might have damaged the sinusoidal core of the neighboring element.

4.5.3 Step 3: Damage quantification

Assuming that each damaged element has the same damage severity, the DSCF method can be further applied for damage quantification using the DLF and DSCF derived in Step 2. The damage quantification results of the DSCF-based damage identification method are presented in Fig. 4.8. The damage quantification results, based on the four damaged elements indicated in Damage Stages 1 and 2, are marked in black in Fig. 4.8. The damage quantification result, based on the six damaged elements indicated in Damage Stage 3, is also shown in gray for comparison in Fig. 4.8. As shown in Fig. 4.8, the damage severity increases consistently with the Damage Stages 1 to 3.

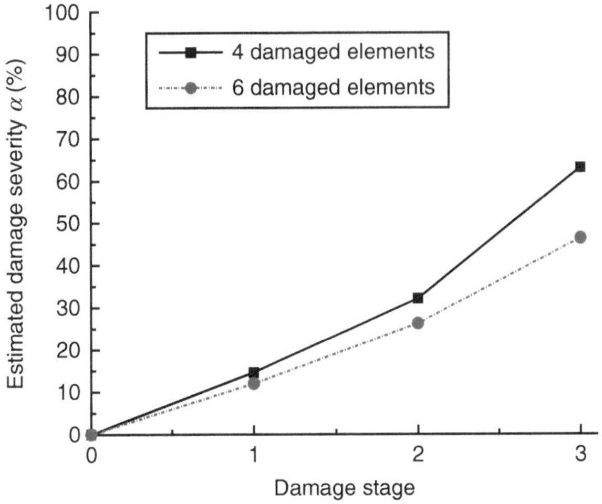

4.8 Damage quantification of the FRP sandwich deck panel at three damage stages.

4.6 Using numerical modal analysis to identify damage

After the experimental modal testing of the full-size FRP sandwich panel, a numerical modal analysis of the FRP sandwich panel is conducted to verify the experimental results. The DSCF-based damage identification method is also applied to the numerical data for damage localization and quantification.

4.6.1 Modeling of face sheets and the sinusoidal core

The lay-up of the face sheets and sinusoidal core is shown in Fig. 4.9, and corresponding layer material properties are listed in Table 4.2. The face laminate includes five ChopSM bonding layers (ChopSM308) and four bi-directional combination mat layers (CDM3208), while the microstructure of core walls consists of two layers of ChopSM (ChopSM308).

To predict the equivalent properties of the face laminates from the fiber and matrix material properties, a micro–macro mechanics approach (Davalos *et al.*, 2001; Qiao and Wang, 2005) is adopted. First, the properties of each layer in the face sheet can be computed using periodic microstructure (PM) approaches (micromechanics) (Barbero, 1999). Then, based on ply properties and lay-up, the apparent stiffnesses of the face laminate can be predicted using classical lamination theory (macromechanics). The equivalent analysis of core is performed using the methods of homogenization and

Table 4.2 Layer material properties from micromechanics models

Ply	Orientation	E_1 (GPa)	E_2 (GPa)	G_{12} (GPa)	Thickness t (mm)	ρ_{ply} (g/cm^3)	Fiber volume fraction V_f
ChopSM 308	Random	10.67	10.67	3.86	1.668	1.37	0.164
CDM 3208	0°	34.38	12.08	4.25	0.49	1.74	0.424
	90°	34.45	12.10	4.26	0.55	1.74	0.425
	Random	19.59	19.59	7.04	0.25	1.70	0.396

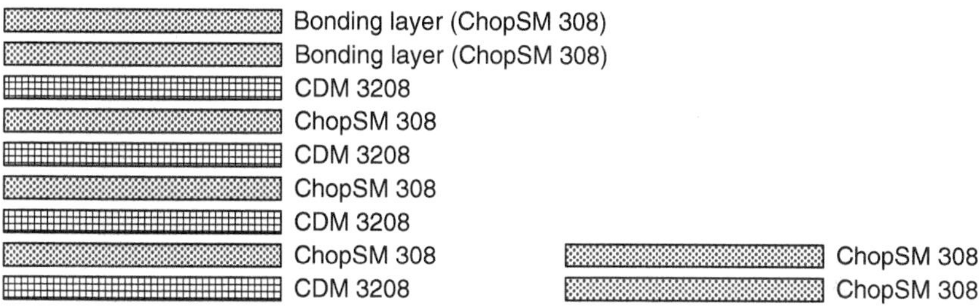

(a) Face laminate lay-up (b) Honeycomb core lay-up

4.9 Lay-up of face laminates and core.

mechanics of materials approaches (Davalos *et al.*, 2001; Qiao and Wang, 2005). The derived equivalent stiffness properties of face laminates and core walls are listed in Table 4.3.

4.6.2 Finite element modal analysis of FRP sandwich panel

The commercial finite element analysis package ABAQUS is used to conduct an eigenvalue analysis of the FRP sandwich deck panel. A detailed model of the full-size sandwich panel is constructed in ABAQUS, as shown in Plate IV (in the color section between pages 172 and 173). The face sheets and the sinusoidal core walls are modeled as four-node first-order plate elements S4 and three-node first-order plate elements S3. The corresponding stiffness properties of face sheets and core walls are given in Table 4.3. The boundary conditions are considered to be simply supported at both the transverse edges.

The FEA is conducted on both the health and damaged sandwich panel. Three damaged sandwich panel models are set up to simulate the FRP

Table 4.3 Equivalent properties of face laminates and core

	ρ (g/cm^3)	E_1 (GPa)	E_2 (GPa)	G_{12} (GPa)	v_{12}
Face laminate	1.506	15.30	15.06	4.21	0.29
Sinusoidal core	1.37	10.67	10.67	3.86	0.38

sandwich deck panels with artificially induced damage in the three damage stages (see Fig. 4.4c). The debonding between the face sheets and core is simulated by detaching the node/element connection between the face sheets and core elements. The cuts in the sinusoidal core are similarly simulated. The simulated mode shapes of the FRP sandwich panel are shown in Plate IV. The displacement mode shapes of the healthy and damaged panel are very close to each other; thus, only the displacement mode shapes of the healthy one are shown in Plate IV. The strain mode shape can be then extracted from the 19 × 15 nodes corresponding and identical to the 19 × 15 PVDF sensors in the experimental test, as shown in Plates V and VI (see in the color section between pages 172 and 173), respectively, for the first and third bending modes.

4.7 Damage identification using numerical data

Similarly, the proposed three-step DSCF-based damage identification method is applied to the numerical finite element modal analysis data, and their results and comparison to the damage identification outcomes with experimental data are presented and discussed in the following subsections.

4.7.1 Step 1: Mode sensitivity

The modal frequencies of the healthy sandwich panel from the FEA are 62.983 and 418.6 Hz, which shows close agreement with the experimental test data of 61.08 and 403.00 Hz for the first and third bending modes, respectively. The result confirms that the boundary conditions at the two transverse supporting ends should be modeled as simply supported, and the free vibration modes of the FRP sandwich panel can be accurately predicted by the detailed finite element (FE) model. The modal eigenvalue change ratios at three damage stages from the FEA and experimental test are compared in Table 4.4. The FEA results also show that both the modes (first and third) are sensitive to the damage, so both can be used for damage identification. It can also be noticed that the FEA tends to underestimate the modal eigenvalue change ratios for the first mode but over-predict the ratios for the third mode.

Table 4.4 Modal eigenvalue change ratios from FE and experiment

Vibration modes		1st stage (%)	2nd stage (%)	3rd stage (%)
1st mode	FEA	0.044	0.095	0.862
	Experiment	0.57	0.91	3.59
3rd mode	FEA	0.925	2.356	8.635
	Experiment	0.71	1.55	4.07

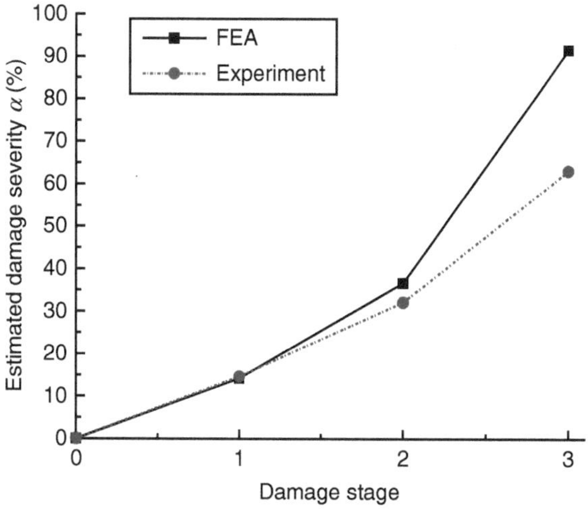

4.10 Damage quantification of FRP sandwich plate.

4.7.2 Step 2: Damage localization

The DSCF-based damage identification method is applied to the curvature mode shapes for damage localization. The damage indices using the modal strain energy from the first and third longitudinal bending modes are shown in Plate VII (in the color section between pages 172 and 173). The results show the same indication of damaged elements in the first and second stages as the results from experimental test data (see Plates IIIa and IIIb). In the third stage, the FE results indicate damaged elements at the two sides of the damaged core, which corresponds to the saw cut in the third damage stage.

4.7.3 Step 3: Damage quantification

Assuming that the damaged elements have the same damage severity, the DSCF method can be further applied for damage quantification using the FEA data. The damage quantification results, based on the damaged elements indicated in Step 2, is shown in Fig. 4.10. As shown in Fig. 4.10, the damage quantification results from both the FEA and experimental data show the same trend, i.e., the damage severity increase consistently with the Damage Stages 1 to 3. The damage quantification results based on the experimental data are also shown for comparison. It can be seen that in Damage Stages 1 and 2, both the experimental and numerical damage quantification results show close agreement with each other; while in Damage Stage 3 the results show considerable discrepancy due to their differences in the damaged element indication and modal eigenvalue change ratios. Based on the FEA data, the core crashing can induce over 90% stiffness reduction in the damaged elements.

4.8 Conclusions

In this chapter, a DSCF-based damage identification method is proposed, and it is then applied to the experimental and numerical modal analysis data for damage localization and quantification of an as-manufactured FRP sandwich deck panel. Both the experimental and numerical finite element modal analyses are adopted to examine the dynamic vibration modes of the FRP sandwich deck panel. The concluding remarks on this study are summarized as follows:

1. The DSCF-based damage identification method can be easily and effectively applied to the experimental and numerical FE modal analysis data for both damage localization and quantification.
2. It is also shown that using the partial strain energy in the dominant direction of a pure bending mode can be used with the DSCF-based damage identification. It thus makes the implementation of surface-based piezoelectric sensors and their strain mode shape measurement along only one direction more feasible.
3. The surface-bonded PVDF sensor network and impact hammer system can be effectively used to capture modal frequencies and curvature mode shapes for the DSCF-based damage identification method.
4. The detailed FE model can be used to accurately predict the free vibration of the FRP sandwich deck panel with sinusoidal core. The shear bolt connections at two transverse supporting ends can be simplified as simply supported boundary conditions.
5. In damage quantification, to avoid non-uniqueness, the assumption of uniform distribution of damage severity on damaged elements is made to reduce the number of unknown parameters in this study. Further study is needed for the damage severity distribution pattern.
6. Studies toward the development of a real-time structural damage identification system using the DSCF-based damage identification method

and piezoelectric sensor array for *in situ* application are promising, and they should be the subjects of future research.

4.9 Acknowledgments

This study is partially supported by Alaska University Transportation Center (AUTC) and US Department of Transportation (Contract/Grant Number: DTRT06-G-0011). The authors also want to acknowledge the support from Dr Jerry Plunkett of Kansas Structural Composites, Inc., who fabricated the FRP sandwich deck panels used in this study.

4.10 References

Barbero, E.J. (1999), *Introduction to Composite Materials Design*. Taylor & Francis.

Cawley, P. and R.D. Adams (1979), The location of defects in structures from measurements of natural frequencies. *Journal of Strain Analysis*, **14**(2): p. 49–57.

Choi, S., S. Park, S. Yoon and N. Stubbs (2005), Nondestructive damage identification in plate structures using changes in modal compliance. *NDT & E International*, **38**(7): p. 529–540.

Choi, S., S. Park, N.H. Park and N. Stubbs (2006), Improved fault quantification for a plate structure. *Journal of Sound and Vibration*, **297**(3–5): p. 865–879.

Davalos, J.F., P.Z. Qiao, X.F. Xu, J. Robinson and K.E. Barth (2001), Modeling and characterization of fiber-reinforced plastic honeycomb sandwich panels for highway bridge applications. *Composite Structures*, **52**(3–4): p. 441–452.

Doebling, S.W., C.R. Farrar, M.B. Prime and D.W. Shevitz (1996), Damage identification and health monitoring of structural and mechanical systems from changes in their vibration characteristics: a literature review. Los Alamos National Laboratory Report, 1996(LA-13070-MS).

Fan, W. (2011), Health Monitoring and Damage Identification of Composite Structures. Ph.D. dissertation, Department of Civil and Environmental Engineering, Washington State University, Pullman, WA, June 2011.

Fan, W. and P.Z. Qiao (2011), Vibration-based damage identification methods: a review and comparative study. *Structural Health Monitoring-an International Journal*, **10**(1): p. 83–111.

Fan, W. and P.Z. Qiao (2012), A strain energy-based damage severity correction factor method for damage identification in plate-type structures. *Mechanical Systems and Signal Processing*, **28C**: p. 660–678.

Farrar, C.R. and D. Jauregui (1996), Damage detection algorithms applied to experimental and numerical modal data from the I-40 bridge. In Technical Report LA-13074-MS. Los Alamos National Laboratory: LA.

Hadjileontiadis, L.J., E. Douka and A. Trochidis (2005), Fractal dimension analysis for crack identification in beam structures. *Mechanical Systems and Signal Processing*, **19**(3): p. 659–674.

Law, S.S., Z.Y. Shi and L.M. Zhang (1998), Structural damage detection from incomplete and noisy modal test data. *Journal of Engineering Mechanics-ASCE*, **124**(11): p. 1280–1288.

Qiao, P.Z. and J.L. Wang (2005), Mechanics of composite sinusoidal honeycomb cores. *Journal of Aerospace Engineering, ASCE*, **18**(1): p. 42–50.

Ratcliffe, C.P. (1997), Damage detection using a modified Laplacian operator on mode shape data. *Journal of Sound and Vibration*, **204**(3): p. 505–517.

Salawu, O.S. (1997), Detection of structural damage through changes in frequency: A review. *Engineering Structures*, **19**(9): p. 718–723.

Shi, Z.Y. and S.S. Law (1998), Structural damage localization from modal strain energy change. *Journal of Sound and Vibration*, **218**(5): p. 825–844.

Shi, Z.Y., S.S. Law and L.M. Zhang (2000), Structural damage detection from modal strain energy change. *Journal of Engineering Mechanics-ASCE*, **126**(12): p. 1216–1223.

Stubbs, N., J.T. Kim and C.R. Farrar (1995), Field verification of a nondestructive damage localization and severity estimation algorithm. *Proceedings of 13th International Modal Analysis Conference*, **1**: p. 210–218.

Stubbs, N. and J.T. Kim (1996), Damage localization in structures without baseline modal parameters. *AIAA Journal*, **34**(8): p. 1644–1649.

Wang, B.T. (1998), Structural modal testing with various actuators and sensors. *Mechanical Systems and Signal Processing*, **12**(5): p. 627–639.

Wang, J.L. and P.Z. Qiao (2007), Improved damage detection of beam-type structures using a uniform load surface. *Structural Health Monitoring, An International Journal*, **6**(2): 99–110.

Zhang, Z. and A.E. Aktan (1998), Application of modal flexibility and its derivatives in structural identification. *Research in Nondestructive Evaluation*, **10**(1): p. 43–61.

5

Large rupture strain (LRS) fibre-reinforced polymer (FRP) composites for seismic retrofit of reinforced concrete (RC) piers

J. G. DAI and Y. L. BAI, The Hong Kong Polytechnic University, Hong Kong, China

DOI: 10.1533/9780857097019.1.117

Abstract: Large rupture strain (LRS) fibre-reinforced polymer (FRP) composites are usually made of polyethylene naphthalate (PEN) or polyethylene terephthalate (PET) fibres. They have a relatively low elastic modulus but a much larger rupture strain (usually greater than 5%) as compared to conventional FRP composites (i.e. carbon FRP, glass FRP and aramid FRP), and therefore provide a cheaper but more effective solution for the seismic retrofit of reinforced concrete (RC) columns. This chapter first discusses the strength and ductility of LRS FRP-confined concrete subjected to monotonic and cyclic compressive loading and then discusses the behaviour and design of RC piers seismically retrofitted with LRS FRP composites under cyclic lateral loading.

Key words: fibre-reinforced polymer, large rupture strain, reinforced concrete (RC) pier, seismic retrofit, confinement.

5.1 Introduction

Many existing RC piers do not satisfy the strength or/and ductility demands for seismic loading as they were designed based on outdated codes of practice. Due to the lack of sufficient transverse reinforcement and/or appropriate seismic detailing, these substandard RC piers may encounter unductile shear failure (before or after the flexural yielding of longitudinal reinforcement), confinement failure of the flexural plastic hinge region and lap splice debonding of the longitudinal reinforcement during a major earthquake (Seible *et al.*, 1997). As a result, they need large strength and ductility improvements in order to meet the requirements of modern earthquake-resistance regulations. In recent years, the use of FRP composites as external reinforcement has become a very effective solution for the strengthening of existing RC structures, due to the high strength-to-weight ratio, non-corrosive property and the ease of construction of FRP composites. One of the most promising applications is to use FRP composites as a jacketing material with the main fibres running in the hoop direction for the seismic retrofit of RC piers. Because FRP jackets facilitate additional shear capacity as well as significant lateral confinement to concrete, the seismic resistance of RC piers can be significantly improved.

Carbon FRP (CFRP), glass FRP (GFRP) and aramid FRP (AFRP), which are classified as conventional FRPs in this chapter, are among the most commonly used FRPs for the seismic retrofit applications. The stress–strain behaviour of conventional FRPs is linearly elastic, with an ultimate tensile rupture strain of around 1.5% for carbon FRP (CFRP), 2% for glass FRP (GFRP) and 3% for aramid FRP (AFRP). It should be noted that for seismic retrofit applications an FRP jacket with a larger rupture strain is generally beneficial, as it leads to more ductile behaviour and a greater energy absorption capacity with the same degree of strength enhancement (Dai *et al.*, 2011, 2012; Dai and Ueda, 2012). In recent years, a new category of FRP composites has emerged as an alternative to conventional FRP composites. These new FRP composites are made of polyethylene naphthalate (PEN) fibres or polyethylene terephthalate (PET) fibres, which are recognized as having a large rupture strain (LRS) (usually larger than 5%) but a relatively low elastic modulus. Compared to conventional FRPs, LRS FRPs are cheaper and more environmentally friendly since they are often made from recycled plastic (e.g. PET bottles). This chapter focuses on the behaviour and modelling of LRS FRP-confined concrete subject to monotonic and cyclic compressive loading as well as LRS FRP-jacketed RC piers under cyclic lateral loading.

5.2 Properties of large rupture strain (LRS) fibre-reinforced polymer (FRP) composites

Both PET and PEN belong to polyester, a category of polymers which contact the ester functional group in their main chain. PET is the most common polyester used for producing fibres as well as soft drink bottles. Nowadays, it is also very common to recycle PET bottles after use by remelting them and extruding them as fibres, to save valuable petroleum raw materials, reduce energy consumption and eliminate solid waste sent to landfills. PEN, a new generation polymer, is a high performance member of the polyester family. Its unique chemical structure renders it useful for fibres, packaging and films. The two condensed aromatic rings of PEN improve strength and modulus, chemical and hydrolytic resistance, gaseous barrier, thermal and thermo-oxidative resistance and ultraviolet (UV) light barrier resistance as compared to PET. Traditionally, significant commercial markets have been developed for the application of PET and PEN in textile and industrial fibres, films and foamed articles, containers for carbonated beverages, water

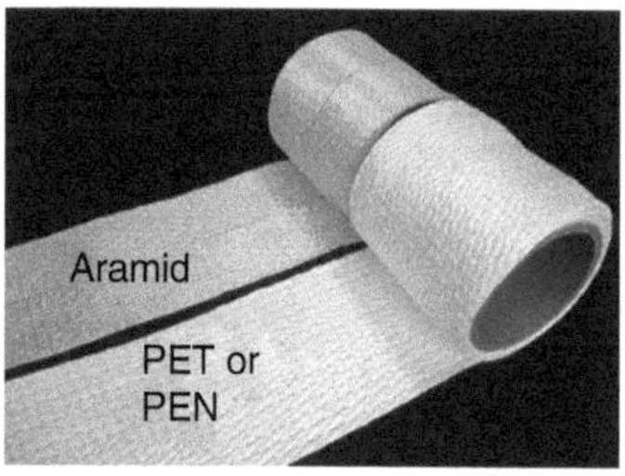

5.1 PET or PEN fibre sheets.

Table 5.1 Properties of different fibres for structural strengthening applications

Fibre	Fibre type	Tensile strength (MPa)	Young's modulus (GPa)	Rupture strain (%)	Density (g/mm³)
Carbon (PAN)	High strength type	3790	242	1.55	1.81
	High modulus type	2300	370	0.62	1.90
Carbon (Pitch)	Ordinary type	1100	41	2.80	1.65
	High strength and modulus type	3210	940	0.36	2.20
Aramid	High strength type	2350	79	3.0	1.40
	High modulus type	2060	118	1.8	1.45
Glass	E-glass	1720	72	2.4	2.6
	S-glass	2530	87	2.9	2.5
PEN		790	15 ± 2	>5.0	1.41
PET		740	10 ± 1	>7.0	1.44

PAN, Polyacrylonitrile

and other liquids, and thermoformed applications. In recent years, Japanese manufacturers have produced PET or PEN in a sheet form (Fig. 5.1) as a type of external reinforcement for the strengthening of existing concrete structures.

Table 5.1 summarizes the typical material properties of LRS fibres (i.e. PET and PEN fibres) as compared to those of conventional ones. It is seen that the tensile strengths of PET and PEN fibres are similar, about one-fourth to one-third of that of carbon fibres. The elastic modulus of PEN is higher than that of PET but is still less than one-fifteenth of that of carbon fibres. The elongation of PET is higher than that of PEN and is about four times that of carbon fibres. It should be noted that in order to achieve a certain level of tensile stiffness for structural strengthening applications, the aerial fibre weight of one layer of LRS fibre sheet should be four or five times that of the commonly used carbon fibre sheet (i.e. 200~300 g/m^2). As a result, the nominal thickness of one layer of LRS fibre sheet is usually about 1.0 mm.

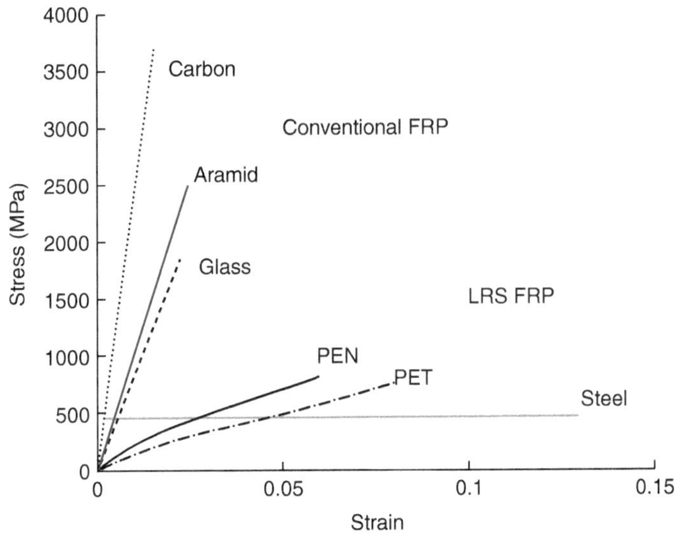

5.2 Tensile stress–strain relationships of conventional and LRS fibres.

As shown in Fig. 5.2, the tensile stress–strain relationships of conventional fibres are linearly elastic to the point of rupture, while LRS fibres all exhibit some material nonlinearity. Such nonlinear stress–strain behaviour is caused by the motion of amorphous phases and the sliding or failing of macromolecular chains in PET and PEN fibres (Lechat *et al.*, 2006, 2011). The initial elastic modulus of LRS fibres is usually higher than that given in Table 5.1, in which a secant modulus of elasticity is given for conservative design purposes.

5.3 LRS FRP-confined concrete under monotonic compressive loading

To achieve a reliable design for FRP strengthening of RC columns under the action of axial loading or combined axial and lateral loading, an in-depth understanding of the compressive stress–strain behaviour of FRP-confined concrete is important. A large number of experimental tests have been conducted on FRP-confined circular concrete columns. Design-orientated and analysis-orientated models (e.g. Lam and Teng, 2003; Jiang and Teng, 2007) have been developed to predict the ultimate strength and strain as well as the full-range stress–strain behaviour of FRP-confined concrete. However, studies on the compressive behaviour of concrete confined with LRS FRPs have been very limited.

5.3 Typical failure of FRP-confined concrete.

Dai *et al.* (2011) conducted compressive tests on a total of 42 cylindrical concrete specimens, comprising 36 FRP-wrapped specimens and six control specimens. Among the 36 specimens, nine specimens were wrapped with AFRP FRP jackets, another nine specimens were wrapped with PEN FRP jackets and the rest were wrapped with PET FRP jackets. It was found that the typical failure of LRS FRP-confined concrete was still due to the hoop tensile rupture of the FRP jacket outside the overlapping zone as observed in conventional FRP-confined concrete (Fig. 5.3). But the failure at the ultimate state was in a quiet manner, unlike the explosive one observed in AFRP-confined concrete. The average hoop rupture strain of FRP jackets from these compression tests was found to be 3.0%, 4.5% and 7.5% for AFRP, PEN FRP and PET FRP, respectively, while most existing tests on concrete confined with GFRP and CFRP exhibited hoop rupture strains of around 2% and 1%, respectively (Lam and Teng, 2004). Similar to those of conventional FRP-confined concrete with a sufficient level of FRP confinement, the stress–strain curves of LRS FRP-confined concrete also exhibit a monotonically ascending bilinear shape with rapid softening in a transition zone around the stress level of unconfined concrete strength. Both the compressive strength and the ultimate axial strain are significantly enhanced (Fig. 5.4).

An analysis-orientated model was also proposed by Dai *et al.* (2011) to predict the experimentally observed stress–strain relationships of LRS FRP-confined circular concrete. The model was based on the Jiang and Teng (2007) model and hence called the modified J&T's model. In principle, an analysis-orientated model is applicable to concrete confined by all types of FRP materials once the relationship between the lateral strain and the longitudinal strain is established. In the analysis-orientated model, a generic

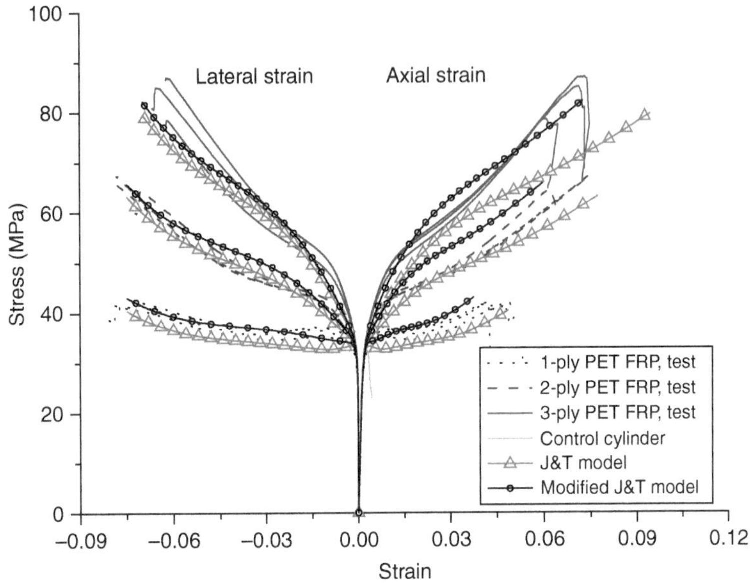

5.4 Typical compressive stress–strain relationship of LRS FRP-confined concrete. J & T: Jiang and Teng (2007).

relationship between the confining pressure and the lateral strain of an FRP jacket is given by:

$$\sigma_l = \frac{E_{\mathrm{FRP}(\varepsilon_l)} \cdot t_{\mathrm{FRP}} \cdot \varepsilon_l}{R} \quad [5.1]$$

where: σ_l is the confining pressure (or lateral stress) of the FRP jacket; E_{FRP} is the secant modulus of FRP, which is a constant for conventional FRP jackets with a linear elastic stress–strain response but is dependent on the lateral strain (i.e. ε_l) of the FRP jacket for LRS FRPs with a nonlinear stress–strain response (Fig. 5.2); t_{FRP} is the nominal thickness of the FRP jacket; and R is the radius of the concrete core. In Jiang's and Teng's (2007) model, the following equation, which was originally proposed by Popovics (1973) and later employed in the model of Mander *et al.* (1988) for steel-confined concrete, is adopted to predict the stress–strain curves of actively-confined concrete:

$$\frac{\sigma_c}{f_{cc}'^*} = \frac{(\varepsilon_c/\varepsilon_{cc}^*)m}{m-1+(\varepsilon_c/\varepsilon_{cc}^*)^m} \quad [5.2]$$

where $f_{cc}^{\prime *}$ and $\varepsilon_{cc}{}^{*}$ are the peak axial compressive stress and the corresponding axial compressive strain of concrete respectively under a specific constant confining pressure; and σ_c and ε_c are the compressive stress and strain of FRP-confined concrete, respectively. The constant m is given by:

$$m = \frac{E_c}{E_c - f_{cc}^{\prime *} / \varepsilon_{cc}^{*}} \qquad [5.3]$$

where E_c is the elastic modulus of unconfined concrete. The peak stress on the stress–strain curve of the actively-confined concrete is given by:

$$f_{cc}^{\prime *} = f_{co}^{\prime} + 3.5\sigma_l \qquad [5.4]$$

while the axial strain corresponding to the peak compressive stress is given by:

$$\varepsilon_{cc}^{*} = \varepsilon_{co}\left[1 + 17.5\left(\frac{\sigma_l}{f_{co}^{\prime}}\right)^{1.2}\right] \qquad [5.5]$$

where $f_{co}^{\prime}$ is the compressive strength of unconfined concrete and ε_{co} is the compressive strain corresponding to the compressive strength of unconfined concrete. Based upon Dai *et al.*'s (2011) test results, the lateral-to-axial strain relationship for LRS FRP-confined concrete can be expressed as follows:

$$\frac{\varepsilon_c}{\varepsilon_{co}} = \left(1.0 + 8.0\frac{\sigma_l}{f_{co}^{\prime}}\right) \cdot \left[1.024\left(\frac{\varepsilon_l}{\varepsilon_{co}}\right)^{0.350} + 0.089\left(\frac{\varepsilon_l}{\varepsilon_{co}}\right)\right] \qquad [5.6]$$

Through a combined use of Equations [5.1], [5.2] and [5.6] and an incremental analysis, the full-range compressive stress–strain curve of LRS FRP-confined concrete can be predicted. It was found that the modified J&T's model performs satisfactorily, while the original J&T's model is able to predict closely the compressive strength of LRS FRP-confined concrete but usually tends to overestimate its ultimate axial strain (Fig. 5.4). This weakness was believed to be attributed to the fact that the lateral strain-to-axial strain relationship, which was originally proposed based on the test results of concrete confined with conventional FRPs (Teng *et al.*, 2007), needs to be improved for high strain range. The modified J&T's model was also proven to be applicable for both conventional and LRS FRP-confined concrete with sufficient accuracy (Fig. 5.5).

Based upon the proposed analysis-orientated model, comparisons among concrete confined with FRP jackets using five different fibre sheets (CFRP,

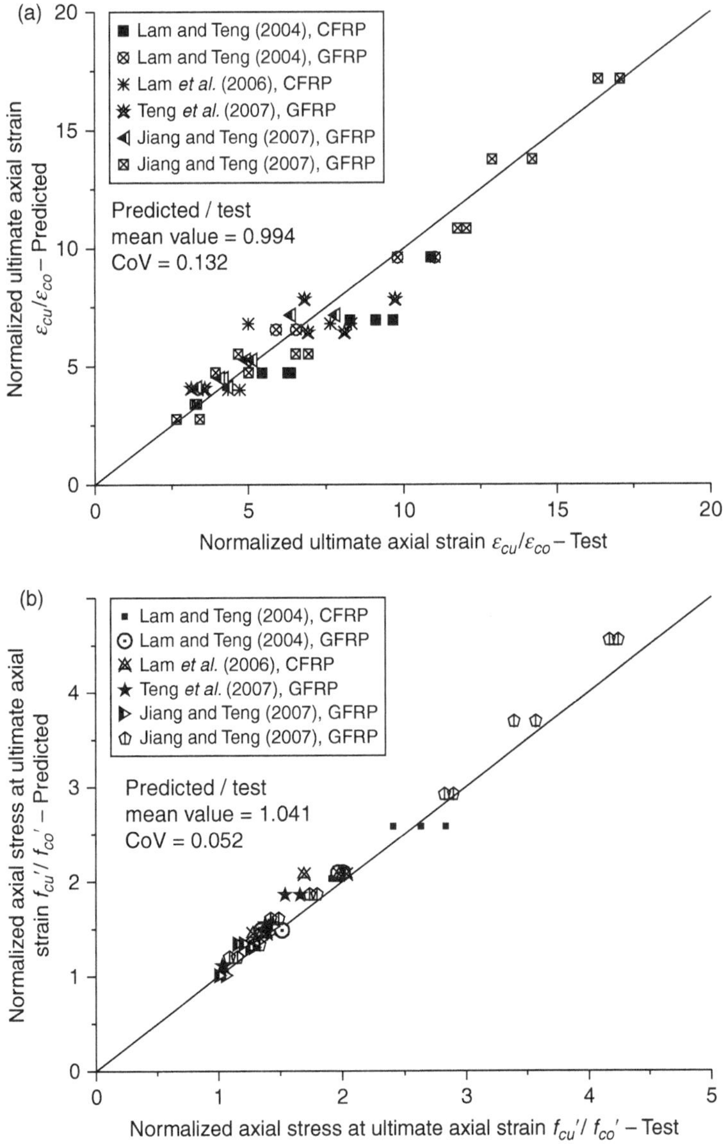

5.5 Performance of the modified Jiang and Teng (2007) model in predicting the ultimate condition for FRP-confined concrete: (a) ultimate strain; (b) ultimate stress. CoV, coefficient of variance.

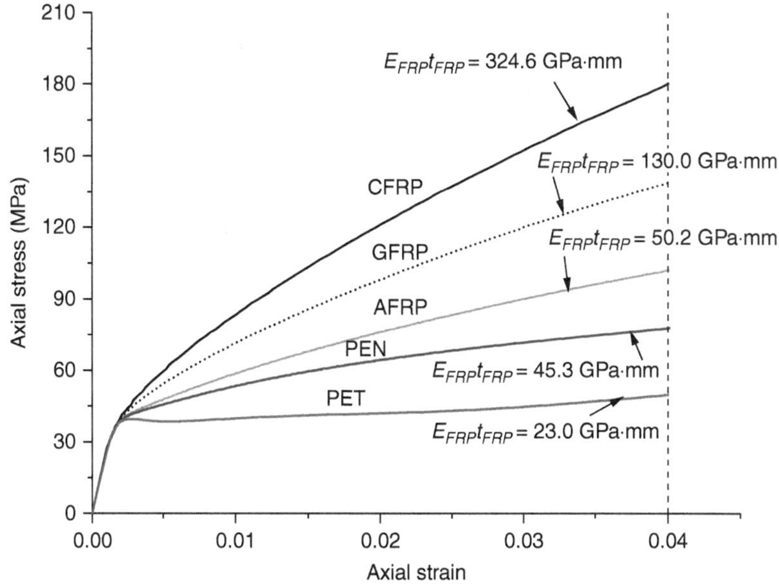

5.6 Comparison between conventional and LRS FRP-confined concrete given the same displacement ductility.

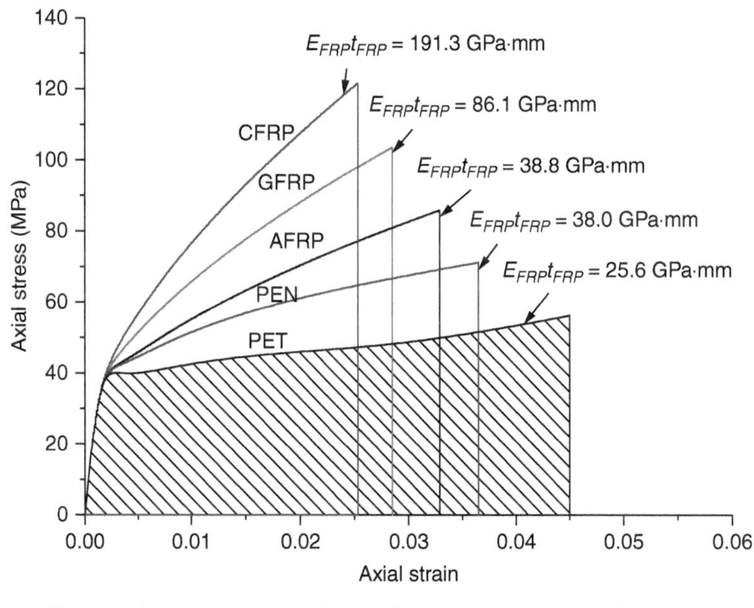

5.7 Energy absorption comparison of conventional and LRS FRP-confined concrete.

GFRP, AFRP, PEN and PET) were made. The results of comparisons are presented in Figs 5.6 and 5.7, in which all FRP-confined concretes are assumed to achieve the same displacement ductility and the same energy absorption at the rupture of the fibre sheet. The energy absorption is evaluated using the area underneath the compressive stress–strain curve. A concrete cylinder with an unconfined concrete strength of 38 MPa was assumed for such a comparison and the rupture strain of fibres used in the analysis were assumed to be 0.095%, 1.7%, 3.0%, 4.5% and 7.5%, which are typical values for CFRP, GFRP, AFRP, PEN FRP and PET FRP, respectively. It is shown for both cases that the needed tension stiffness of PEN FRP and PET FRP jackets is significantly lower than that of conventional FRP jackets. It is therefore very clear that LRS FRP jackets lead to a more economic and attractive confinement solution for seismic retrofit purposes.

5.4 LRS FRP-confined concrete under cyclic compressive loading

Compared with the large number of studies on the monotonic stress–strain behaviour of FRP-confined concrete, only a few studies exist on FRP-confined concrete under cyclic axial compression (Lam *et al.*, 2006; Shao *et al.*, 2006; Lam and Teng, 2009; Wang and Li, 2009; Ozbakkaloglu and Akin, 2012). No study has been conducted on the cyclic stress–strain behaviour of concrete confined with LRS FRP. In addition, only three theoretical models (Shao *et al.*, 2006; Lam and Teng, 2009; Wang and Li, 2009) have been proposed to predict such behaviour in the open literature.

Bai *et al.* (2012, 2013) conducted tests on a total of 12 circular concrete cylinders (152 mm in diameter and 305 mm in height), comprising eight FRP-confined concrete cylinders and four unconfined control cylinders. All the tests were carried out under cyclic concentric axial compression. The average unconfined cylinder compressive strength was 35.6 MPa. The test parameters included the type of FRP materials (i.e. PEN and PET) and the number of layers of fibre sheets (i.e. one or two layers). For each specimen configuration, two identical specimens were prepared, but they were tested with different loading schemes. The two major loading schemes adopted in the tests are shown in Fig. 5.8: the first scheme consists of one single full loading/unloading cycle at each prescribed displacement (Fig. 5.8a); the second scheme consists of three full loading/unloading cycles at each prescribed displacement, with some partial unloading cycles and partial reloading cycles incorporated (Fig. 5.8b).

The typical axial stress–strain curves of LRS FRP-confined concrete are shown in Fig. 5.9. It can be seen that, similar to conventional FRP-confined concrete, the unloading path of LRS FRP-confined concrete exhibits a

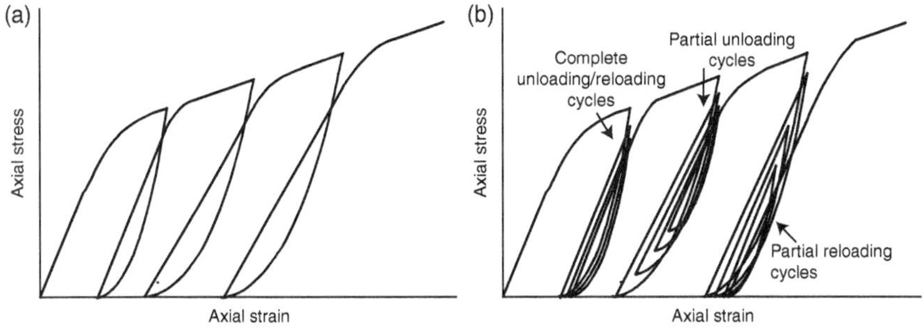

5.8 Loading schemes for LRS FRP-confined concrete: (a) loading scheme A; and (b) loading scheme B.

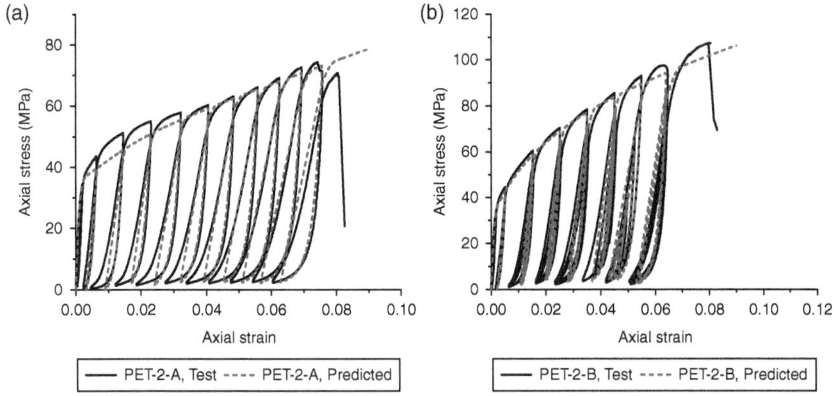

5.9 Typical cyclic stress–strain relationship of LRS FRP-confined concrete under compression: (a) two layers of PET fibre sheets; (b) two layers of PEN fibre sheets.

polynomial shape. The unloading stiffness of the stress–strain curve is initially large but decreases gradually with the axial strain, with the rate of decrease becoming much larger when the axial stress approaches zero. A similar trend is seen for the reloading paths. As the axial stress increases from zero, the reloading stiffness gradually increases until it reaches a nearly constant value. As a result, the major part of the reloading path is a nearly straight line, which is followed by a smooth parabolic portion before it returns to the envelope curve. The energy consumed in each unloading and reloading cycle increases with the plastic axial strain in concrete.

The prediction of the cyclic stress–strain relationships of LRS FRP-confined concrete involves two steps of work: the first is to predict the envelope curve; the second is to predict the loading/unloading paths. A basic hypothesis in studies on the cyclic stress–strain behaviour of unconfined, and FRP or steel-confined, concrete is that an envelope exists and this envelope

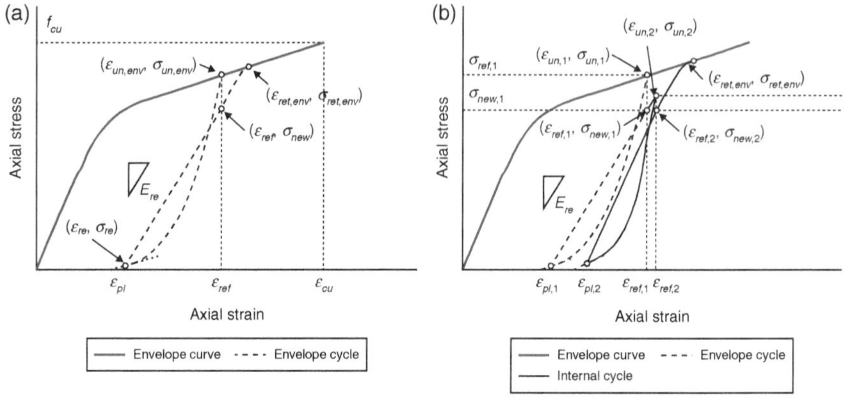

5.10 Cyclic stress–strain model of FRP-confined concrete under compression: (a) envelope cycle and (b) internal cycle.

curve is approximately the same as the stress–strain curve of the same concrete under monotonic loading (Lam *et al.* 2006). Such an envelope curve can be considered as the upper boundary of the response of the concrete subjected to various cyclic loadings in the stress–strain domain. This hypothesis has been verified by many studies (Ilki and Kumbasar, 2002, 2003; Lam *et al.*, 2006; Ozbakkaloglu and Akin, 2012). The above discussed analysis-orientated stress–strain model can be used for predicting the envelop curve of LRS FRP-confined concrete under monotonic loading to account for the nonlinear properties of LRS FRP.

It is worth noting that Lam and Teng (2009) developed a model to define unloading and reloading paths of CFRP-confined concrete under cyclic compressive loading. In their model (Fig. 5.10a and 5.10b), the unloading path from the unloading strain to the reloading strain is described by a polynomial equation, while the reloading path is divided in two parts: the first part is a linear curve extending from the reloading strain to the reference strain; the second part is a parabola extending to the envelope curve (or the next unloading strain). For use of Lam's and Teng's (2009) model with its envelope curve being given by the incremental analysis-orientated model, the determination of the intersection point between the reloading path and the envelope involves a double iteration process. This is more involved than the original Lam's and Teng's (2009) model, for which the envelope curve has a constant slope in the second portion (normally the much larger portion) of the stress–strain curve. For LRS FRP-confined concrete, this constant slope assumption is no longer valid and the slope predicted by analysis-orientated model may vary significantly over the entire strain range (Fig. 5.9). More detail on the determination of the loading and unloading paths can be found in Bai *et al.* (2012).

Figure 5.9a and 5.9b present the comparisons between experimental cyclic stress–strain curves and predictions based on the proposed model, which is a combination of the analysis-orientated model for the envelope curve by Dai *et al.* (2011) and the unloading and reloading paths of Lam's and Teng's (2009) model with slight modifications. Only two specimens are presented as examples. For the predictions, the unloading strains, reloading stresses and actual jacket rupture strains of LRS FRP-confined concrete obtained from the tests were used as the input parameters. The unloading stress, the plastic strain and the new stress at the unloading strain, as well as the whole loading/unloading history, can all be predicted using the proposed model. The predicted curve terminates when the hoop strain of the jacket reaches its rupture strain as found from the confined cylinder test.

Overall, very close agreement between the test results and the predictions is seen in Fig. 5.9. When the axial strain is relatively large, the initial part of the experimental reloading curve has a parabolic shape, which is different from the straight line predicted by Lam's and Teng's (2009) model. This discrepancy may be explained by noting that LRS FRP-confined concrete usually has a large ultimate axial strain, and the degree of damage experienced by the confined concrete in a high axial strain domain is generally larger than that in a small axial strain range. As a result, at the beginning of reloading at a large axial strain, the heavily damaged concrete undergoes a process of re-integration to regain its stiffness. To further evaluate the performance of the proposed cyclic compressive stress–strain model, Fig. 5.11 presents

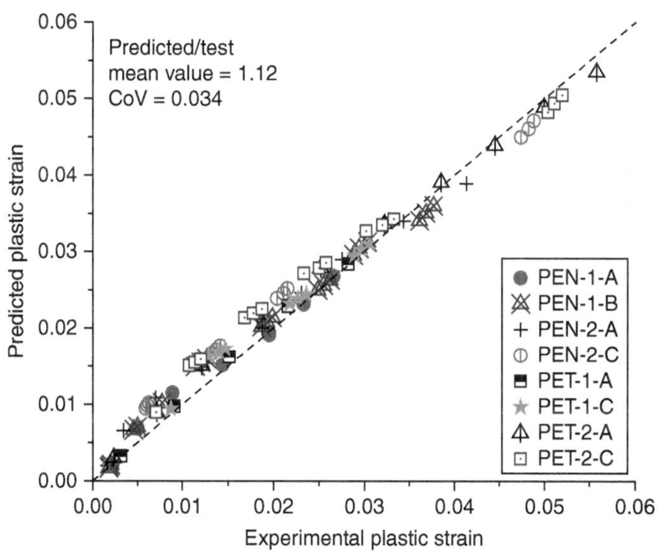

5.11 Comparison between the predicted and experimental plastic strains.

the experimentally observed plastic strain of each unloading cycle in comparison with the predicted results. In Fig. 5.11, the plastic strain of concrete is defined as the residual axial strain of concrete when it is unloaded to zero stress. Since many of the unloading curves were not terminated exactly at zero stress during the test, the plastic strain was determined by tangentially extending the unloading path to zero stress (Fig. 5.10). It is seen in Fig. 5.11 that the average ratio of the predicted to experimental plastic strain is 1.12 with a coefficient of variation (CoV) of 0.034, further demonstrating the applicability of the proposed cyclic stress–strain model.

5.5 Seismic retrofit of reinforced concrete (RC) piers using LRS FRP composites

Two major issues are needed for consideration for the seismic retrofit of RC piers: one is shear enhancement and the other is ductility enhancement. FRP jackets provide additional shear capacity, which reduces the risk of a brittle shear failure, and also increase the strain capacity of concrete in compression, leading to significant enhancement in the member ductility as described above. For the purpose of shear enhancement, the strain capacity of FRP may not be a concern. In fact, it is usually required to control the dilation of RC piers by limiting the strain in FRP under a certain value, e.g. 0.4% (Priestley *et al.*, 1996), to prevent the shear deterioration of concrete. LRS FRP composites have a relatively low modulus of elasticity. In order to satisfy a shear demand, a large amount of fibre may be needed. Therefore, it may not be efficient to use LRS FRP composites for the shear strengthening purpose. On the other hand, the inelastic deformation capacity of flexural plastic hinge regions can be increased by the confinement of RC piers using LRS FRP composites. It is noted that the ultimate state of a conventional FRP-confined RC pier is usually caused by fibre rupture within its plastic hinge zone, particularly at the corner of a rectangular section, leading to an explosive failure of conventional FRP composites. LRS FRP composites are thus an excellent alternative to conventional FRP composites as a confinement material of the plastic hinge zone of RC piers to avoid such an explosive failure manner.

To combine the advantages of both conventional and LRS FRP composites, a hybrid jacketing system has been suggested for the seismic retrofit of RC piers. In this system two types of FRP composites are used: one is the conventional FRP composite used outside the plastic hinge zone for shear strengthening purposes, and the other is the LRS FRP composite used within the plastic hinge zone for confinement purposes (Fig. 5.12). The plastic hinge zone can be assumed to have a length of

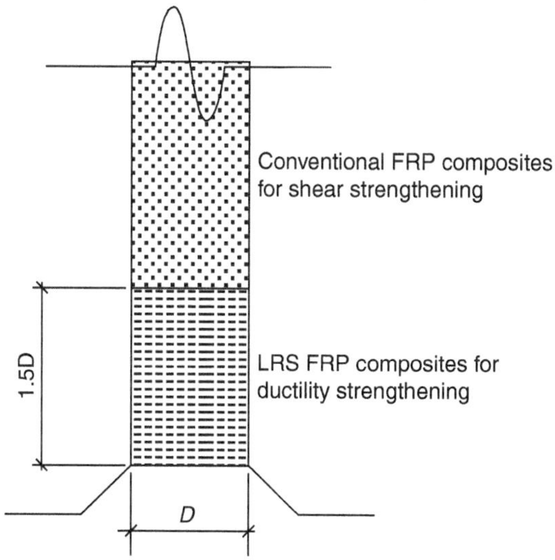

5.12 Hybrid jacketing system for seismic retrofit of RC piers.

1.5D above the pier base for simplicity, in which D is the diameter of a circular pier section or the dimension of the bending side of a rectangular pier section.

Figure 5.13 shows the procedure for the seismic retrofit of an RC pier using the above-described hybrid jacketing method. The first step is a failure mode check to guarantee the pier to fail in flexure. The second step involved is a ductility check. The two-step check can be executed independently for a rectangular/square pier section, since the influence of FRP confinement on the sectional moment capacity is marginal and can be ignored (Nakai *et al.*, 2005; Dai *et al.*, 2012). However, for a circular pier section, the improved sectional moment capacity due to FRP confinement may need to be considered based on sectional analysis and accounting for the stress–strain behaviour of FRP-confined concrete. Thus, after the ductility strengthening design, the shear capacity may need to be re-checked as shown in Fig. 5.13. It should be also noted that, for bridge RC piers, usually the above two-step checks are needed to be conducted for both the longitudinal and transverse directions of a bridge.

The shear capacity of FRP-confined rectangular RC piers can be calculated as follows (JSCE, 2001):

$$V_{fyd} = V_{cd} + V_{sd} + V_{\mathrm{FRP}} \qquad [5.7]$$

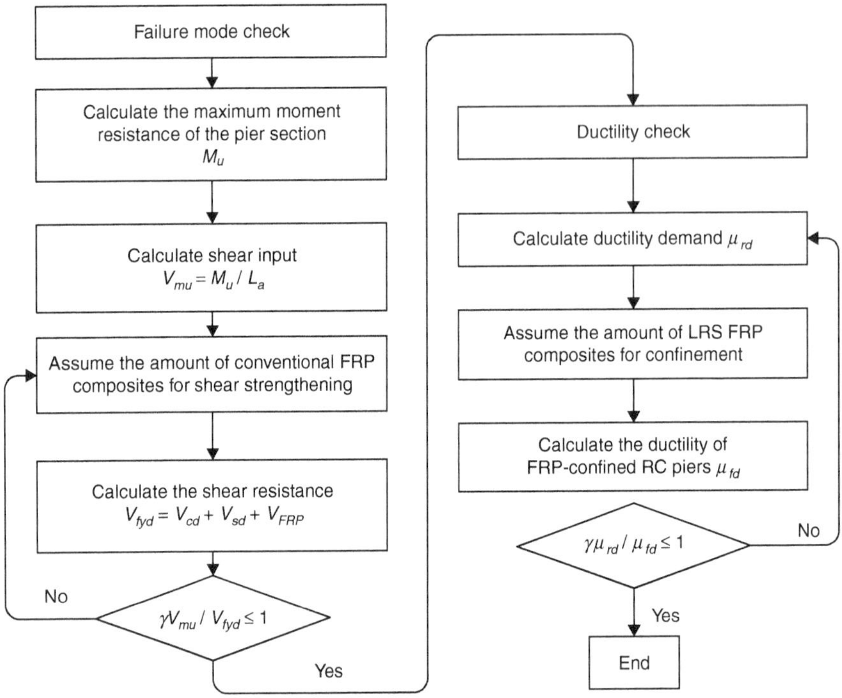

5.13 Flow chart for seismic retrofit design of RC piers using the hybrid FRP jacketing method.

where: V_{fyd} is the shear strength of FRP-strengthened RC piers; V_{cd} is the shear contribution of concrete; V_{sd} is the shear contribution of transverse reinforcement and V_{FRP} is the shear contribution of FRP sheets.

$$V_{cd} = \beta_d \cdot \beta_p \cdot f_{vcd} \cdot b_w \cdot d$$

$$V_{sd} = \left[\frac{A_w \cdot f_{wyd} \cdot (\sin\alpha_s + \cos\alpha_s)}{s_s} \right] \cdot z$$

where: $f_{vcd} = 0.2\sqrt[3]{f'_{cd}} \leq 0.72\,(\text{MPa})$; $\beta_d = \sqrt[4]{1/d} \leq 1.5$; $\beta_p = \sqrt[3]{100\rho_s}$; b_w = member width (mm); d = member effective depth (mm); $\rho_s = \sqrt[4]{A_s/(b_w \cdot d)}$; A_s = section area of flexural reinforcement (mm^2); f_{cd}' = cylinder compressive strength of concrete (N/mm^2); A_w = the cross-sectional area of transverse steel reinforcement within a distance s_s (mm^2); f_{wyd} = yielding strength of transverse steel reinforcement (N/mm^2); α_s = angle between transverse steel reinforcement and member longitudinal axis; s_s = spacing of transverse steel reinforcement (mm) and $z = d/1.15$ (the arm length of flexural reinforcement).

Regarding V_{FRP}, for conventional FRP sheets:

$$V_{FRP} = K\left[\frac{A_{FRP} \cdot f_{FRP} \cdot (\sin\alpha_{FRP} + \cos\alpha_{FRP})}{s_{FRP}}\right] \cdot z$$

$$K = 1.68 - 0.67R, \ (0.4 \le K \le 2.0)$$

$$R = (\rho_{FRP} E_{FRP})^{1/4}\left(\frac{f_{FRP}}{E_{FRP}}\right)^{2/3}\left(\frac{1}{f'_{cd}}\right)^{1/3}, \ (0.5 \le R \le 2.0)$$

where: A_{FRP} = the total cross-sectional area of transverse FRP sheets within a distance s_{FRP} (mm²); f_{FRP} = tensile strength of FRP sheets (N/mm²); E_{FRP} = elastic modulus of FRP sheets (N/mm²); α_{FRP} = angle between fibre direction and member longitudinal axis; s_{FRP} = spacing of FRP sheets (mm) and $\rho_{FRP}\,(= 2t_{FRP}/b_w)$ is the volumetric ratio of FRP sheets, where t_{FRP} is the thickness of FRP sheets.

Very limited work has been conducted on the shear contribution of LRS FRP sheets (The Japan Railway Research Institute, 2006; Senda and Ueda, 2009). The Japan Railway Research Institute (2006) recommended a preliminary design equation for predicting the shear contribution of LRS FRP sheets as follows:

$$V_{FRP} = E_{FRP,d} \cdot \varepsilon_{FRP,d}\left[\frac{A_{FRP} \cdot (\sin\alpha_{FRP} + \cos\alpha_{FRP})}{s_{FRP}}\right] \cdot z$$

$$\varepsilon_{FRP,d} = \frac{0.12(0.5V_{cd} + V_{sd})}{\left[V_{mu} \cdot (E_{FRP,d} \cdot \rho_{FRP})^{1/3}\right]}, \quad (0.014 \le \varepsilon_{FRP,d} \le 0.04 \quad \text{and} \quad \varepsilon_{FRP,d} \le 0.5\varepsilon_{FRP,u})$$

where: $\varepsilon_{FRP,d}$ is the design strain of LRS FRP for shear strength calculation; $E_{FRP,d}$ is the design elastic modulus of LRS FRP (N/mm²) and $\varepsilon_{FRP,u}$ is the rupture strain of LRS FRP.

The ductility of RC piers confined with LRS FRP composites under seismic loading can be determined according to the moment-drift ratio envelope curve as shown in Fig. 5.14. Anggawidjaja *et al.* (2006) proposed the following way to calculate the maximum drift ratio of LRS FRP-jacketed RC piers:

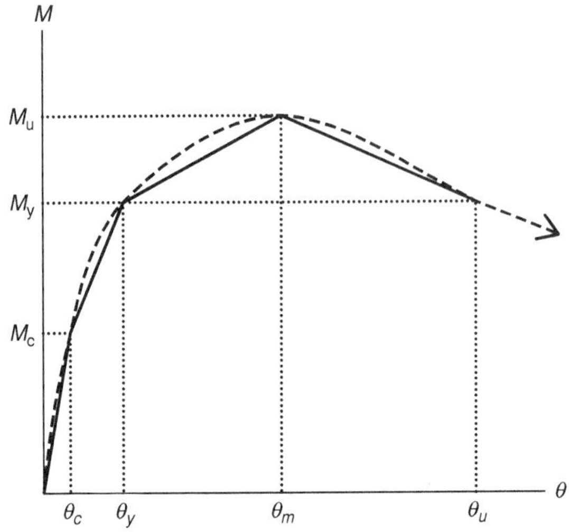

5.14 Moment-drift ratio envelop curve.

$$\theta_u = \theta_m + \eta\left(1 - \frac{M_y}{M_u}\right) \qquad [5.8]$$

$$\theta_m = \theta_{me} + \theta_{\text{pull-out}} + \theta_{\text{shear}} + \theta_{mp}$$

$$\theta_{mp} = \frac{0.021k_s K_o \rho_w + 0.013}{0.79\rho_t + 0.153}$$

$$\eta = 1.22\ K_u + 0.04$$

$$K_o = 0.0019E_{\text{FRP}} \cdot \rho_{\text{FRP}} + 3.65$$

$$K_u = \frac{V_w + V_c}{V_{mu}} \cdot \frac{E_{\text{FRP}} \cdot \rho_{\text{FRP}}}{E_w \cdot \rho_w}$$

where: θ_u = the drift ratio of the member at the ultimate state; θ_m = the drift ratio of the member at the maximum load; θ_{mp} = the drift ratio of the member due to the plastic member rotation at the maximum load; θ_{me} = the drift ratio of the member due to elastic flexural member rotation at the maximum load; $\theta_{\text{pull-out}}$ = the drift ratio of the member due to the rotation caused by the pull-out slips of longitudinal reinforcement at the footing-to-pier interface; θ_{shear} = the drift ratio of the member due to the shear deformation in the plastic hinge zone; M_y = the moment at the yielding load; M_u = the moment at the peak load; ρ_w = the ratio of lateral reinforcement (%); ρ_t = the ratio of tensile reinforcement (%); η = a factor to take into account the slope of the linear descending branch of the load-displacement responses; K_o = a factor representing the confinement effects of FRP jackets; k_s = a factor to take into account the strength of transverse reinforcement and K_u = a factor to represent the FRP strengthening ratio. In Equation [5.8], the drift ratio, θ_{me}, due to elastic flexural member rotation at the maximum load is calculated using Euler's beam theory and Branson's equation to account for the effective moment of inertia of cracked RC columns (Anggawidjaja *et al.*, 2006). The drift ratio, $\theta_{pull\text{-}out}$, is calculated using Ishibashi *et al.* (2002)'s approach, which was also adopted by the JSCE Standard Specifications for Seismic Performance Verification of Concrete Structures (JSCE, 2005). In Anggawidjaja *et al.* (2006)'s model, the drift ratio, θ_{shear}, is estimated to be 5~12% of the total drift ratio at the ultimate state.

Different from the above approach, The Japan Railway Research Institute (2006) also proposed a simple empirical equation accounting for all the deformation components at the ultimate state based upon the same test data as reported by Anggawidjaja *et al.* (2006):

$$\theta_u = \frac{5.03\times10^{-3}\times L_a^{0.651}\times\left(E_{FRP,d}\cdot\rho_{FRP}\right)^{0.0321}}{d^{0.801}\times\rho_s^{0.515}} \qquad [5.9]$$

where L_a = the shear span of the RC pier (mm).

A significant difference between the ultimate state of LRS FRP-confined RC piers and that of conventional FRP-confined ones is that the fibre sheets hardly rupture at the ultimate state in the former case (Fig. 5.15). On the contrary, the ultimate state of conventional FRP (particularly CFRP)-confined RC piers is usually caused by the sudden rupture of fibres (e.g. at the corner of a rectangular pier section) followed by the complete loss of concrete cover in an explosive manner. The use of LRS FRP composites allows progressive bulking of longitudinal steel reinforcement while maintaining the integrity of concrete cover (Dai *et al.*, 2012), so that a sudden loss of the load-carrying capacity can be avoided. Such a progressive and ductile failure mode is highly desired for structures in earthquakes. Figure 5.16 shows a practical application in which LRS FRP sheets were used for the seismic retrofit of a railway bridge pier. Given the same ductility performance, the

5.15 Typical failure of RC piers jacketed with LRS FRP composites within the hinge zone at the ultimate state.

5.16 Seismic retrofit of a railway bridge pier using PET FRP composites.

use of PET FRP sheets as an alternative to AFRP sheets led to a decrease of construction cost by approximately 70% (Fig. 5.17), demonstrating that LRS FRP composites are a more economic and effective seismic retrofit solution for RC piers.

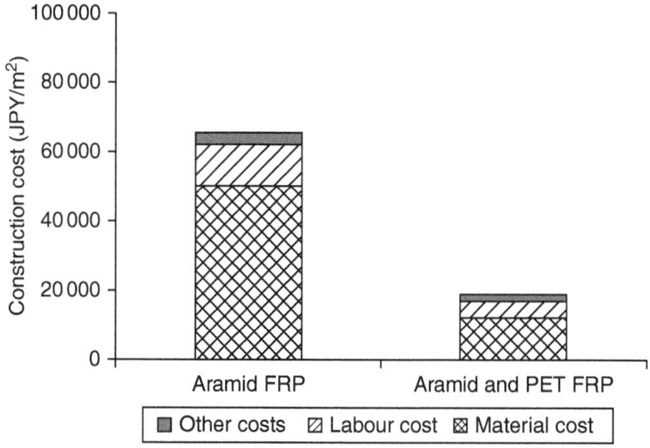

5.17 Cost performance comparison.

5.6 Acknowledgements

The authors are grateful for the financial support received from The Hong Kong Polytechnic University (Project Code: A-PC1L). Thanks also go to Prof. T. Ueda, Mr H. Jaqin, Mr D. Anggawidjaja and Mr M. Senda from Hokkaido University, Japan and Prof. J. G. Teng from The Hong Kong Polytechnic University for their kind help and collaboration towards this research project. The authors are particularly thankful to Mr H. Nakai, Maeda Kosen Co., Japan for providing photos and design guides on LRS FRP composites for the seismic retrofit of RC piers.

5.7 References

Anggawidjaja, D., Ueda, T., Dai, J.G. and Nakai, H. (2006), Deformation capacity of RC columns wrapped by new fibre-reinforced polymer with large fracture strain, *Cement and Concrete Composites*, **28**(10), 914–927.

Bai, Y.L., Dai, J.G. and Teng, J.G. (2012), Behavior of concrete confined with large rupture strain FRP under cyclic axial compression, *Proceedings of International Conference on Earthquake Engineering Research Challenges in the 21st Century – Commemorating the 100th Anniversary of the Birth of Professor Huixian LIU*, Paper No. ICEER-002, CD-ROM, 19–21 May 2012, Harbin, China.

Bai, Y.L., Dai, J.G. and Teng, J.G. (2013), Cyclic compressive behavior of concrete confined with large rupture strain FRP composites, *Journal of Composites for Construction, ASCE*, http://ascelibrary.org/doi/. pdf/10.1061/(ASCE)CC.1943-5614.0000386.

Dai, J.G., Bai, Y.L. and Teng, J.G. (2011), Behavior and modeling of concrete confined by large rupture strain FRP composites, *Journal of Composites for Construction, ASCE*, **15**(6), 963–973.

Dai, J.G., Lam, L. and Ueda, T. (2012), Seismic retrofit of square RC columns with polyethylene terephthalate (PET) fibre reinforced polymer composites, *Construction and Building Materials*, **27**(1), 206–217.

Dai, J.G. and Ueda, T. (2012), Strength and deformability of concrete members wrapped with fibre reinforced polymer composites with a large rupture strain, in Fardis Michael, N. (ed.), *Innovative Materials and Techniques in Concrete Construction*, ISBN 978-94-007-1996-5, Springer Publisher, Chapter 14, 225–241.

Ilki, A. and Kumbasar, N. (2002), Behavior of damaged and undamaged concrete strengthened by carbon fiber composite sheets, *Structural Engineering and Mechanics*, **13**(1), 75–90.

Ishibashi, T., Tsuyoshi, T., Kobayashi, K., Yoshida, T. and Umihara, T. (2002), Study on ductility evaluations on reinforced concrete columns subjected to reversed cyclic loading with large deformations, *Proceedings of JSCE*, **56**(711), 79–96.

Jiang, T. and Teng, J.G. (2007), Analysis-oriented stress-strain models for FRP-confined concrete, *Engineering Structures*, **29**, 2968–2986.

Japan Railway Research Institute (2006), *Guidelines of Design and Construction Seismic Retrofit of Railway RC Bridge Piers Using A&P Methods* (in Japanese).

JSCE 292 Committee on Concrete Structures with Externally Bonded Continuous Fibre Reinforcing Materials (2001), *Recommendation for Upgrading Concrete Structures with Use of Continuous Fibre Sheets*, Concrete Engineering Series 41, JSCE.

JSCE Subcommittee on English Translation of the Standard Specifications (2005), *Standard Specifications for Concrete Structures – 2002 'Seismic Performance Verification'*.

Lam, L. and Teng, J.G. (2003), Design-oriented stress-strain model for FRP-confined concrete. *Construction and Building Materials*, **17**(6–7), 471–489.

Lam, L. and Teng, J.G. (2004), Ultimate condition of fibre reinforced polymer-confined concrete. *Journal of Composites for Construction, ASCE*, **8**(6), 539–548.

Lam, L., Teng, J.G., Cheung, C.H. and Xiao, Y. (2006), FRP-confined concrete under axial cyclic compression, *Cement and Concrete Composites*, **28**(10), 949–958.

Lam, L. and Teng, J.G. (2009), Stress-strain model for FRP-confined concrete under cyclic axial compression, *Engineering Structures*, **31**(2), 308–321.

Lechat, C., Bunsell, A.R. and Davies, P. (2011), Tensile and creep behavior of polyethylene terephthalate and polyethylene naphthalate fibres, *Journal of Materials Science*, **46**(2), 528–533.

Lechat, C., Bunsell, A.R., Davies, P. and Piant, A. (2006), Mechanical behavior of polyethylene terephthalate and polyethylene naphthalate fibres under cyclic loading, *Journal of Materials Science*, **41**(6), 1745–1756.

Mander, J.B., Priestley, M.J.N. and Park, R. (1988), Theoretical stress-strain model for confined concrete, *Journal of Structural Engineering, ASCE*, **114**(8), 1804–1826.

Nakai, H., Sato, H., Dhannyanto, A. and Ueda, T. (2005), Seismic retrofit of RC piers using large rupture strain FRP composites, *Proceedings of the Japan Concrete Institute*, **27**(2), 1081–1086.

Ozbakkaloglu, T. and Akin, E. (2012), Behavior of FRP-confined normal- and high-strength concrete under cyclic axial compression, *Journal of Composites for Construction, ASCE*, **16**(4), 451–463.

Priestley, M.J.N., Seible F. and Calvi G.M. (1996), *Seismic Design and Retrofit of Bridges*. John Wiley, New York.

Popovics, S. (1973), A numerical approach to the complete stress-strain curve of concrete, *Cement and Concrete Research*, **3**(5), 583–599.

Seible, F., Priestley, M.J., Hegemier, G.A. and Innamorato, D. (1997), Seismic retrofit of RC columns with continuous carbon fibre jackets, *Journal of Composites for Construction, ASCE*, **1**(2), 52–62.

Senda, M. and Ueda, T. (2009), Experimental study on shear strengthening by continuous fibre with high fracturing strain, *Proceedings of the Concrete Structure Scenarios, JSMS*, **8**, 249–256 (in Japanese).

Shao, Y., Zhu, Z. and Mirmiran, A. (2006). Cyclic modeling of FRP-confined concrete with improved ductility, *Cement and Concrete Composites*, **28**(10), 959–968.

Teng, J.G., Huang, Y.L., Lam, L. and Ye, L.P. (2007), Theoretical model for fiber reinforced polymer-confined concrete, *Journal of Composites for Construction, ASCE*, **11**(2), 201–210.

Wang, Z.Y. and Li, H.P. (2009), Loading and reloading criteria of FRP-confined concrete under cyclic compression, *Building Structure*, **39**(7), 100–104 (in Chinese).

Part II

Applications

6
All-composite superstructures for accelerated bridge construction

H. S. JI, Daewon University College, Republic of Korea and
W. SONG and Z. J. MA, University of Tennessee at Knoxville, USA

DOI: 10.1533/9780857097019.2.143

Abstract: This chapter presents the design, manufacture, installation, and in-service structural performance of an all-composite sandwich bridge superstructure for accelerated bridge construction (ABC). The superstructure is made of fiber-reinforced polymer (FRP). The discussions about the manufacture and installation include construction details when the FRP superstructure is implemented in practice. The assessment of the in-service structural performance may provide baseline data for the capacity rating and serve as part of the long-term performance monitoring of the FRP superstructure. Finally, this chapter explains why all-composite superstructures for ABC can be economically competitive in the future.

Key words: accelerated bridge construction, capacity rating, fiber-reinforced polymer (FRP) composites, long-term performance, sandwich bridge superstructure.

6.1 Introduction

Recently, a main concern in bridge engineering had been the rehabilitation and reconstruction of deteriorated bridge members. Accelerated bridge construction (ABC) can overcome the shortcomings of traditional construction methods, which are typically labor-intensive and time-consuming. ABC can minimize disruption to public communities and save indirect costs associated with detouring traffic. Concrete members prefabricated in factory are often utilized in ABC, where the connections of prefabricated concrete members have been extensively studied in recent years (Li *et al.* 2010a, 2010b; Ma *et al.* 2012a, 2012b; Zhu *et al.* 2012a, 2012b). As alternatives to prefabricated concrete members, bridge superstructures entirely made of advanced composites or fiber-reinforced polymer (FRP) are also potential solutions to ABC, due to the following advantages of the materials:

- ***Light weight*** – FRP superstructures weigh about 25% or less of a conventional reinforced concrete superstructure. Therefore, FRP composites reduce the dead load of a superstructure significantly. Reduction of dead load can increase the rating factor when capacity rating of a bridge is performed. Since capacity rating of a bridge is proportional to the rating factor, the load carrying capacity of a superstructure reconstructed or rehabilitated with FRP composites may be re-rated to its original design load.

- ***Durability*** – Corrosion of steel reinforcement in concrete is the leading cause of deterioration of conventional concrete superstructures. Steel reinforcement is susceptible to oxidation when exposed to deicing salt and other chlorides. Compared to conventional materials, FRP composites provide high resistance to corrosion. Therefore, less maintenance may be required for FRP superstructures.
- ***Rapid Installation*** – Factory fabrication and modular construction of FRP superstructures allows rapid installation on construction sites. Installation of FRP superstructures in bridges with high traffic volume can substantially reduce the construction time of superstructures to half when compared to the labor-intensive and time-consuming construction process of a conventional cast-in-place concrete superstructure. As a result, disruption to communities, and indirect costs associated with detouring traffic, can be minimized by FRP superstructures.
- ***High Specific Stiffness/Strength*** – Advanced composites offer the advantages of high stiffness-to-weight ratios (specific stiffness) and strength-to-weight ratios (specific strength) when compared to conventional construction materials. Therefore, they can provide good combinations of strength/stiffness and weight for bridge superstructures.
- ***Lower Life-cycle Cost*** – A life-cycle cost usually consists of an initial construction cost and maintenance cost. Since FRP superstructures are expected to have good long-term durability and require little maintenance, they can be more cost-effective in service life than bridge superstructures with reinforced concrete.

In the United States, there are more than 40 bridge superstructures with FRP composites (Triandafilou and O'Connor, 2009). No-Name Creek Bridge built in Russell, Kansas, was the first all-FRP bridge superstructure in the US (Plunkett, 1997). The FRP superstructure had a sandwich construction with honeycomb cores. Tech 21, the Smith Road Bridge built in Butler County, Ohio, was the third all-FRP superstructure in the US (Farhey, 2005). New York State Department of Transportation built an FRP superstructure in a demonstration project that had a cell core system providing stiffness in two directions (Alampalli, 2006). In other countries, all-composite superstructures have also been utilized in bridge construction. An FRP superstructure was installed and opened to the public in Yeongwol, Gangwon-Do, Korea in 2002 (Ji *et al.*, 2007). The bridge superstructure was a sandwich structure with corrugated cores that could contribute significantly to the stiffness in the selected direction compared with other cores.

The projects above demonstrate how advanced composites or FRP materials were implemented in new bridge construction, or accelerated the rehabilitation of deteriorated bridges. Research on FRP superstructures indicates that their design should be controlled by stiffness rather than strength (Triandafilou and O'Connor, 2009). Although the advantages of FRP composites are recognized, their application in bridges needs further investigation. In fact, a number of technical challenges concerning design methods, connection details, dynamic effects, and long-term performance still remain unsolved. This chapter demonstrates structural analysis and design, manufacture and installation, and in-service structural evaluation of an FRP superstructure to consider these challenges. Finally, costs of the FRP superstructure are compared to those of reinforced concrete bridge superstructures.

6.2 Structural analysis and design

When designing all-composite bridge superstructures, designers can tailor FRP materials and structural configurations of superstructures to meet design requirements. As an example, the design of a two-lane FRP superstructure is discussed in this section. The FRP superstructure is now in service in South Korea, and has a sandwich construction (Ji *et al.*, 2010). Corrugated cores are sandwiched between top facing and bottom facing. The sandwich construction in this study can reduce weight and provide significant stiffness in the selected direction. The discussion in this section includes design requirements, material selection, preliminary design based on design requirements, and detailed design of the cross-section.

6.2.1 Design requirements

The design of FRP superstructures is stiffness-oriented (Federal Highway Administration, 2011). Although no official specifications are available for the deflection limit of FRP superstructures, general design guidelines for conventional bridge superstructures are provided by the AASHTO LRFD Bridge Design Specifications (AASHTO, 2010). The AASHTO's requirement for the stiffness of bridge superstructures with conventional materials is that the maximum deflection due to vehicular live loads should not exceed 1/800 of the bridge span length.

The criterion above was used to design the FRP superstructure in this chapter. The FRP superstructure in this chapter was expected to be in service in South Korea. As a result, the design vehicular live load in this study

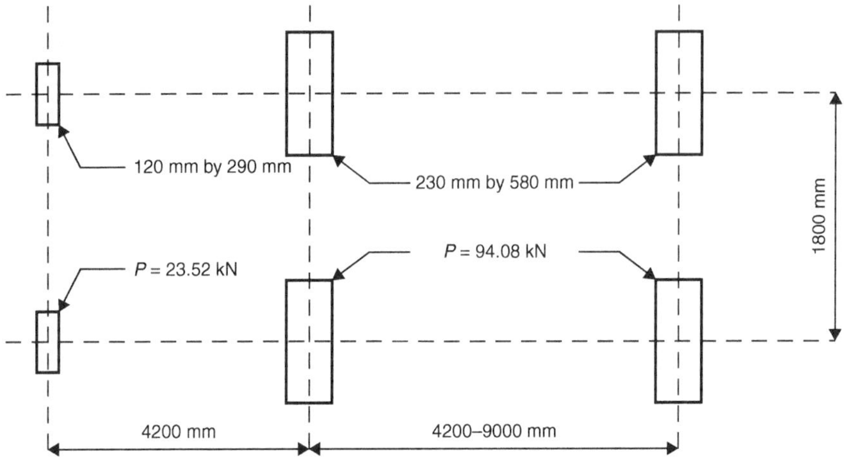

6.1 DB-24 truck load and tire contact areas.

was consistent with Ministry of Construction and Transportation (MOCT) (2000a). As for the strength requirements, the maximum strain under service load should be limited to 20% of the ultimate strain. Laminate failure prediction was based on the first-ply failure criterion.

For a simply-supported bridge with span less than 44 m, the standard design truck load DB-24 produces higher moment and deflection than the design lane load (MOCT, 2000a). A DB-24 truck load is approximately 1.3 times heavier than the HS-20 truck load specified in AASHTO LRFD Bridge Design Specifications (2010). One standard design truck load DB-24 with tire contact areas, as shown in Fig. 6.1 (MOCT, 2000b), was assumed to be the design live load. The weight of a DB-24 truck is 423 kN.

6.2.2 Materials and lamina design

Superstructures with FRP composites including the one in this study can be characterized at three levels: the lamina level (a single ply), the laminate level (multiple plies), and the structural level (Altenbach *et al.*, 2004). This section focuses on the materials and lamina design. The design at the latter two levels will be addressed in the following sections.

The primary constituents of FRP composites are fibers and resin matrices. In FRP bridge superstructures, glass fibers are mainly used as fiber reinforcement, and polyester or vinyl ester is the most widely used resin matrix. Their combinations can be the most cost-effective in current practice. In this study, E-glass fibers and vinyl ester resin were chosen as the main constituents for the all-FRP superstructure for ABC. Their material properties are given in Table 6.1 (Ji *et al.*, 2010).

Table 6.1 Properties of constituent materials

Material	E (GPa)	G (GPa)	υ	ρ (kN/m³)
E-glass fiber	72.4	27.6	0.22	25.4
Vinyl ester resin	3.91	1.38	0.37	12.4

E is the Young's modulus; *G* is the shear modulus; υ is the Poisson's ratios; ρ is the density.

To design a superstructure with FRP composites, it is necessary to know the mechanical properties of a single ply or a lamina made of fiber reinforcement and resin matrix. The properties of a lamina with unidirectional fibers can be predicted from properties of fibers and resin matrix according to micro-mechanics models based on several simple rules of mixtures. Most calculations about the properties of a unidirectional lamina depend on the volume fraction of fibers and resin matrix. The fiber and matrix volume fractions, V_f and V_m, respectively, for a unidirectional lamina based on their weight fractions and densities are calculated in Equation [6.1].

$$V_f = \frac{w_f/\rho_f}{(w_f/\rho_f)+(w_m/\rho_m)} \quad \text{and} \quad V_m = 1 - V_f \qquad [6.1]$$

where w_f and w_m are the weight fractions of fibers and resin matrix, ρ_f and ρ_m are the densities of fibers and resin matrix.

The mechanical properties of a unidirectional lamina with known fiber and matrix properties and their volume fractions can be calculated by simple rules of mixtures as shown in Equations [6.2]–[6.5].

$$E_1 = \eta_0 E_f V_f + E_m (1 - V_f) \qquad [6.2]$$

$$E_2 = \frac{1}{(V_f/E_f)+((1-V_f)/E_m)} \qquad [6.3]$$

$$G_{12} = \frac{1}{(V_f/G_f)+((1-V_f)/G_m)} \qquad [6.4]$$

$$v_{12} = v_f V_f + v_m (1 - V_f) \qquad [6.5]$$

where η_0 is the efficiency factor of fibers – for unidirectional fibers, η_0 is equal to 1.0; E_f and E_m are the Young's moduli of fibers and resin matrix, respectively; G_f and G_m are the shear moduli of fibers and resin matrix, respectively, and; υ_f and υ_m are the Poisson's ratios of fibers and resin matrix, respectively.

Table 6.2 Mechanical properties of a unidirectional FRP lamina

	E_1 (GPa)	E_2 (GPa)	G_{12} (GPa)	υ_{12}
Predicted values	30.8	6.22	2.20	0.323
Experimental values	36.9	10.1	3.34	0.325

For the FRP sandwich superstructure in this study, the unidirectional laminae in it have w_f equal to 57% and w_m equal to 43%. Assuming that there are no voids in the FRP laminae, their mechanical properties predicted using Equations [6.1]–[6.5] are shown in Table 6.2 (Ji *et al.*, 2010).

The FRP laminae discussed here, with various orientations, were used to construct the laminates for the facings and core walls of the FRP superstructure. The details about the lay-ups of the laminates were determined based on the preliminary and detailed design, and will be discussed in the following sections. In design, it is often necessary to have a preliminary estimation of the thickness of a laminate. The thickness that can be achieved is related to the mass of the materials used in the laminate. The thickness of a laminate can be predicted using a thickness constant in Equation [6.6]:

$$\text{Thickness of a laminate (mm)} = \text{material mass (kg/m}^2) \times \text{thickness constant} \qquad [6.6]$$

The thickness constant can be calculated from a given density in Equation [6.7]:

$$\text{Thickness constant} = \frac{1}{\text{density}\left(\text{kg}/\text{m}^3\right)} \times 10^3 \qquad [6.7]$$

6.2.3 Preliminary design

At the preliminary design stage, the required material and section properties for the FRP sandwich superstructure were determined based on the design load mentioned in Section 6.2.1. They should be obtained to guarantee that the midspan deflection Δ under the maximum live-load effect would not exceed 1/800 of the bridge span length. The FRP superstructure discussed here was a simply-supported bridge with a single span and two traffic lanes. It was 10 m long by 8 m wide. The placement of three wheel loads from the DB-24 truck is shown in Fig. 6.2. It was determined by the influence line to maximize the moment effect at midspan (Ji *et al.*, 2010). The loads in Fig. 6.2 are the wheel loads multiplied by the impact factor of 1.3.

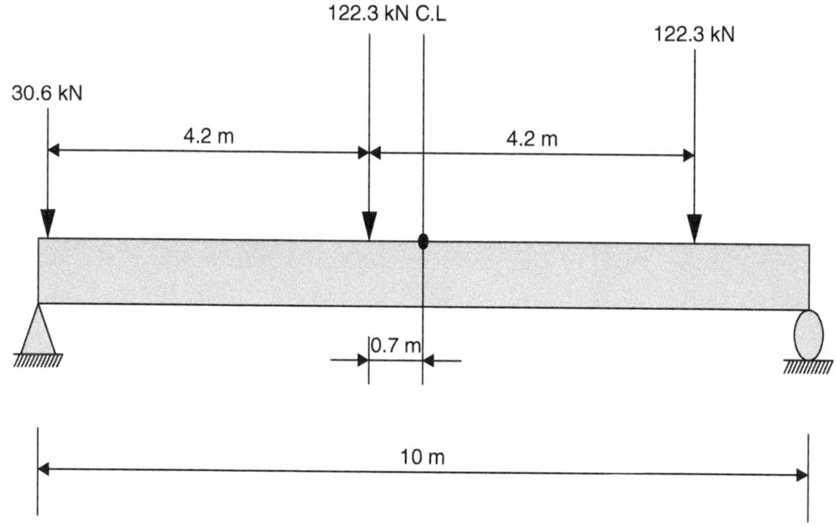

6.2 DB-24 truck load's placement location. (C.L. refers to center line.)

At the preliminary design stage, it is desirable to perform the design without invoking finite element analysis (FEA) (Triandafilou and O'Connor, 2009). The properties of the top and bottom facings and the depth of the FRP superstructure could be designed by the following two simplified structural analysis methods without using FEA. In this study, the subscripts x and y denote the transverse direction and span direction, respectively.

Principle of virtual work

According to the principle of virtual work, the required stiffness of the cross-section E_yI times the midspan deflection Δ for the loading case shown in Fig. 6.2 could be calculated in Equation [6.8] (Ji *et al.*, 2010). It should be noted that E_yI is constant along the span length.

$$\Delta E_y I = \int_0^L M_L M_U \mathrm{d}y \overset{\Delta=\frac{L}{800}}{\Rightarrow} \left(E_y I\right)_{\min} = \frac{800}{L}\int_0^L M_L M_U \mathrm{d}y \qquad [6.8]$$

where M_U is the moment caused by the unit load acting at the midspan, M_L is the moment caused by the actual loads, E_y is the Young's modulus of the cross-section in the longitudinal direction, L is the bridge span length, and I is the moment of inertia of the superstructure's cross-section.

The FRP superstructure in this study has a sandwich construction. If the contribution of core walls to the moment of inertia I is neglected and the

Young's moduli of top and bottom facings in the longitudinal direction are assumed to be the same, the expression for E_yI is given in Equation [6.9].

$$E_yI = E_y^f\left[\frac{Wt_{tf}^3}{12} + Wt_{tf}\left(h - \frac{t_{tf}}{2} - z_c\right)^2 + \frac{Wt_{bf}^3}{12} + Wt_{bf}\left(z_c - \frac{t_{bf}}{2}\right)^2\right] \quad [6.9]$$

where h is the depth of the superstructure, W is the width of the superstructure's cross-section, t_{tf} and t_{bf} are the thickness of the top and bottom facing respectively, z_c is the distance between the centroid of the superstructure's cross-section and the extreme bottom fiber of bottom facing, and E_y^f is the Young's modulus of the top and bottom facing in the longitudinal direction.

In order to satisfy the deflection limit for the FRP superstructure, the actual E_yI should be larger than the required one $(E_yI)_{\text{min}}$. This relationship is indicated by Equation [6.10]. In practice, the material properties and thickness of the laminated facings and the depth of the FRP superstructure can be varied to satisfy Equation [6.10].

$$\frac{(E_yI)_{\text{actual}}}{(E_yI)_{\text{min}}} \geq 1.0 \quad [6.10]$$

Equivalent strip width

The preliminary design could be further refined by using the equivalent strip width for FRP superstructures (Song, 2012). The FRP superstructure in this study was directly treated as a 1-D beam on the principle of virtual work, so that its actual width was utilized in the previous calculation. This assumption should be modified because the FRP superstructure might behave like a 2-D plate instead of a 1-D beam. Further, Equations [6.8] and [6.9] neglected the deflection of the FRP superstructure due to out-of-plane shear deformation. Considering these problems, the design at the preliminary design stage could be improved by using the equivalent strip width to design the superstructure as a 1-D Timoshenko beam.

Typical FRP superstructures can be idealized as single-layer specially orthotropic plates (Davalos *et al.*, 1996, 2001; Zhou, 2002). An example of the idealization from previous research is given in Fig. 6.3 (Zhou, 2002). At the preliminary design stage, the equivalent material properties of idealized orthotropic plates, including the one in this study, are not known. As is evidenced in Fig. 6.3, the equivalent material properties are influenced by the properties of FRP laminae and laminates, and structural configuration and

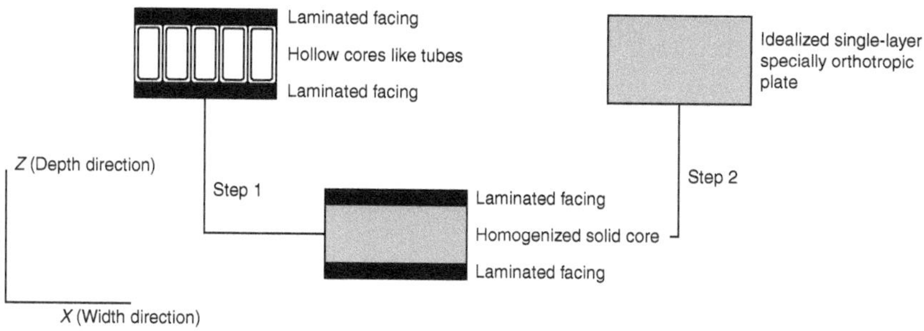

6.3 Equivalent single-layer orthotropic plate (Zhou, 2002).

geometry of FRP superstructures. Few of these factors are available at the preliminary design stage. However, reasonable assumptions can be made about the equivalent material properties of idealized orthotropic plates.

FRP superstructures in practice are typically less than 18.3 m and their width-to-span ratios are less than 1.5. Most of them have one lane or two lanes and equivalent material properties described below. The ratio of the two equivalent Young's moduli E_x/E_y varies between 0.1 and 0.9. The equivalent major Poisson's ratio υ_{yx} changes from 0.1 to 0.4. The ratio of the equivalent in-plane shear modulus G_{yx} to the equivalent Young's modulus E_x is between 0.1 and 0.4. The ratio of the equivalent out-of-plane shear modulus G_{xz} to the equivalent Young's modulus E_x, and the ratio of the equivalent out-of-plane shear modulus G_{yz} to the equivalent Young's modulus E_y, are at least 0.025. The combination of E_x, E_y, G_{yx}, and υ_{yx} for an FRP superstructure usually satisfies the relationship in Equation [6.11].

$$\left(\frac{E_y}{E_x}\right)^{0.5} > \upsilon_{yx} + 2\left[1 - \upsilon_{yx}^2\left(\frac{E_x}{E_y}\right)\right]\left(\frac{G_{yx}}{E_x}\right) \qquad [6.11]$$

Considering the discussions above about practical FRP superstructures, the equivalent strip width for the preliminary design of FRP superstructures is given in Equations [6.12]–[6.19] (Song, 2012). It should be mentioned that in essence Equations [6.12]–[6.19] are simplifications of 2-D plate analysis. Therefore, each load on an FRP superstructure may correspond to an equivalent strip width. The equivalent strip width here is used to facilitate the midspan deflection calculation by Timoshenko beam theory, in which only loading conditions in the longitudinal or span direction are required. In the transverse direction, it is assumed that design live loads have symmetric distribution with respect to the centerlines of FRP superstructures. The symmetry can be utilized to reduce the number of equivalent strip width calculations at the preliminary design stage.

$$EW = \frac{w_p (b/L)\left[3 - 4(b/L)^2\right]}{384 \sin(b\pi/L) w'} \qquad [6.12]$$

$$n_1 = E_y/E_x,\ n_2 = L/W,\ s_1' = \sqrt{(0.5n_1^{0.5} + 0.4)},\ s_2' = \sqrt{(0.5n_1^{0.5} - 0.4)} \qquad [6.13]$$

$$C_1' = \frac{4s_1's_2' \cosh(\pi s_1'/2n_2)\cos(\pi s_2'/2n_2) + \sinh(\pi s_1'/2n_2)\sin(\pi s_2'/2n_2)}{\left(1 + 1/\sqrt{n_1}\right)s_2' \sinh(\pi s_1'/n_2) + \left(1 - 1/\sqrt{n_1}\right)s_1' \sin(\pi s_2'/n_2)} \qquad [6.14]$$

$$C_2' = \frac{4s_1's_2' \sinh(\pi s_1'/2n_2)\sin(\pi s_2'/2n_2) - \cosh(\pi s_1'/2n_2)\cos(\pi s_2'/2n_2)}{\left(1 + 1/\sqrt{n_1}\right)s_2' \sinh(\pi s_1'/n_2) + \left(1 - 1/\sqrt{n_1}\right)s_1' \sin(\pi s_2'/n_2)} \qquad [6.15]$$

$$w' = \sum_{m=1}^{29} [k_1' + k_2' + k_3'] \frac{\sin(w_p m\pi/2W)\sin(am\pi/W)(1 - \upsilon_{yx}^2/n_1)}{m\pi^5\left[(mn_2)^4/n_1 + 1.4(mn_2)^2/n_1 + 1\right]} \qquad [6.16]$$

$$k_1' = C_1'\left[\frac{(mn_2)^3}{n_1} + \frac{1.4mn_2}{n_1}\right]\cos\left(\pi s_2'\left(\frac{x}{L} - \frac{0.5}{n_2}\right)\right)\cosh\left(\pi s_1'\left(\frac{x}{L} - \frac{0.5}{n_2}\right)\right) \qquad [6.17]$$

$$k_2' = C_2'\left[\frac{(mn_2)^3}{n_1} + \frac{1.4mn_2}{n_1}\right]\sin\left(\pi s_2'\left(\frac{x}{L} - \frac{0.5}{n_2}\right)\right)\sinh\left(\pi s_1'\left(\frac{x}{L} - \frac{0.5}{n_2}\right)\right) \qquad [6.18]$$

$$k_3' = \sin\left(\frac{m\pi x}{W}\right) \qquad [6.19]$$

where EW is the equivalent strip width corresponding to a specific load, w_p is the width of a pressure in the transverse direction, b is the distance from the center of a pressure to the closest support in the span direction, and a is the position of a pressure's center in the transverse direction. L is the span length and W is the width of the superstructure. x is the coordinate of a point in the transverse direction at midspan. For one-lane FRP superstructures, x is suggested to be either $W/2$ or a. For multi-lane FRP superstructures, x is suggested to be either a or the coordinate of the center of each lane in the transverse direction.

Equations [6.12]–[6.19] only include three key parameters n_1, n_2, and υ_{yx}, which have significant influences on the equivalent strip width (Song, 2012). The actual values about E_x, E_y, G_{yx}, G_{xz}, and G_{yz}, which are not known and difficult to assume at the preliminary design stage, are not required for the calculation of the equivalent strip width. Consequently, these equations can avoid inconvenient 2-D plate design based on first-order shear deformation theory (FSDT) when the equivalent strip width from them is directly used with Timoshenko beam theory. It should be emphasized that E_x, E_y, G_{yx}, G_{xz}, and G_{yz} are the equivalent material properties of an idealized single-layer specially orthotropic plate. They are generally not equal to the material properties of a lamina or a laminate in an FRP superstructure, as is implied in Fig. 6.3. Their calculation may be referred to several research studies conducted previously (Davalos *et al.*, 1996, 2001).

The equivalent strip width discussed here could be used to perform the preliminary design of the FRP sandwich superstructure in this study. Considering the properties of the FRP laminae in Table 6.2, n_1 should be 6 or less if the contribution of core walls were neglected. At the beginning of the preliminary design, E_y was arbitrarily assumed as 6.0 GPa. υ_{yx} might be assumed as 0.3. The span-to-depth ratio was considered as 12 (Federal Highway Administration, 2011). To use Timoshenko beam theory in design, another assumption about the out-of-plane shear modulus G_{yz} should also be made, although this assumption had no impact on the equivalent strip width calculation. In this study, G_{yz} was $E_y/40$.

The midspan deflections at ($W/4$, $L/2$) predicted by Timoshenko beam theory with the equivalent strip width at several n_1 are shown as the analytical results in Fig. 6.4 and Table 6.3. Correspondingly, the midspan deflections at ($W/4$, $L/2$) from FEA using the commercial package ABAQUS are also shown in Fig. 6.4. The maximum deflections from tire contact areas are shown in Table 6.3 to investigate the local deformation within these regions.

When the results in Fig. 6.4 and Table 6.3 were predicted, the placement of DB-24 truck loads in the span direction was the same as that in Fig. 6.2. In the transverse direction, two DB-24 trucks were considered, and each truck was symmetric with respect to the centerline of each lane. The modeling in FEA was consistent with FSDT. In FEA, G_{yx}/E_x was equal to 0.25 and G_{xz}/E_x was equal to 0.025. These two ratios did not affect the equivalent strip width and Timoshenko beam theory but did affect FSDT (Song, 2012).

Figure 6.4 shows that the equivalent strip width together with Timoshenko beam theory can be used to predict midspan deflections and perform the preliminary design of the FRP sandwich structure in this study. Table 6.3 indicates that the local deformation at tire contact areas is not significant enough to invalidate the 1-D Timoshenko beam design with the equivalent strip width discussed here. The maximum analytical deflection from

Table 6.3 Deflections at (*W*/4, *L*/2) and tire contact area (unit: mm)

n_1	1	2	3	4	5	6
Analytical results	7.629	7.927	7.995	8.015	8.020	8.021
FEA results	8.144	8.214	8.271	8.321	8.365	8.406

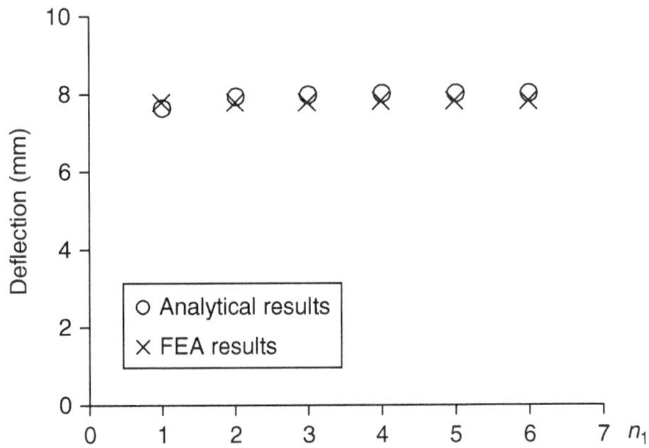

6.4 Deflections at the position (*W*/4, *L*/2).

Table 6.3 is about 1/1250 of the span length of the FRP sandwich superstructure. As a result, a cross-section whose Young's modulus in the span direction E_y and depth h satisfy the relationship in Equation [6.20] is necessary to the FRP superstructure in this study. According to Fig. 6.3, E_y is a function of laminated facings and core walls. It also depends on the depth of the cross-section. For the FRP superstructure in this study, the material properties and thickness of the laminates in facings and cores and the depth can be varied to satisfy Equation [6.20].

$$Ey \cdot h^3 \geq (3840) \cdot (10\,000/12)^3 \text{ (N} \cdot \text{mm)} \qquad [6.20]$$

It should be mentioned that the equivalent strip width from Equations [6.12]–[6.19] is also suitable to perform parametric study and design optimization. The procedure is explained as follows. The first two steps in the procedure are utilized in the example above. The remaining steps should be followed if parametric study and design optimization at the preliminary design stage are conducted without using FEA. Although Equations [6.12]–[6.19] appear to be complex, they can be conveniently utilized in a spreadsheet. Once a spreadsheet is set up, minor modifications are required to cover different cases in the parametric study.

1. Calculate the aspect ratio n_2 from the geometry of an FRP superstructure and make an assumption about the depth. The span-to-depth ratio is suggested to be equal to 12.
2. Make an initial assumption about n_1 and υ_{yx}. A reasonable assumption about n_1 is 10, and υ_{yx} may be taken as 0.3. Then calculate the equivalent strip width for each load.
3. Assume that G_{yz}/E_y is 1/40, and then calculate the required E_y by Timoshenko beam theory based on the argument that the total midspan deflection under design live loads is equal to its limit.
4. Based on the calculated E_y, tailor FRP materials and design cross-sections.
5. Idealize the FRP superstructure as a single-layer specially orthotropic plate and update the equivalent properties E_y, E_x, υ_{yx}, G_{yx}, G_{yz}, and G_{xz}.
6. Based on the actual equivalent properties of the cross-section, update the equivalent strip width.
7. Modify laminates in components of the FRP superstructure and/or reduce the depth if predicted midspan deflections with the updated equivalent strip width are not close to the limit.
8. Repeat the process from Steps 5 to 7 until satisfactory results are achieved.

6.2.4 Detailed design

The FRP superstructure in this study has a sandwich structure with corrugated core walls. Besides the parameters influencing the relationship in Equation [6.10] or [6.20], several geometric factors of corrugated cores also affect the behavior of the superstructure. These geometric factors are the corrugation angle, the ratio of the core depth to core wall's thickness, and the ratio of the pitch to core depth as shown in Fig. 6.5 (Libove and Hubka, 1951). All the parameters were determined based on a detailed design by FEA using the LUSAS software. In the detailed design, DB-24 trucks multiplied by the impact factor of 1.3 were considered as design live loads. The dead loads, including the self-weight of the superstructure and asphalt wearing surface with a thickness of 50 mm, were also considered in the design.

Based on the FEA, the final proposed section of the FRP sandwich superstructure with corrugated cores for manufacturing is shown in Fig. 6.6 (Ji *et al.*, 2007, 2010). The FRP superstructure is 10 m in length, 8 m in width, and 0.94 m in depth. The superstructure's cross-section is constant along the whole span. The lay-ups of the FRP laminae in Section 6.2.2 in the laminates for facings and core walls are shown in Table 6.4. The corrugation angle, the ratio of the core depth to core wall's thickness, and the ratio of

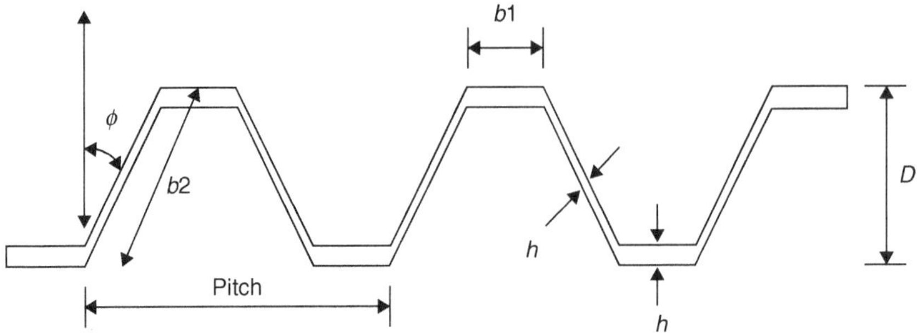

6.5 Configuration of hat-type corrugation.

the pitch to core depth are selected to be 70°, 22, and 1.6, respectively. If the contribution of the core walls to flexural stiffness is neglected, it is straightforward to show that the geometry and construction in Table 6.4 can satisfy the relationship in Equations [6.10] and [6.20]. Considering the material properties in Table 6.2, the E_y^f and E_y in Equations [6.9] and [6.20] can be calculated according to Equations [6.21]–[6.23] (Bank, 2006; Barbero, 2010). The proposed section in Fig. 6.6 also satisfied the deflection limit in the FEA.

$$A_{11} = \sum_{k=1}^{60} \frac{E_y^k t_k}{1 - \upsilon_{xy}^k \upsilon_{yx}^k}, \quad A_{22} = \sum_{k=1}^{60} \frac{E_x^k t_k}{1 - \upsilon_{xy}^k \upsilon_{yx}^k}, \quad A_{12} = \sum_{k=1}^{60} \frac{\upsilon_{yx} E_x^k t_k}{1 - \upsilon_{xy}^k \upsilon_{yx}^k} \quad [6.21]$$

$$D_{11} = 2\sum_{k=1}^{60} \frac{E_y^k}{1 - \upsilon_{xy}^k \upsilon_{yx}^k}\left(z_k^2 t_k + \frac{t_k^3}{12}\right), \quad D_{22} = 2\sum_{k=1}^{60} \frac{E_x^k}{1 - \upsilon_{xy}^k \upsilon_{yx}^k}\left(z_k^2 t_k + \frac{t_k^3}{12}\right),$$
$$D_{12} = 2\sum_{k=1}^{60} \frac{\upsilon_{yx} E_x^k}{1 - \upsilon_{xy}^k \upsilon_{yx}^k}\left(z_k^2 t_k + \frac{t_k^3}{12}\right) \quad [6.22]$$

$$E_y^f = \frac{A_{11}A_{22} - A_{12}^2}{t_f A_{22}}, \quad E_y = \frac{12(D_{11}D_{22} - D_{12}^2)}{h^3 D_{22}} \quad [6.23]$$

where k denotes the kth lamina in the laminate for a facing, t_k is the thickness of the kth lamina, z_k is the distance between the centroid of the kth lamina and the centroid of the cross-section of the FRP superstructure, and $E_y{}^k$, $E_x{}^k$, and υ_{yx}^k are E_1, E_2, and υ_{12} for the kth lamina if its orientation is 0°, and are E_2, E_1, and υ_{21} for the kth lamina if its orientation is 90°.

Table 6.4 Lay-up of the laminates for facings and core walls

Components	Lamina lay-up	Ply thickness (mm)	No. of plies	Total thickness (mm)
Top facing	$[0°/90°/90°/0°]_{15}$	0.5	60	30
Core wall	$[0°/45°/-45°/90°/mat]_{16}$	0.5	80	40
Bottom facing	$[0°/90°/90°/0°]_{15}$	0.5	60	30

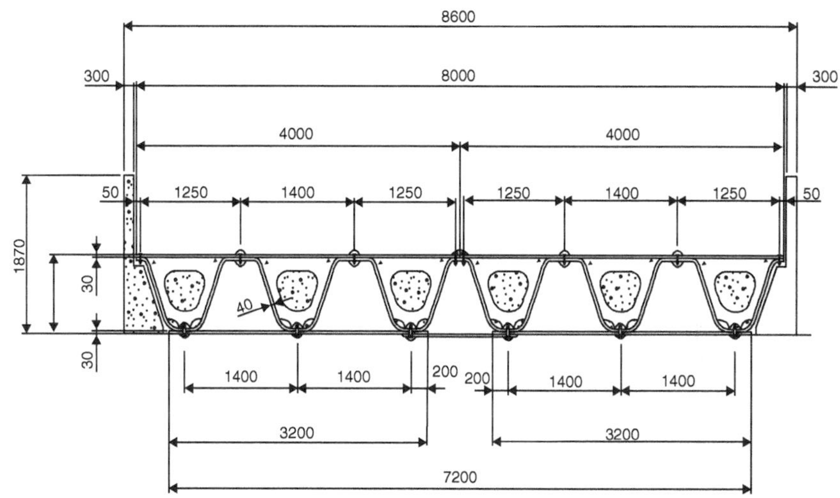

6.6 Cross-section of the FRP bridge superstructure (all dimensions in mm).

The vehicle traffic barrier system should also be designed for the FRP superstructure at this stage. Many demonstration projects show that concrete barriers are mainly used to contain vehicles on FRP bridge decks. The embedded steel bars and grout are usually applied to construct and attach reinforced concrete barriers to FRP bridge decks. As a result, the components of the FRP decks, such as flanges and webs, are often partly damaged. In this study, an improved vehicle traffic barrier system separating barriers and the FRP superstructure, which is shown in Fig. 6.7 (Ji *et al.*, 2010), was proposed for the FRP bridge construction. This barrier system was attached to the concrete abutment instead of attaching to the FRP superstructure. Consequently, it does not cause any damage to the FRP superstructure, such as local tearing of fibers and resin matrix. It will also make future barrier upgrading possible without damaging the FRP superstructure.

Currently, there is no design guidance for the barrier system used for this study. As a result, specifications from AASHTO LRFD (2010) were chosen to design it. The design force for the barrier system is related to the Test Levels 1–6 specified in AASHTO LRFD (2010). The choice of test level depends on several factors, such as bridge alignment, and traffic volume and speed. For the FRP superstructure in this study, Test Level 1 was chosen for the design of the barrier system. The load capacity for the barrier system in this study is 675 kN, which is larger than the 120 kN required by Test Level 1 (Ji *et al.*, 2010). Consequently, this barrier system can be used for the FRP superstructure.

6.3 Manufacture and installation

The FRP sandwich superstructure with corrugated cores was manufactured by commercial fabricators according to the detailed design drawing shown in Fig. 6.6. It consisted of two modular pieces, each 10 m long, 4 m

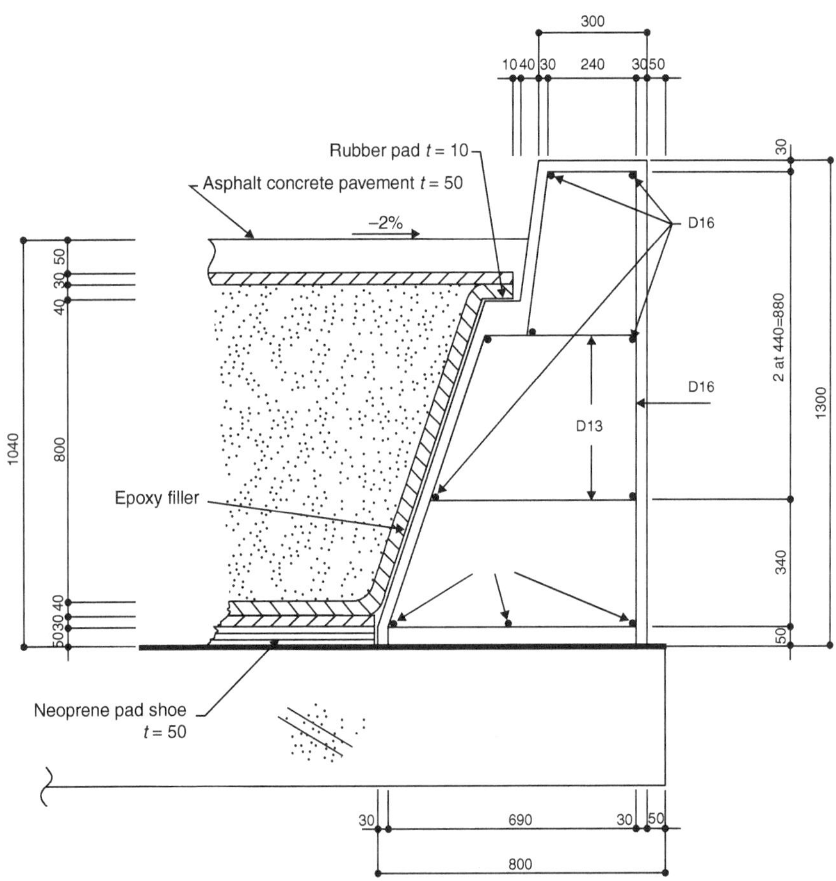

6.7 Barrier details (units: mm).

wide, and 0.94 m deep. The modular pieces were the largest transportable units, due to shipping constraints. The FRP superstructure was completed by joining the two pieces on the field application site. Each modular piece consists of three components: top facing, corrugated cores, and bottom facing. All components were manufactured by a hand lay-up method. Facings were hand-laid and their thickness was approximately 30 mm according to Table 6.4. The corrugated core's walls were hand-laid at room temperature on wood molds. Each wall contained 80 plies and the thickness was 40 mm. Then, internal transverse FRP frames were coated with adhesives and bonded to the inside surfaces of corrugated cores to provide stability and to maintain wall thickness. The transverse frames were also applied at the superstructure's end supports to provide end rigidity for load transfer from the superstructure to the abutments. After their fabrication, the corrugated cores were hoisted onto bottom facings by a light duty crane. Then, M20 steel bolts with the spacing of 32 mm were used to fasten the facings to the corrugated cores. Figure 6.8 shows one finished modular piece of the FRP superstructure (Ji *et al.*, 2010).

The two modular pieces were prefabricated in a factory and assembled on the construction site. Their top and bottom facings were bolted to FRP cover plates in the field. Figure 6.9 shows the connections of the modular pieces. Bolt diameter and bolt spacing were designed according to the EURO design code for bolting in FRP members (Eurocomp, 1996).

6.8 An FRP modular piece.

(a)

(b)

6.9 Connections of the modular pieces: (a) front view; (b) side view.

Figure 6.10 shows the entire FRP bridge superstructure (Ji *et al.*, 2010). In order to prevent the local deflection of the top facings between the corrugated cores, the hollow cells of the superstructure were later filled with expandable polyurethane foams.

6.10 The entire FRP superstructure.

In order to protect FRP superstructures from moisture and contamination, wearing surfaces such as polymer concrete and asphalt concrete could be utilized in practical construction. In the FRP superstructure of this study, asphalt concrete was used as the wearing surface. Epoxy-coated fine aggregate overlay was applied on the surface of the superstructure to avoid delamination between the asphalt wearing surface and the top facing. The epoxy-coated fine aggregate overlay was 0.88 mm thick. The asphalt concrete wearing surface was 50 mm thick. In terms of thermal behavior, the FRP materials could resist a temperature up to 175°C. There was no deformation over the superstructure's surface due to the heat of the asphalt concrete.

In this study, the FRP superstructure was easily placed on the concrete abutments using a light duty crane. The bridge seats were 50 mm thick neoprene pad shoes. In the actual construction, the neoprene shoes were adjusted to accommodate the unevenness of the superstructure's bottom facing. Holes in the bottom facing of the superstructure were made to fit over two stainless steel threaded rods on each concrete abutment. The threaded rods acted as anchors to tie the superstructure to the concrete abutments and prevent uplift at the bridge supports. Expansion joints which were 10 mm wide were provided at the supports. Finally, the reinforced concrete barriers were built after the first field load testing of the FRP superstructure. The completed bridge is shown in Fig. 6.11 (Ji *et al.*, 2010).

6.11 Completed FRP bridge.

6.4 In-service structural performance evaluation

The primary goals of the in-service structural performance evaluation were to investigate if the FRP superstructure could satisfy the design requirements for stiffness and strength in Section 6.2.1. The results from the in-service structural performance evaluation could be utilized to investigate dynamic effects on the FRP superstructure and to perform capacity rating of the FRP superstructure as well.

6.4.1 Field testing

The field testing was conducted in 2002, 2004, and 2011. The bridge with the FRP superstructure was tested using one or two dump trucks, like the one shown in Fig. 6.12. The wheel weights of one dump truck used for the field testing in 2011 are shown in Fig. 6.13. The axle weights of the dump trucks for the field testing in 2002 and 2004 are given in previous research (Ji *et al.*, 2007). Both static and dynamic tests were conducted in the field load testing. In static tests, the truck or trucks were placed at pre-marked positions shown in Fig. 6.13 to maximize the bending moment at midspan. The dynamic tests were conducted with the vehicle traveling over the bridge at speeds of 10–60 km/h. Static and dynamic tests consisted of four loading

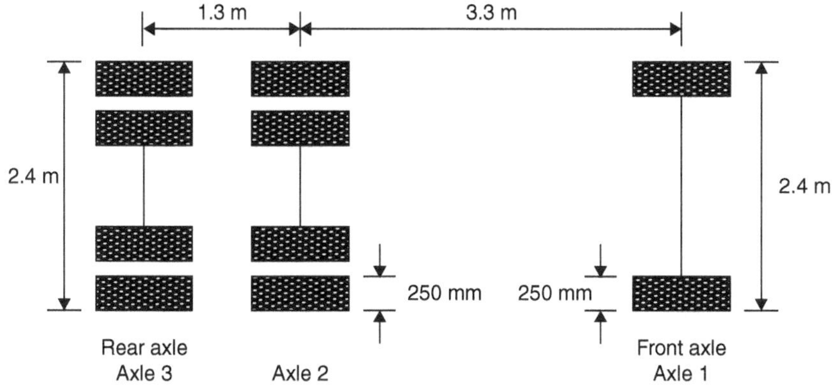

6.12 Dimensions of the test vehicle.

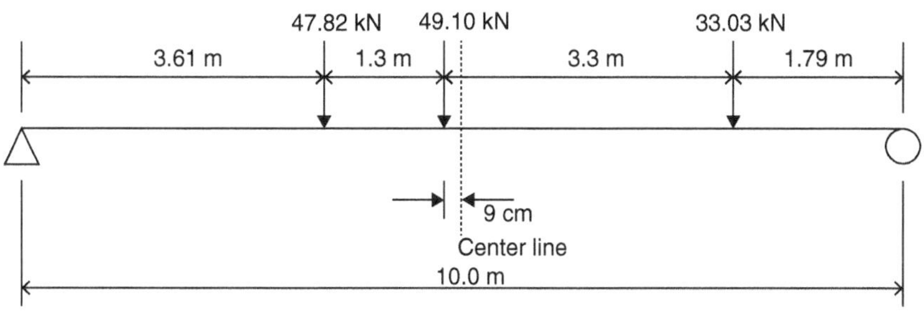

6.13 Test vehicle's placement location.

cases in which the truck or trucks are on the right lane (Load case 1 (LC1)), left lane (LC2), center lane (LC3), and two lanes (LC4) of the bridge.

The FRP superstructure was instrumented with conventional strain gages and linear variable differential transformers (LVDTs) on its bottom surface. The numbering of the positions where strain gages and/or LVDTs might be applied is shown in Fig. 6.14. The positions were chosen to investigate responses at several points at midspan and quarterspan. Further, an accelerometer was instrumented at the Position 5 for the field load testing. Figure 6.15 illustrates the instrumentation of the accelerometer. Figures 6.16 and 6.17 show the results of static and dynamic displacements under the LC3 from the field load testing conducted in 2011. It is noteworthy that the letter 'D' before a number denotes the deflection at the point corresponding to the number.

6.4.2 Results and discussion

In order to investigate if the FRP superstructure satisfied the design requirements for stiffness and strength, the results from the field testing in 2002 should be invoked and compared with the deflection and strength limit. The field testing in 2002 was conducted before the FRP superstructure had the

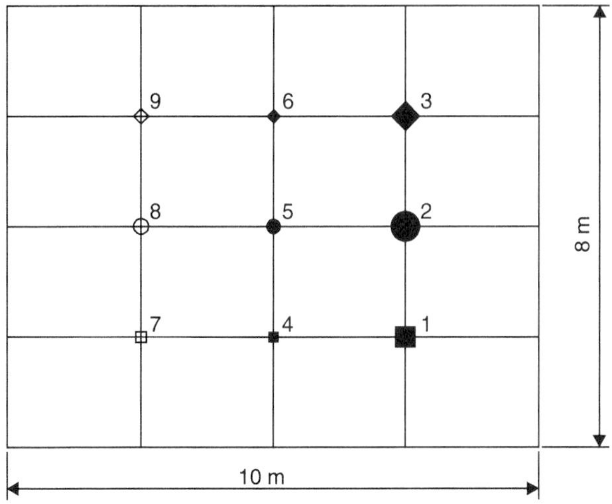

6.14 Numbering of the positions for strain gages and LVDTs.

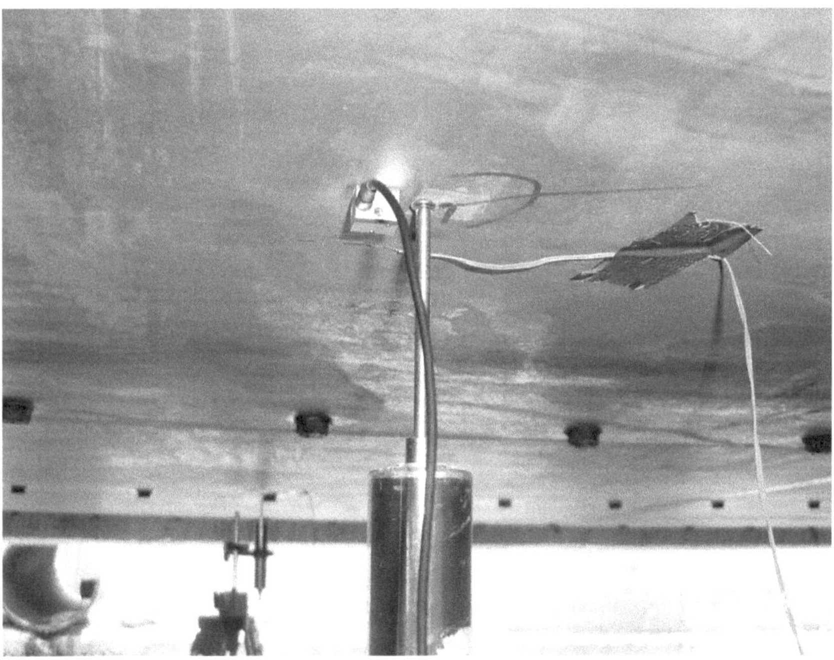

6.15 LVDT and accelerometer for the field testing in 2011.

concrete barrier system and was opened to the public. The static responses measured in the LC3 by a loaded truck 'crawling' across the FRP bridge at about 5 km/h were used as a reference. The measured deflection at midspan was 4.72 mm, less than its design limit $L/800$, which was 12.5 mm in this

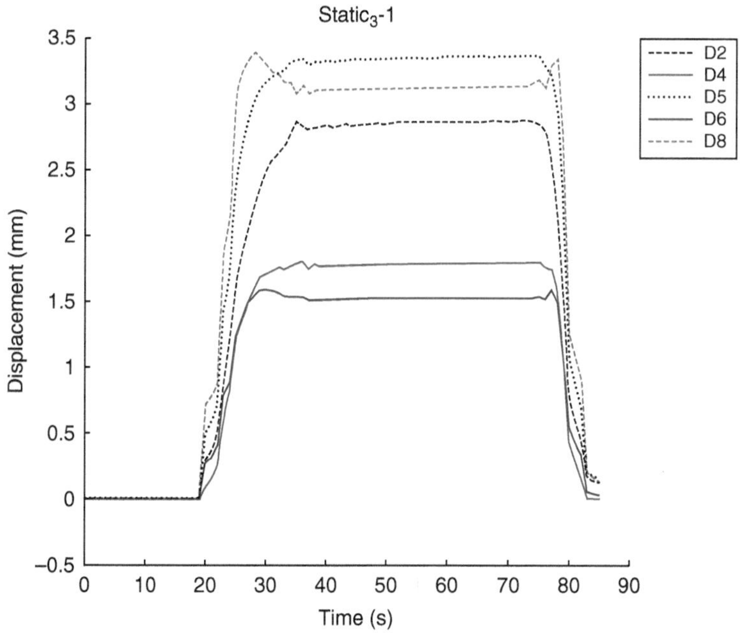

6.16 Results of static displacements at midspan (LC3).

study. The midspan bending strain in tension was 101.72 με at a truck load of 250.56 kN. It was less than the allowable strain, which was 20% of the failure strain 19 317 με obtained from the manufacturers' data. The results above also verified that the FRP superstructure design should be controlled by stiffness rather than strength.

To study dynamic effects on the FRP superstructure, the data taken from the dynamic tests in 2002 under the LC3 at six speeds (10–60 km/h) were analyzed. The maximum dynamic deflection was measured to be 4.78 mm at the speed of 40 km/h. This value was just a little more than 4.72 mm, which was from the static test under the same loading case. The largest impact factor was 0.016, and it was relatively small compared to the value of 0.33 for the cases of conventional materials (AASHTO LRFD, 2010). The results above showed that the dynamic effects induced by the passage of the truck at various speeds were not of much significance. This conclusion was attributed to the difference between the natural frequency of the FRP superstructure and the forced frequency from the passing truck.

The damped natural frequencies from the test data obtained in 2002, 2004, and 2011 were calculated by using the fast Fourier transformation (FFT) analysis of the natural vibration part of acceleration components. An example of the FFT analysis using the data obtained in 2011 is shown in Fig. 6.18. Table 6.5 shows the comparison of the damped frequencies from the experimental results and FEA.

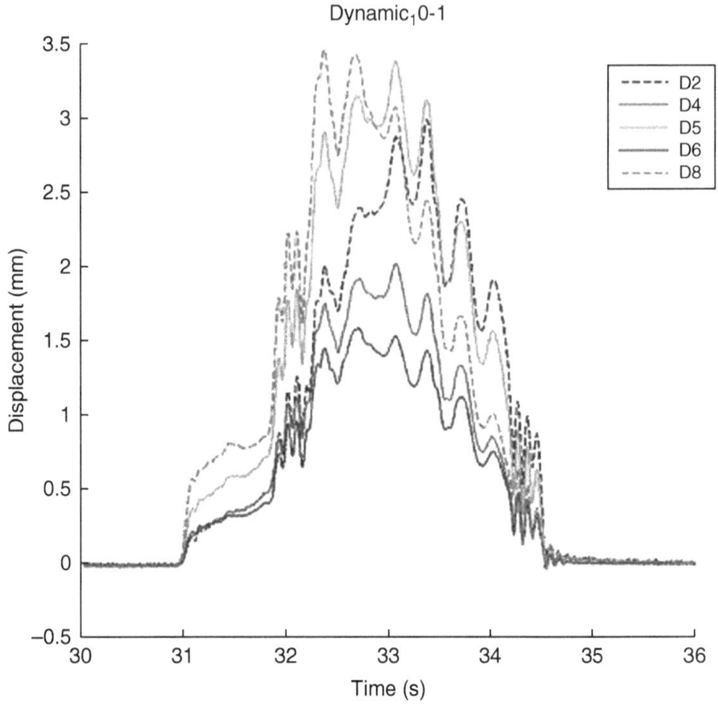

6.17 Results of dynamic displacements at midspan (LC3).

The recorded damped frequencies of the FRP superstructure are close to the predicted value of 10 Hz from $f = 82L^{-0.9}$, where L is 10 m (Hayes *et al.*, 2000). They are also close to the value calculated by the FEA shown in Fig. 6.19. The FEA in Fig. 6.19 was conducted by ANSYS software (ANSYS, 2005). Nine-node thick shell elements were used to model the FRP laminates in the FRP superstructure. In the finite element modeling, the material properties and fiber orientation of each lamina that constitutes a laminate were input for these shell elements. The boundary conditions were modeled as a pin and a roller at the end supports.

In 2002, the shear strength of the interfaces between the asphalt wearing surface and the surface of the superstructure was measured from shear tests. The shear strength values were within the range 0.35 to 0.61 MPa. They exceeded the required shear strength of 0.15 MPa for the interfaces between steel deck plates and asphalt wearing surfaces according to the specification for highway bridges in Korea (MOCT, 2000b). The measurement indicated that there was a good interface bond between the asphalt concrete wearing surface and the surface of the superstructure at that time. According to the field inspection in 2011, cracks between the asphalt concrete pavement and the FRP superstructure have been found. These cracks might be caused by

Table 6.5 Damped frequency comparisons

Load case	Speed (km/h)	Field measurement (Hz)			FEA (Hz)
		2002	2004	2011	
LC3	Crawl	–	–	–	11.117
	10	9.9945	12.0056	11.5500	
	20	9.6802	12.6709	10.4444	
	30	10.0614	10.6567	11.2941	
	40	10.4340	11.7676	11.7059	
	50	9.4238	10.6567	10.6098	
	60	9.8813	11.4258	13.6184	
	Max	–	12.0728	11.0880	

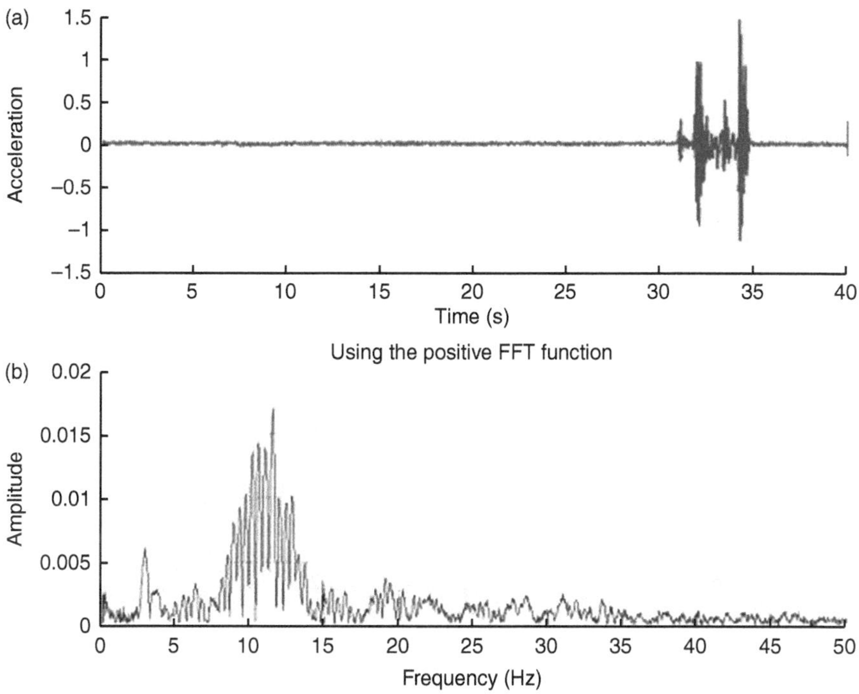

6.18 Dynamic response of the FRP bridge in adjusted time: acceleration vs. frequency.

decreased shear strength between the asphalt concrete pavement and the FRP superstructure.

6.4.3 Capacity rating

Recently, interest in field load testing of highway bridges has partly resulted from the large number of deteriorated bridges with posted load limits that

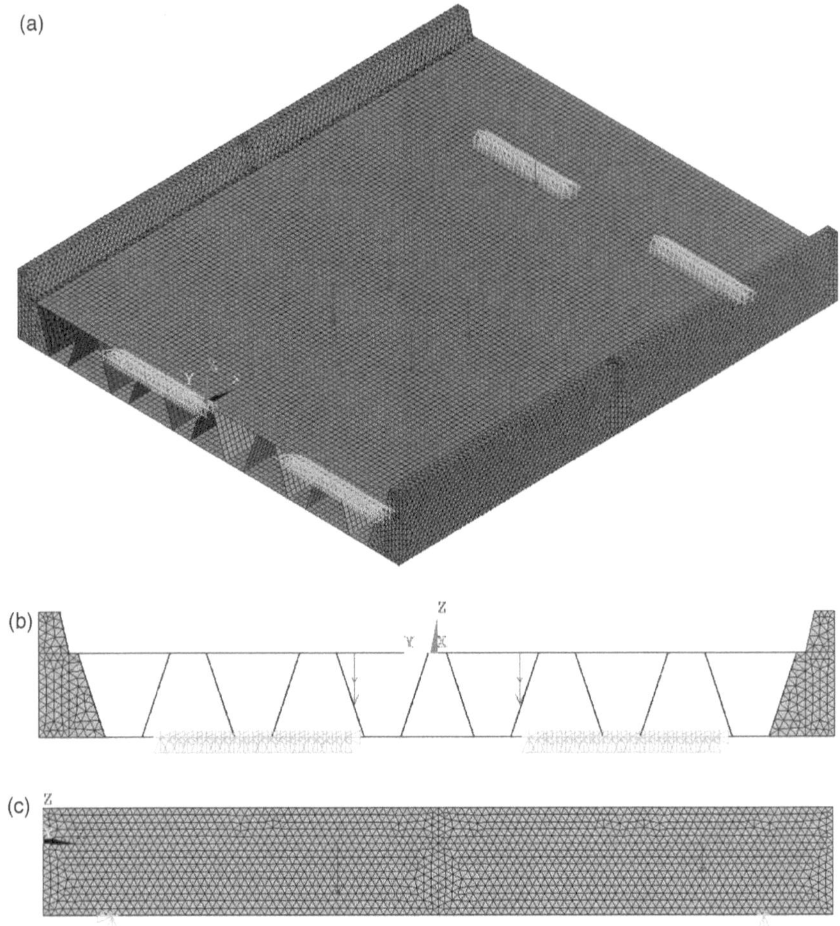

6.19 Finite element model of the superstructure: (a) overview; (b) front view; (c) side view.

are below standard truck weights. Field load testing is an attractive tool to perform capacity rating of a bridge (Stallings and Yoo, 1993). The results of the field testing in this study are used to refine bridge rating calculations.

The formula for capacity rating of an in-service bridge with conventional materials is given in Equation [6.24].

$$P = \mathrm{RF} \times k_s \times P_r \tag{6.24}$$

where RF, k_s, and P_r denote the rating factor, stress modification coefficient, and test truck loading, respectively, and P is the capacity of the bridge.

Table 6.6 Summary of RFs

Classification	f_d (MPa)	f_l (MPa)	RF
LC1 + self-weight	2.205	9.192	8.78
LC2 + self-weight	2.205	8.928	9.05
LC3 + self-weight	2.087	8.340	9.69
LC4 + self-weight	2.205	8.477	9.53

For the FRP superstructure in this study, the rating factor (RF) in Equation [6.24] could be determined based on the allowable stress method. Its calculation is given in Equation [6.25].

$$\mathrm{RF} = \frac{f_a - f_d}{f_l(1+i)} \quad [6.25]$$

where f_a is the allowable stress, f_d is the stress due to dead loads, f_l is the maximum live-load stress produced by the standard vehicle, and i is the impact factor that accounts for dynamic loading effects. The f_a based on experimental data is 107.16 MPa in this study. The f_d and f_l could be determined from FEA. According to MOCT (2000a), the impact factor i could be calculated based on Equation [6.26]. L in Equation [6.26] is the span length in meters. A summary of the RF obtained based on Equation [6.25] and Fig. 6.13 is shown in Table 6.6.

$$i = \frac{15}{40+L} \le 0.3 \quad [6.26]$$

The stress modification coefficients k_s in Equation [6.24] can be calculated using Equation [6.27a] and/or Equation [6.27b]

$$k_s = \frac{\varepsilon_{\mathrm{cal}}}{\varepsilon_{\mathrm{mea}}} \times \frac{1+i_{\mathrm{cal}}}{1+i_{\mathrm{mea}}} \quad [6.27a]$$

$$k_s = \frac{\delta_{\mathrm{cal}}}{\delta_{\mathrm{mea}}} \times \frac{1+i_{\mathrm{cal}}}{1+i_{\mathrm{mea}}} \quad [6.27b]$$

where ε and δ are the strain and deflection of the bridge, respectively. The subscript 'mea' denotes the measured values and the subscript 'cal' denotes the calculated value from FEA.

Table 6.7 Response ratios of static deflections from LC1

Location	Field measurement (mm)			FEA (mm)	Response ratios		
	2002	2004	2011		2002	2004	2011
1	–	1.21	1.66	1.62	–	1.34	0.97
2	–	1.27	1.62	0.60	–	0.47	0.97
3	–	–	–	0.03	–	–	–
4	4.5	1.95	1.97	2.29	0.51	1.17	1.16
5	1.12	1.54	1.76	1.56	0.53	0.38	0.34
6	–	0.24	0.63	0.06	–	0.25	0.09
7	3.68	–	–	1.78	0.48	–	–
8	1.23	1.58	1.72	0.33	0.27	0.21	0.19
9	0.25	–	–	0.06	–	–	–

The response ratios based on static deflections or strains will be defined as $\delta_{cal}/\delta_{mea}$ or $\varepsilon_{cal}/\varepsilon_{mea}$. According to Equation [6.27], they are required to calculate the stress modification coefficients k_s. The response ratios from various loading cases based on the field testing in 2002, 2004, and 2011 are shown in Tables 6.7–6.14.

Finally, the capacity rating of the FRP bridge in 2002, 2004, and 2011 was calculated using the minimum RFs and stress modification coefficients k_s based on the tables discussed above. The minimum RF could be obtained from Table 6.6 directly. According to Equation [6.27], the minimum k_s should be the product of the minimum response ratios and impact factor ratios, defined as $(1+i_{cal})/(1+i_{mea})$. The minimum response ratios could be obtained from Tables 6.7 to 6.14. The impact factor ratio defined above should not exceed 1.0. Once the minimum RF and k_s were obtained, the capacity rating could be conducted according to Equation [6.24]. The results are summarized in Table 6.15.

The capacity rating for the FRP bridge turned out to be DB24.00 (total weight of 432.645 kN) in 2002, DB24.99 (total weight of 441.10 kN) in 2004, and DB23.88 (total weight of 429.39 kN) in 2011. There was little difference in the capacity rating between 2002, 2004, and 2011. Since the rated capacity of the FRP superstructure in this study was close to the design live load, it might be concluded that the FRP superstructure had good in-service performance in the past 9 years.

6.5 Construction time and costs

The construction of the FRP superstructure was quite efficient in time and labor. It was completed in only 8 hours, with a crane and four workers on the construction site. If the fabrication period in the factory was also included, the FRP superstructure was completed in 30 days. The total construction

Table 6.8 Response ratios of static deflections from LC2

Location	Field measurement (mm)			FEA (mm)	Response ratios		
	2002	2004	2011		2002	2004	2011
1	–	0.19	0.47	0.04	–	0.21	0.09
2	–	1.23	1.39	0.42	–	0.34	0.30
3	–	–	–	1.68	–	–	–
4	1.07	0.23	0.52	0.07	–	0.30	0.14
5	1.22	1.45	1.57	1.41	0.29	0.48	0.41
6	–	1.87	2.01	2.26	–	1.21	1.12
7	0.92	–	–	0.06	–	–	–
8	2.09	1.48	1.76	0.66	0.32	0.45	0.38
9	7.29	–	–	1.68	0.23	–	–

Table 6.9 Response ratios of static deflections from LC3

Location	Field measurement (mm)			FEA (mm)	Response ratios		
	2002	2004	2011		2002	2004	2011
1	–	0.67	1.59	0.67	–	1.0	0.42
2	–	2.57	2.97	2.67	–	1.04	0.90
3	–	–	–	0.78	–	–	–
4	1.66	0.79	1.77	0.94	–	1.19	0.53
5	4.70	3.24	3.46	3.10	0.60	0.87	0.82
6	–	0.68	1.79	0.97	–	1.43	0.54
7	1.21	–	–	0.75	–	0.62	–
8	3.67	3.26	3.52	2.61	0.71	0.80	0.74
9	1.69	–	–	0.69	0.41	–	–

Table 6.10 Response ratios of static deflections from LC4

Location	Field measurement (mm)			FEA (mm)	Response ratios		
	2002	2004	2011		2002	2004	2011
1	–	1.42	2.22	1.67	–	1.17	0.75
2	–	2.77	3.21	1.02	–	0.37	0.32
3	–	–	–	1.71	–	–	–
4	6.95	2.18	2.56	2.36	0.34	1.08	0.92
5	6.87	6.48	7.04	5.32	0.27	0.38	0.35
6	–	2.14	2.64	2.32	–	1.08	0.88
7	4.87	–	–	1.85	0.38	–	–
8	4.01	3.29	3.63	0.99	0.25	0.30	0.27
9	6.31	–	–	1.74	0.28	–	–

Table 6.11 Response ratios of static strains from LC1

Location	Field measurement (με)			FEA (με)	Response ratios		
	2002	2004	2011		2002	2004	2011
1	–	9.2	27.3	59.6	–	6.48	2.18
2	–	19.3	15.8	18.3	–	0.95	1.16
3	–	5.4	10.7	3.2	–	0.59	0.30
4	–	51.2	54.3	101.7	–	1.99	1.87
5	41.1	37.8	31.4	26.8	0.65	0.71	0.85
6	–	3.6	7.5	6.5	–	1.81	0.87
7	162.5	26.7	67.0	76.1	0.47	2.85	1.14
8	–	24.9	28.1	21.8	–	0.88	0.78
9	4.4	6.3	11.2	13.1	2.98	2.08	1.17

Table 6.12 Response ratios of static strains from LC2

Location	Field measurement (με)			FEA (με)	Response ratios		
	2002	2004	2011		2002	2004	2011
1	–	9.8	9.5	4.1	–	0.42	0.43
2	–	19.4	19.1	19.3	–	0.99	1.01
3	–	1.8	35.0	71.3	–	39.61	2.03
4	–	6.3	11.7	7.3	–	1.56	0.62
5	35.6	30.4	31.6	25.6	0.72	0.84	0.81
6	–	56.7	55.0	99.7	–	1.76	1.81
7	1.3	6.6	12.9	14.9	–	2.26	1.16
8	–	23.2	27.2	19.8	–	0.85	0.73
9	194.3	39.2	52.3	58.3	0.30	1.49	1.11

Table 6.13 Response ratios of static strains from LC3

Location	Field measurement (με)			FEA (με)	Response ratios		
	2002	2004	2011		2002	2004	2011
1	–	24.9	31.0	24.3	–	0.98	0.78
2	–	26.6	38.9	31.3	–	1.18	0.80
3	–	16.2	32.1	38.2	–	2.36	1.19
4	–	18.9	37.0	42.9	–	2.27	1.16
5	91.9	76.7	81.6	60.4	0.66	0.79	0.74
6	–	11.9	33.8	46.3	–	3.89	1.37
7	44.7	21.7	55.6	31.9	0.71	1.47	0.57
8	–	49.9	58.8	40.4	–	0.81	0.69
9	45.1	23.5	36.6	27.1	0.60	1.15	0.74

Table 6.14 Response ratios of static strains from LC4

Location	Field measurement (με)			FEA (με)	Response ratios		
	2002	2004	2011		2002	2004	2011
1	–	15.8	41.2	63.6	–	4.03	1.54
2	–	36.4	46.1	37.6	–	1.03	0.82
3	–	4.3	47.4	74.5	–	17.33	1.57
4	–	51.7	58.3	108.9	–	2.11	1.87
5	90.3	70.2	76.3	52.5	0.58	0.75	0.69
6	–	57.9	66.3	106.3	–	1.84	1.60
7	197.5	28.7	82.0	90.9	0.46	3.17	1.11
8	–	51.1	64.5	41.6	–	0.81	0.64
9	200.6	39.6	66.7	71.4	0.36	1.80	1.07

Table 6.15 Assessment of capacity rating

Year	Minimum RF	Minimum k_s			Capacity rating
		Response ratio		Impact factor ratio	
		Deflection	Strain		
2002	8.78	0.23	0.30	0.77	$1.56 \times P_r$
2004	8.78	0.20	0.32	1.00	$1.76 \times P_r$
2011	8.78	0.19	0.30	1.00	$1.66 \times P_r$

cost of the bridge was \$140 050. The total cost covered both the FRP superstructure (\$84 030) and concrete abutments (\$56 020). The estimated construction cost of the FRP sandwich superstructure with corrugated cores was \$1750/m^2, while an estimated cost for a concrete bridge was \$1400/m^2. The construction cost of the FRP superstructure was 25% higher than that for conventional concrete bridges. With further design optimization and efficient manufacturing methods, the costs of bridges with FRP superstructures could be decreased in future applications. Therefore, bridges with FRP superstructures like the one in this study may be able to compete with those with conventional materials, especially when shorter on-site construction time and less required maintenance are considered in the future.

6.6 Conclusions

The design, fabrication, construction, and in-service structural performance of an FRP superstructure for ABC are demonstrated in this chapter. Based on the previous discussions, the following conclusions can be drawn:

1. The FRP sandwich superstructure discussed can effectively meet the requirements for stiffness and strength. Connection details are also presented and found working well.
2. The preliminary design approach explained here is effective for estimating the behavior of the FRP sandwich superstructure. It allows the preliminary design to be performed without applying more complicated methods like FEA.
3. The dynamic effects in this study are of less significance. The dynamic responses are close to the static ones, partly due to the difference between the natural frequency of the FRP superstructure and the forced frequency from the passing truck.
4. The construction of the bridge with the FRP superstructure is very efficient in time and labor. The proposed system is considered as a highly competitive short-span bridge alternative for ABC.
5. In the FRP superstructure, there is no corrosion, or damage such as local tearing of the fibers and resin matrix, around the connections of the FRP components. This observation could lead to reduced maintenance strategies compared to other superstructures with conventional materials. However, some maintenance about the wearing surface might be required in the future.
6. To date, a key issue is the lack of design standards relevant to long-term performance of bridges with FRP superstructures. The results presented in this chapter should provide baseline data for future capacity rating assessment, and serve as a part of the long-term performance monitoring of the FRP superstructure.

6.7 Acknowledgment

The work presented in this chapter was partially sponsored by National Science Foundation under Grant No. 0550899 awarded to the third author. The assistance and support from the Korea Land Corporation, Won-chang Technology Ltd, and Seong-won Construction Ltd are also gratefully acknowledged.

6.8 References

AASHTO LRFD (2010), *AASHTO LRFD Bridge Design Specifications*, 5th Ed., American Association of State Highway and Transportation Officials, Washington, DC.

Alampalli, S. (2006), 'Field performance of an FRP slab bridge', *Composite Structures,* **72**(4), 494–502.

Altenbach, H., Altenbach, J. and Kissing, W. (2004), *Mechanics of Composite Structural Elements*, Springer-Verlag, New York, 468.

ANSYS Inc. (2005), *ANSYS 10.0 Documentation*, Canonsburg, PA, USA.

Bank, L. C. (2006), *Composites for Construction: Structural Design with FRP Materials*, John Wiley & Sons, Inc., 551.

Barbero, E. J. (2010), *Introduction to Composite Materials Design*, 2nd Ed., CRC Press, 520.

Davalos, J. F., Salim, H. A., Qiao, P., Lopez-Anido, R. and Barbero, E. J. (1996), 'Analysis and design of pultruded FRP shapes under bending', *Composites Part B: Engineering*, **27**(3–4), 295–305.

Davalos, J. F., Qiao, P., Xu, X. F., Robinson, J. and Barth, K. E. (2001), 'Modeling and characterization of fiber-reinforced plastic honeycomb sandwich panels for highway bridge applications', *Composite Structures*, **52**(3-4), 441–452.

Eurocomp (1996), *Structural Design of Polymer Composites, Eurocomp Design Code and Handbook*, E&FN Spon, London.

Farhey, D. N. (2005), 'Long-term performance monitoring of the Tech 21 all-composite bridge', *Journal of Composites for Construction*, **9**(3), 255–262.

Federal Highway Administration (FHWA) (2011), 'Current practices in FRP composites technology FRP bridge decks and superstructures', <http://www.fhwa.dot.gov/bridge /frp/deckprac.cfm>.

Hayes, M. D., Lesko, J. J., Haramis, J., Cousins, T. E., Gomez, J. and Masarelli, P. (2000), 'Laboratory and field testing of composite bridge superstructure', *Journal of Composite for Construction*, **4**(3), 120–128.

Ji, H. S., Son, B. J. and Chang, S. Y. (2007), 'Field testing and capacity-ratings of advanced composite materials short-span bridge superstructure', *Composite Structures*, **78**(2), 299–307.

Ji, H. S., Song, W. and Ma, Z. J. (2010), 'Design, test and field application of a GFRP corrugated-core sandwich bridge', *Engineering Structures*, **32**, 2814–2824.

Li, L., Ma, Z. J., Griffey, M. E. and Oesterle, R. G. (2010a), 'Improved longitudinal joint details in decked bulb tees for accelerated bridge construction: Concept development,' *Journal of Bridge Engineering*, **15**(3), 327–336.

Li, L., Ma, Z. J. and Oesterle, R. G. (2010b), 'Improved longitudinal joint details in decked bulb tees for accelerated bridge construction: Fatigue evaluation', *Journal of Bridge Engineering*, **15**(5), 511–522.

Libove, C. and Hubka, R. E. (1951), 'Elastic constants for corrugated-core sandwich plates', *National Advisory Committee for Aeronautics Technical Note 2289*, 105.

LUSAS version 13.2 (1999), Finite Element Analysis System, FEA Ltd. London.

Ma, Z. J., Cao, Q., Chapman, C., Burdette, E. and French, C. (2012a), 'Longitudinal joint details with tight bend diameter U-bars,' *ACI Structural Journal*, **109**(6), 815–824.

Ma, Z. J., Lewis, S., Cao, Q., He, Z., Burdette, E. and French, C. (2012b), 'Transverse joint details with tight bend diameter U-bars for accelerated bridge construction,' *Journal of Structural Engineering*, **138**(6), 697–707.

Ministry of Construction and Transportation (MOCT) (2000a), *Standards Specifications for Highway Bridges*, 2nd Ed., Korea.

Ministry of Construction and Transportation (MOCT) (2000b), *Design Manuals for Highway Bridges*, 2nd Ed., Korea.

Plunkett, J. D. (1997), *Fiber-Reinforced Polymer Honeycomb Short Span Bridge for Rapid Installation,* Kansas Structural Composites, Inc., Russell, Kansas, USA.

Song, W. (2012), 'Stiffness-driven design and interface debonding study of FRP sandwich structures for bridges', Ph.D. Dissertation, The University of Tennessee, Knoxville, TN, 111.

Stallings, J. M. and Yoo, C. H. (1993), 'Tests and ratings of short-span steel bridges', *Journal of Structural Engineering,* **119**(7), 2150–2168.

Triandafilou, L. and O'Connor, J. (2009), 'FRP composites for bridge decks and superstructures, state of the practice in the U. S.', *Proceedings of FRP Composites for Infrastructure Applications*, Stockton, CA.

Zhou, A. (2002), 'Stiffness and strength of fiber reinforced polymer composite bridge deck systems', Ph.D. Dissertation, Virginia Polytechnic Institute and State University, Blacksburg, VA, 234.

Zhu, P., Ma, Z. J., Cao, Q. and French, C. (2012a), 'Fatigue evaluation of transverse U-bar joint details for accelerated bridge construction', *Journal of Bridge Engineering*, **17**(2), 191–200.

Zhu, P., Ma, Z. J. and French, C. (2012b), 'Fatigue evaluation of longitudinal U-bar joint details for accelerated bridge construction', *Journal of Bridge Engineering*, **17**(2), 201–210.

7
Engineered cementitious composites for bridge decks

M. LI, University of Houston, USA

DOI: 10.1533/9780857097019.2.177

Abstract: Engineered cementitious composites (ECCs), by reducing the brittle nature of concrete, have opened a new world of possibilities to enhance the safety, durability, and sustainability of civil infrastructure. With a strategically designed microstructure, ECC features incredible ductility under tension, previously only seen in ductile metals. Cost savings can be realized through more efficient design, and less frequent maintenance and repairs. When long-term environmental impacts are accounted for, the ECC advantages over concrete become even more compelling. This chapter presents the ECC fundamentals related to bridge construction, including the basics of material design theory, mechanical and durability characteristics, and application in bridges.

Key words: engineered cementitious composites, micromechanics, ductility, multiple cracking, durability.

7.1 Introduction

In the past decades, great strides have been made in developing high-performance-fiber-reinforced cementitious composites (HPFRCCs). HPFRCCs were first classified by Naaman and Reinhardt[1] as materials that can achieve different degrees of tensile ductility, often accompanied by a macroscopic pseudo-strain-hardening response after first-cracking. Under this definition, HPFRCC is differentiated from normal fiber-reinforced concrete (FRC), which has a tension-softening response (Fig. 7.1).[2] Efforts to design tensile ductility into cementitious materials started in the 1970s, and have mainly focused on using continuous aligned fibers or large volumes of discontinuous fibers. Aveston *et al.*[3] and Krenchel and Stang[4] achieved tensile ductility hundreds of times that of normal concrete using continuous, aligned fibers in cementitious materials. Textile reinforced concrete materials, representing the modern version of continuous FRC, were developed by Curbach and Jesse[5] and Reinhardt *et al.*[6] Pultruded continuous FRC was recently developed by Mobasher *et al.*[7] Additionally, cementitious composite materials using discontinuous fibers at high dosage (4–20%) in cement laminates[8] or in

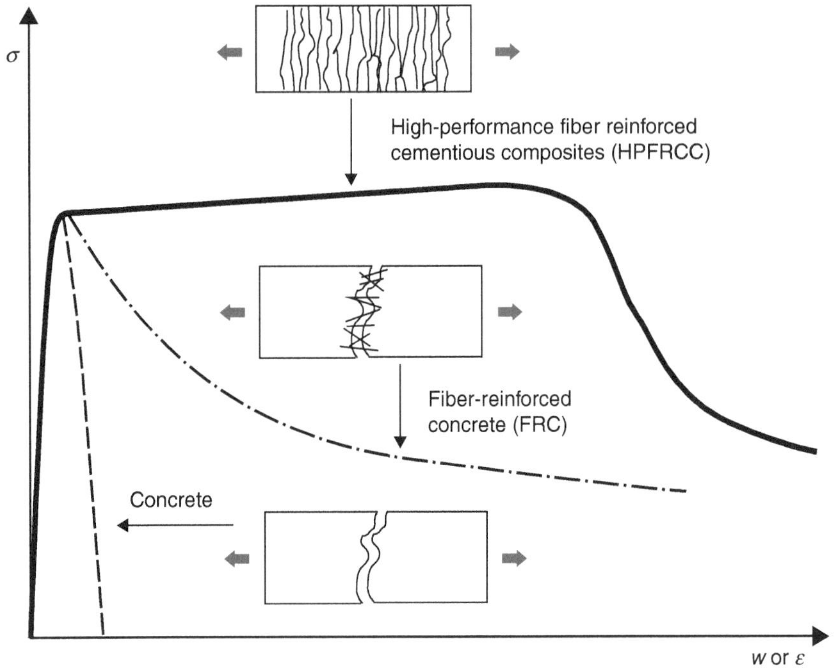

7.1 Uniaxial tensile stress σ vs strain ε (or crack opening w) relation of concrete, FRC, and HPFRCC.

slurry infiltrated fiber concrete (SIFCON)[9,10] attained higher tensile strength and strain capacity than normal concrete, but much lower tensile ductility than continuous fiber and textile reinforced cementitious materials.

In recent years, a new class of HPFRCCs has emerged. Engineered cementitious composites (ECC) represent a family of fiber-reinforced cementitious composite materials that uniquely feature tensile ductility and intrinsic crack width control capacity with a moderate amount (2% by volume) of short discontinuous fibers such as polyethylene, poly-vinyl alcohol, or polypropylene fibers.[11,12] While containing similar ingredients to concrete or FRC, the microstructure of ECC can be deliberately tailored through the use of micromechanical models to achieve tensile strain-hardening behavior and ductility levels approximately 200–600 times that of concrete under tension, thereby leading to delayed fracture localization (Fig. 7.2).[13,14] The fiber/matrix interfacial micromechanical parameters are strategically tailored to allow ECC to dissipate energy through multiple microcracking with crack widths less than 100 μm (Fig. 7.3). The tensile strain-hardening behavior of ECCs significantly differentiates them from other FRCs such as steel, polymeric, glass, and carbon FRCs, which exhibit tension-softening behavior.

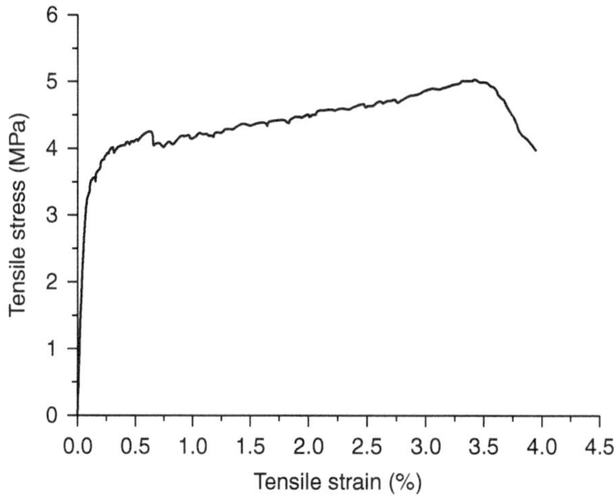

7.2 A typical stress–strain curve of ECC under direct uniaxial tension.[14]

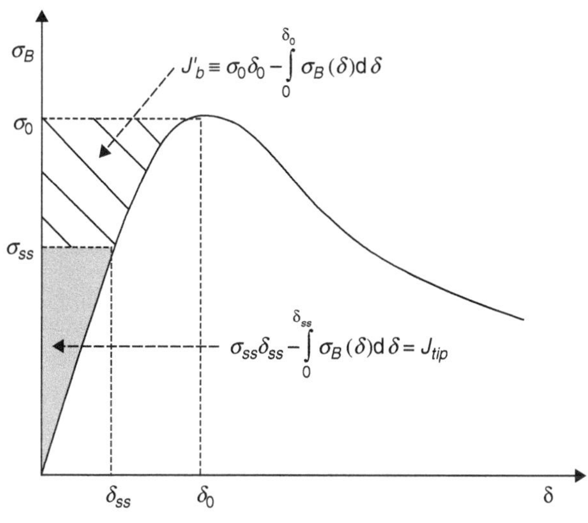

7.3 Typical fiber bridging stress vs crack opening $\sigma(\delta)$ curve for a tensile strain-hardening composite. Hatched area represents complementary energy J'_b. Shaded area represents crack tip toughness J_{tip}.[16,17]

The name ECC was adopted by its original developers.[11] The word 'engineered' emphasizes the micromechanics-based design and manipulation of the material microstructure to achieve specific composite behavior dictated by particular structural applications. In this way, ECC represents a family of materials without fixed ingredients or mixing proportion; more importantly, it also represents a versatile design philosophy. Microstructure design, rheology tailoring, and composite material science together serve as powerful tools to guide ECC material development for targeted composite properties, and enable a meaningful linkage between materials engineering and structural performance design.[15,16] It should be noted that the design of ECC is not dependent on a certain type of fiber, although a large body of literature is on ECC incorporating poly-vinyl alcohol fibers. Nor does ECC rely on increasing fiber volume to achieve strain-hardening behavior. Instead, the ECC design philosophy emphasizes the synergistic interaction between fiber, matrix, and the fiber/matrix interface. An alternative name for ECC is strain-hardening cementitious composite (SHCC), given by RILEM TC HFC Technical Committee in 2006,[17,18] emphasizing its tensile strain-hardening behavior as a constitutive law for structural engineering. Another name for ECC, given by Japan Society of Civil Engineers(JSCE), is multiple fine cracking fiber-reinforced cementitious composites,[19] emphasizing the multiple fine microcracking behavior of ECC and its durability aspects.

The development and understanding of ECC is still evolving. This article presents some basic knowledge of ECC that does not cover the whole range of ECC materials that have been or are being developed. In the next section, the fundamentals of ECC design theory are introduced, followed by a section summarizing the mechanical properties and durability of ECC. The last section describes the implementation of a regular version of ECC, and a high early strength version of ECC in bridge-deck applications.

7.2 Engineered cementitious composites (ECCs) design theory

The design of ECC is based on micromechanics and fracture mechanics theory. Once the target ECC composite property is determined, it can be achieved through microstructure design coupled with rheology tailoring.

7.2.1 ECC design framework: scale linking

The micromechanics-based design theory of ECC was first established in the early 1990s.[20,21] This theory links the measurable micromechanics parameters to the cracking propagation mode, and then to conditions for composite tensile strain-hardening. Scale linking is a fundamental characteristic of the ECC design approach, in which understanding and tailoring of microscale constituent parameters are the keys to achieving target macroscale composite behavior.[16]

As a composite material, ECC contains three main components: fiber, matrix, and fiber/matrix interface. Each component can be defined by a set of micro-parameters. Under uniaxial tension, ECC composites exhibit tensile strain-hardening behavior at the macroscale (mm–cm) through a multiple cracking process. Steady-state crack propagation is a necessary condition to ensure multiple cracking, which is governed by the fiber bridging properties across cracks at the mesoscale (μm–mm). The fiber bridging spring law across a crack, quantified by the fiber bridging stress vs crack opening relationship $\sigma(\delta)$, is the integration of the bridging force contributed by every single fiber bridging the crack. For an individual fiber, its bridging force for a given crack opening is determined by its debonding and pullout behavior from the surrounding matrix, and governed by fiber and interface properties at the microscale (nm–μm), as well as by fiber embedment length and inclination angle between the fiber axis and the crack face normal. In summary, the micromechanics model links microscale constituent parameters to fiber bridging constitutive behavior on the mesoscale; steady-state crack analysis links fiber bridging properties to tensile strain-hardening behavior on the composite macroscale. Once established, the model-based linking provides a systematic framework for optimizing ECC composite tensile properties with the minimum amount of fiber by strategically tailoring the microstructure at the smallest scale.[16]

7.2.2 Conditions for tensile strain-hardening

The tensile strain-hardening behavior of ECC is realized by tailoring the synergistic interaction between the fiber, matrix, and fiber/matrix interface using micromechanics theory. As a fiber-reinforced brittle mortar matrix composite, ECC's pseudo-strain-hardening behavior is achieved through sequential formation of matrix multiple cracking. The fundamental requirement for matrix multiple cracking is that steady-state flat crack propagation prevails under tension, which was first characterized by Marshall and Cox[22] for continuous aligned fiber-reinforced ceramics, and extended to discontinuous fiber-reinforced cementitious composites by Li and Leung.[20] To ensure steady-state cracking, the crack tip toughness J_{tip} must be less than the complementary energy J_b' calculated from the fiber bridging stress σ vs crack opening δ curve, as illustrated in Fig. 7.3 and shown in Equations [7.1] and [7.2].

$$J_{tip} \leq \sigma_0\delta_0 - \int_0^{\delta_0} \sigma(\delta)\mathrm{d}\delta \equiv J_b' \qquad [7.1]$$

$$J_{tip} = \frac{K_m^2}{E_m} \qquad [7.2]$$

where σ_0 is the maximum bridging stress corresponding to the opening δ_0, K_m is the matrix fracture toughness, and E_m is the matrix Young's modulus.

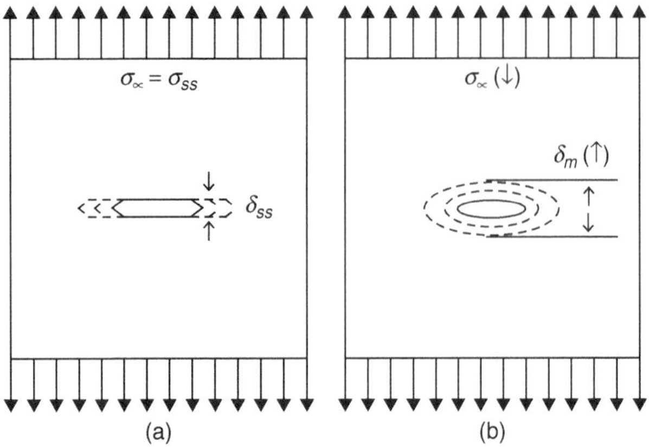

7.4 (a) Steady-state cracking with a constant crack opening δ_{ss} accompanied by a constant ambient load σ_{ss}; (b) Griffith cracking with a widening crack opening δ_m accompanied by a descending ambient load.[16]

Equation [7.1] employs the concept of energy balance during flat crack extension between external work ($\sigma_0\ \delta_0$), crack flank energy absorption through fiber/matrix interface debonding and sliding $\left(\int_0^{\delta_0} \sigma(\delta)\mathrm{d}\delta\right)$, and crack tip energy absorption through matrix breakdown (J_{tip}). This energy-based criterion determines whether the crack propagation mode is steady-state flat cracking or Griffith cracking,[23] as illustrated in Fig. 7.4.

Apart from the energy criterion, another condition for pseudo-strain-hardening is that the matrix tensile cracking strength σ_c must not exceed the maximum fiber bridging strength σ_0.

$$\sigma_c < \sigma_0 \qquad [7.3]$$

where σ_c is determined by the matrix fracture toughness K_m and the pre-existing internal flaw size a_0 and distribution. While the energy criterion governs the crack propagation mode, the strength-based criterion controls the initiation of cracks. Satisfaction of both equations is necessary to achieve ECC strain-hardening behavior; otherwise, the composite behaves as a normal FRC, and results in tension-softening behavior.

7.2.3 Conditions for saturated multiple microcracking

For ECC materials with pseudo-strain-hardening behavior, high tensile strain capacity results from saturated formation of multiple microcracks. Once the steady-state cracking criteria are satisfied, the number of microcracks that can be developed is determined by (i) the maximum fiber bridging stress σ_0, which is determined by fiber/matrix interface properties and fiber properties, and (ii) the matrix properties, in particular the pre-existing flaw-size distribution, and the matrix fracture toughness. Considering that ECC is a non-homogenous brittle-matrix composite, first-cracking strength is determined by the largest flaw size in the section where the first microcrack is initiated. Its ultimate tensile strength is determined by the 'weakest' section, where the fiber bridging capacity (maximum bridging stress σ_0) is the lowest among all sections subjected to the same level of stress. Therefore, the maximum fiber bridging stress σ_0 at the 'weakest' section imposes a lower bound on critical flaw size c_{mc}, so that only those flaws larger than c_{mc} can be activated and contribute to multiple cracking. There also exists a minimum crack spacing controlled by interface properties, which imposes an upper bound for the density of multiple cracking.

Matrix imperfections, for example random distribution of pre-existing flaws, is one cause of the variation in crack spacing and tensile strain capacity. Therefore, a large number of flaws slightly larger than c_{mc} are preferred for saturated multiple cracking and high tensile strain capacity. Flaws much larger than c_{mc} will lead to a reduction in the net cross-section and fiber bridging stress at the crack section. Fiber dispersion non-uniformity is another contributor to the variation in tensile strain capacity and unsaturated multiple cracking. With a fixed fiber volume percentage, the maximum fiber bridging stress σ_0 at the weakest section is determined by the degree of fiber dispersion uniformity in the composite. Fiber dispersion uniformity is directly influenced by the rheology characteristics of the fresh ECC during processing.[24] Ideally, processing of ECC should optimize fiber dispersion to achieve a uniform random distribution state, thereby minimizing the probability of creating 'weak' sections with lower fiber content. By these means, the largest possible tensile strain capacity can be achieved by maximizing multiple cracking behavior.

7.3 ECC mechanical properties and durability

ECC's high tensile ductility, deformation compatibility with existing concrete, and self-controlled microcrack width lead to its superior mechanical behavior and durability under various loading and environmental conditions,[25] as summarized below.

7.3.1 Tensile behavior

The sole method to validate the true ECC tensile strain-hardening behavior is through direct uniaxial tension test. This is because some tension-softening materials can exhibit 'pseudo-deflection hardening' under bending test, depending on the geometry of the specimen. Figure 7.2 shows the typical tensile strain-hardening behavior of ECC under uniaxial tension, with tensile strain capacity approximately 3.5%, which is 350 times that of concrete and FRC. Such strain-hardening and ductile behavior indicates high fracture energy of the material for damage tolerance. Despite the difference in mix design, the ECC tensile stress–strain curve generally follows the same pattern containing three stages:

(1) the initial elastic stage, characterized by Young's modulus;
(2) the strain-hardening stage, accompanied by multiple microcracking formation. During this stage, the width of each microcrack slightly increases up to a strain level of 0.5% approximately, and then remains nearly constant while the increasing applied strain only increases the number of microcracks. This stage is characterized by first-cracking strength (i.e. the stress when the first microcrack occurs), ultimate tensile strength (i.e. the peak stress before localized fracture occurs), and the strain capacity of the material (i.e. the strain corresponding to peak stress). As the uniaxial tension test is conducted under displacement control, small load 'drops' on the curve correspond to the released energy during the formation of each microcrack
(3) The tension-softening stage, accompanied by the formation of a localized fracture at one of the microcracks and the continuous decrease in the ambient load. At this stage, ECC behaves similar to tension-softening FRC materials.

7.3.2 Compressive properties

The compressive strength of ECC typically ranges from 30 to 100 MPa,[26] depending on its mix design. The most commonly studied version of ECC, M45, has a 28-day compressive strength of 50–60 MPa. The Young's modulus of ECC is 20–25 GPa, which is lower than concrete because ECC does not contain any coarse aggregates in general in its mix design. This relatively low elastic modulus of ECC is desirable for limiting the tensile stress built up under restrained shrinkage conditions.[27] The compressive strain capacity of ECC at 0.45–0.65% is slightly higher than concrete; the softening branch of ECC compressive stress–strain curve is longer, allowing for a more 'ductile' compressive failure mode with an approximately 45° final fracture plane, in contrast with brittle splitting failure mode of concrete or high strength concrete (Fig. 7.5).[28]

7.3.3 Inherent crack width control capacity

Concrete cracking is a result of the combined effects of mechanical loading and environmental conditions.[29] Concentrated chloride exposure is one of

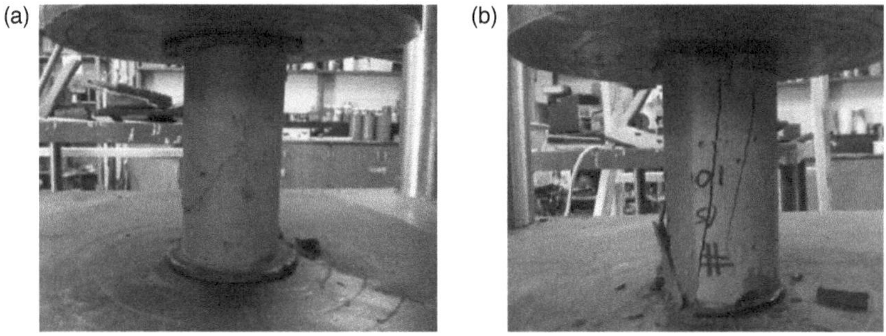

7.5 Compressive failure mode of (a) ECC and (b) concrete.[28]

7.6 Multiple microcracking patterns in ECC during strain-hardening stage. Final failure occurs during the tension-softening stage.[40]

the most severe and aggressive among all possible forms of environmental exposure for concrete structures.[30] Serious concrete deterioration is generally initiated by concrete cracking under restrained shrinkage or stress concentration, which leads to accelerated penetration of aggressive agents, subsequent corrosion of embedded reinforcing steel, and spalling of concrete cover.[31–34] Maximum allowable crack widths are thereby required in various codes and specifications for the design of reinforced concrete structures exposed to aggressive chloride environments. The allowable maximum crack width ranges from 150 to 300 μm,[35–39] with the most stringent requirements specified by JSCE[38] and American Concrete Institute(ACI) 224R.[25] According to ACI 224R,[29] the maximum crack width at the tensile face of reinforced concrete structures is specified as 150 μm for exposure conditions of seawater, seawater spray, wetting, and drying, and 180 μm for deicing chemical exposure. As the microcrack width of ECC formed during its strain-hardening stage is intrinsically limited to below 100 μm (Fig. 7.6),[40] which is independent of structural geometry and applied deformation, steel reinforcement is not required to control crack width even for the most stringent allowable crack width requirement.

7.3.4 Flexural behavior

The popular nickname for ECC – 'bendable concrete' – results from its ductile flexural behavior. Under four-point bending, ECC first undergoes an elastic stage, followed by a deflection-hardening behavior accompanied by the formation of a number of microcracks on the tensile side of the specimen. This allows ECC to undergo a large curvature without fracture failure under excessively applied deformation. The deflection-hardening response of ECC reflects its tensile ductility, and is thus an intrinsic property. This is not the case for tension-softening FRC, for which, as beam specimen height increases, deflection hardening becomes more difficult to attain.[41] Highly deformed ECC coupon and beam specimens are shown in Fig. 7.7.[14]

7.3.5 Abrasion and wear resistance

To evaluate ECC's abrasion and wear resistance for bridge-deck and roadway applications, surface friction and wear track testing according to Michigan

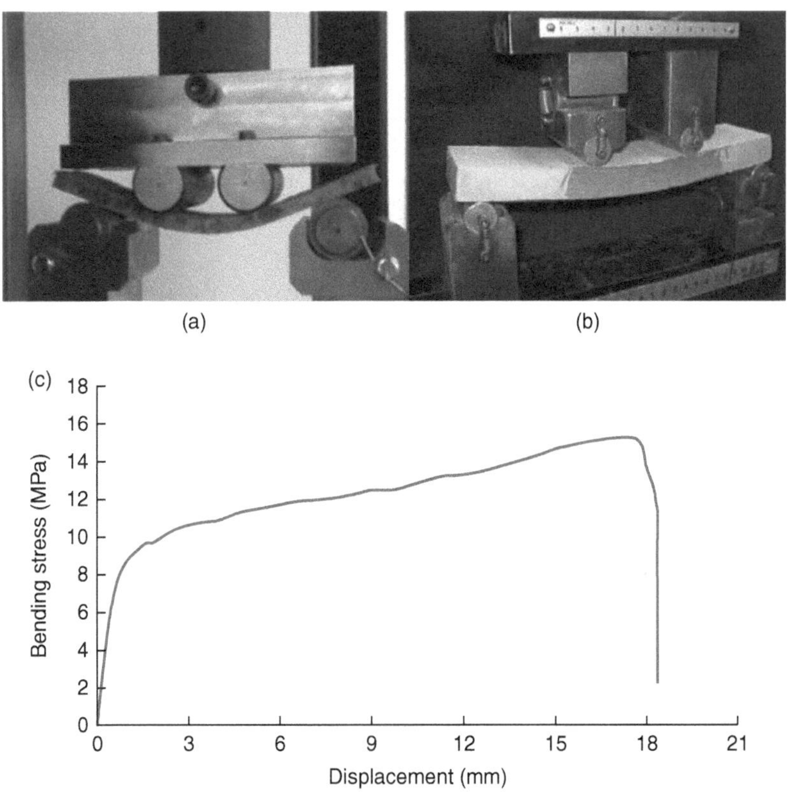

7.7 Flexural behavior of ECC[14]: (a) an ECC plate specimen under four-point bending; (b) an ECC beam specimen under four-point ending; (c) bending stress vs displacement relation.

Test Method 111 was conducted in conjunction with Michigan Department of Transportation (MDOT)[42] using a static friction tester (Fig. 7.8). These tests were conducted on a wet pavement surface with vehicle tires operating at 65 kph (40 mph). Initial friction forces between the tire and the ECC surfaces were determined. The ECC specimens were then subjected to 4 million tire passes to simulate long-term wear. After wearing, friction forces were again determined to evaluate deterioration or surface polishing during wearing. These final measured frictional forces are called the aggregate wear index (AWI). AWI values for the textured ECC samples range from 1.6 to 2.3 kN, which are higher than the required minimum AWI of 1.2 kN for Michigan trunkline road surfaces. The test results show that ECC has good abrasion and wear resistance to heavy traffic.

7.3.6 Freezing and thawing

Saturated concrete exposed to freeze–thaw cycles may experience disintegration. Reduced freeze–thaw resistance has also been found in some

7.8 Driving surface static friction tester.[42]

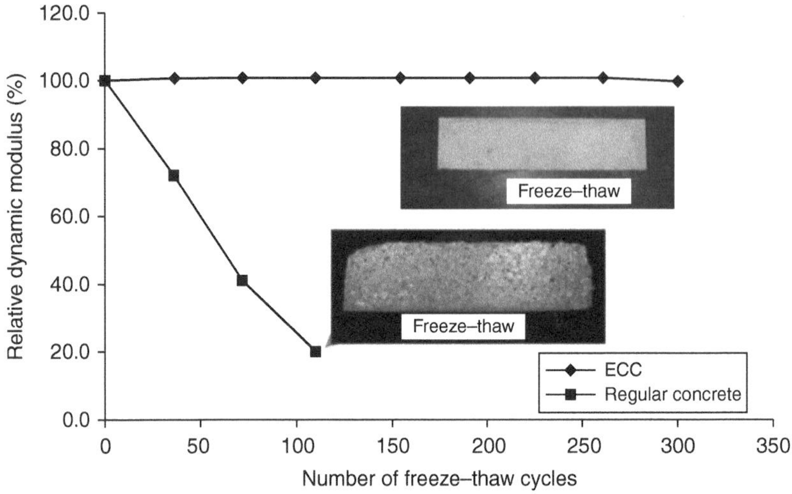

7.9 Relative dynamic modulus vs number of freeze–thaw cycles.[42] Testing in accordance with ASTM C666A. (HES-ECC – high early strength engineered cementitious composites.)

very high early strength concrete mixes,[43] limiting their applications in cold regions. Freezing and thawing testing was conducted on non-air-entrained ECC and normal concrete prisms[42] over 14 weeks based on ASTM C666A.[44] After 5 weeks (110 cycles of freezing and thawing), the non-air-entrained concrete specimens had severely deteriorated and were therefore removed from the test. The non-air-entrained ECC specimens survived 300 cycles with no degradation of dynamic modulus (Fig. 7.9). The freeze–thaw durability factor of ECC was calculated as 100, far larger than 10 for the non-air-entrained concrete. Additionally, uniaxial tension tests were performed on ECC coupons after being exposed to 200 cycles of freezing and thawing, and revealed no significant decrease in tensile strain capacity. The freeze–thaw resistance of non-air-entrained ECC is attributed to its pore structure, which is influenced by (i) the entrapped air size and distribution during processing of the mixture with specific viscosity and fiber content, and (ii) the hydration process. The increased toughness (i.e. reduced brittleness) of ECC also leads to increased resistance to disintegration.

7.3.7 Restrained shrinkage cracking

Shrinkage of concrete, when constrained, often causes early-age cracking in concrete. The tensile strain capacity of ECC is two orders higher than its shrinkage strain. Therefore, ECC can accommodate the restrained shrinkage deformation to suppress localized cracking.[40,45] Shrinkage ring tests were conducted to simulate the shrinkage of a freshly cast ECC layer constrained

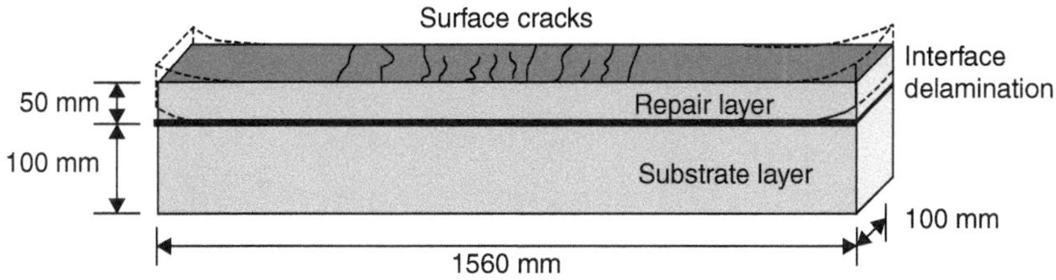

7.10 Layered repair specimen dimensions and potential failure modes.[45]

by a steel ring. In contrast with the localized cracking in the control concrete specimens, whose crack width increased as the shrinkage strain increased with time, the ECC layer exhibited a number of distributed microcracks with crack width under 100 μm. The width of these microcracks did not increase the shrinkage strain increased was time; instead, the number of microcracks increased. The results indicated that ECC does not require steel reinforcement to control shrinkage cracking.

In concrete repairs, shrinkage can induce repair surface cracking and repair/old interface delamination. As shrinkage deformation of the repair layer is restrained by surrounding concrete that has undergone most of its shrinkage, tensile stress can build up in the repair layer and cause surface cracking; a combination of tensile and shear stress can build up at the repair/old interface and lead to interface delamination. Cracking and delamination are the main causes that initiate numerous durability issues in concrete repairs. Li and Li[45] experimentally (Fig. 7.10) and numerically validated that the high tensile ductility of ECC repair can accommodate its shrinkage deformation by forming multiple microcracks with tight crack width below 60 μm. By these means, tensile and shear stresses at the repair/substrate interface were relieved, so that both repair surface cracking and interface delamination were limited to below 70 μm. It was also found that for normal concrete repair, several localized cracks over 450 μm in width formed in the repair layer. For steel FRC repair, several localized cracks formed with widths larger than 120 μm. Since the cracks were bridged by steel fibers and could not open traction-free to accommodate the repair layer's shrinkage deformation, the interface delamination became much larger (280 μm) than concrete and ECC repairs.

7.3.8 Water permeability

The transport properties of concrete cover determine the time needed for the penetration of water or aggressive agents to reach the steel reinforcements. Transport mechanisms include permeation, diffusion, and capillary absorption. Permeability of cracked concrete scales with the third power of

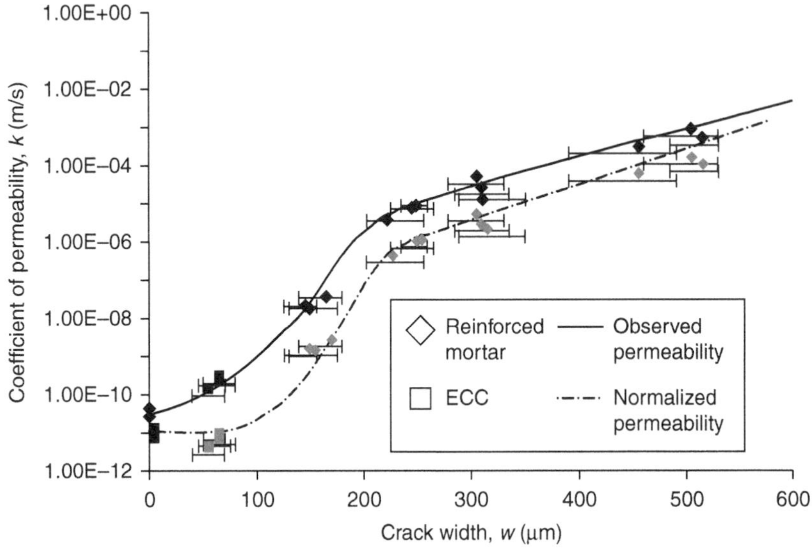

7.11 Permeability of precracked ECC and reinforced mortar.[48]

crack width,[46,47] and a crack with width below 100 μm (50 μm for gas permeability) tends to behave like sound concrete. Wang *et al.*[47] and Lepech and Li[48] found that microcracked ECC exhibited nearly the same permeability as sound concrete, even when loaded to a tensile strain of 1.5% (Fig. 7.11). The low water permeability of ECC is due to its tight self-controlled crack width, which does not rely on the steel reinforcement ratio and is independent of the applied strain level, as long as the strain capacity of the material is not surpassed.

7.3.9 Chloride diffusion

The corrosion of steel in concrete is a major durability concern for reinforced and prestressed concrete structures, demanding significant amounts of repair and rehabilitation.[49–52] When concrete is cast around steel reinforcements, a passivation film forms around the steel bars and protects them from corrosion initiation. However, chloride ions, whether airborne or from deicing salt or seawater, are often present, especially in coastal areas and snow regions where deicing salts are used. Corrosion starts when chloride ions pass through the sound or cracked concrete cover, reach the steel, and accumulate beyond a threshold concentration level. At this point the natural protection layer surrounding the steel becomes depassivated, if oxygen and moisture are present in the steel–concrete interface. The oxidation products generate expansive forces within the concrete, eventually causing

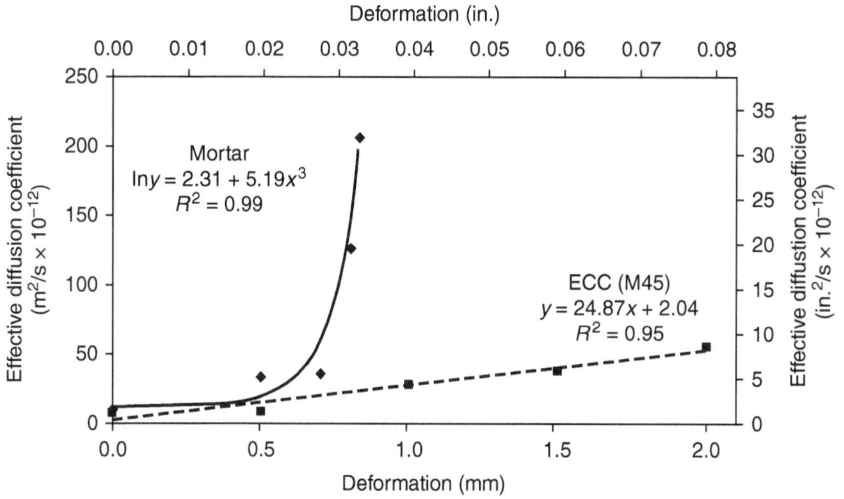

7.12 Diffusion coefficient vs pre-loading deformation level for reinforced ECC and reinforced mortar.[55]

the concrete cover to spall.[53] Moreover, the cross-section of the reinforcing bar or prestressing strand is now diminished, leading to a reduction in the load carrying capacity of the concrete member. It is widely recognized that corrosion of reinforcing steel has led to the premature deterioration of many concrete structures in the United States, especially bridges and marine structures, before their design life was reached. Although corrosion is not the sole cause of all structural deficiencies, it is a significant contributor and has become a major concern.[51,54]

Corrosion initiation time is governed by the chloride diffusion coefficient. Chloride diffusion coefficients of ECC beam specimens were determined and compared with control reinforced mortar specimens (without coarse aggregates).[55] The specimens were loaded under four-point bending to various deformation levels, and then were subjected to salt ponding of 3% chloride solution according to AASHTO T259–80.[56] The chloride diffusion coefficients of unloaded and loaded specimens were determined based on the chloride concentration profile along the depth of the specimen. As shown in Fig. 7.12, the chloride diffusion coefficients were 6.75×10^{-12} m²/s and 10.58×10^{-12} m²/s for unloaded ECC and reinforced mortar, respectively. When the applied bending deformation increased, the chloride diffusion coefficients increased with a significantly faster rate for reinforced mortar than ECC. This was because the increasing deformation of ECC was accompanied by an increasing number of microcracks, while the crack width of each microcrack remained hardly changed; in contrast, the increasing deformation of reinforced mortar was accompanied by increasing crack width. It was further revealed that the chloride diffusion coefficient of ECC increased linearly with the number of cracks, whereas the diffusion coefficient of reinforced mortar was proportional to the square of the crack width. At the same level of applied deformation, the chloride diffusion coefficient of ECC was far lower than reinforced mortar, indicating a significantly longer chloride initiation time.

7.3.10 Corrosion resistance

In addition to prolonged corrosion initiation time, the high tensile ductility of ECC can lead to higher resistance to spalling induced by corrosion, and thus further prolonged service life. Şahmaran *et al.* (2008)[57] conducted an experimental investigation on reinforced (R)/ECC cylinder with a single steel rebar embedded and subjected to accelerated corrosion by electrochemical method (Fig. 7.13). The corrosion-induced crack width of the mortar specimens increased with time as the steel corrosion progressed. Larger crack widths, up to 2 mm wide, were observed in mortar at higher levels of corrosion. On the other hand, crack widths of ECC remained nearly constant (~0.1 mm) with time as corrosion activity progressed, while the number of cracks on the surface of the specimen increased. The results of this study showed that ECC has significant anti-spalling capability as compared to conventional mortar. If 0.3 mm maximum crack width limit for outdoor exposure, as specified by AASHTO (2004),[58] were used to represent the serviceability limit of reinforced concrete structures, the service life of reinforced ECC would be at least 15 times that of the reinforced mortar.

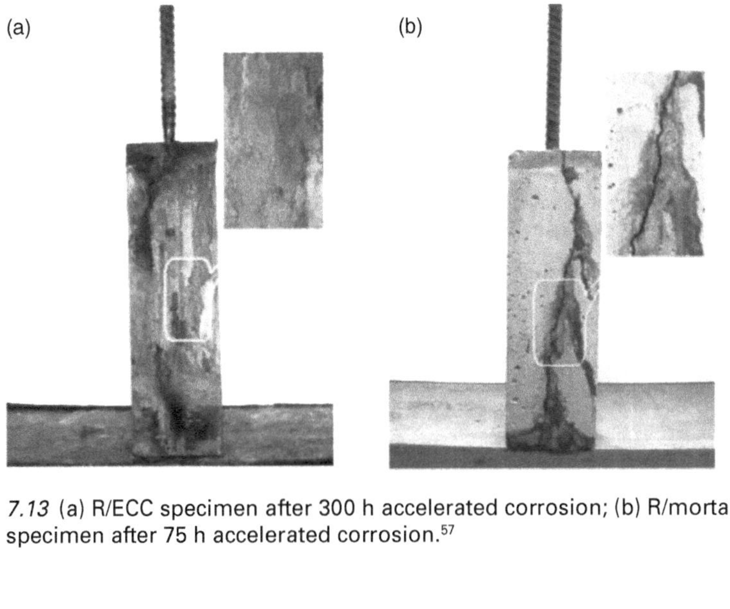

7.13 (a) R/ECC specimen after 300 h accelerated corrosion; (b) R/mortar specimen after 75 h accelerated corrosion.[57]

Reinforcement corrosion in mortar specimens resulted in a marked reduction in stiffness and flexural load capacity, as a result of mass loss of steel.[57] After 25 h of accelerated corrosion exposure, the flexural strength reduced to about 34% of the original flexural capacity of the control mortar beam. In contrast, the ECC specimens after 50 h of accelerated corrosion exposure retained almost 100% of the original flexural capacity. Beyond 50 h, the flexural capacity decreased, but retained over 45% that of the original capacity, even after 300 h of accelerated corrosion exposure. Longitudinal cracks due to expansion of the corrosion products also affected the failure mode of the reinforced mortar under four-point bend loading. However, ECC deterioration due to the corrosion of reinforcement did not modify the ductile failure mode in ECC beams. Overall, the experimental results from this study suggest that the propagation period of corrosion could be safely included in estimating service life of a structure when concrete is replaced by ECC.

7.3.11 Fatigue behavior

ECC has greatly improved fatigue response compared with normal concrete and FRC. Suthiwarapirak *et al.*[59] conducted flexural fatigue tests on ECC, and demonstrated higher ductility and fatigue life compared with polymer cement mortars commonly used in repair applications. Furthermore, fatigue loading does not seem to increase crack width. Kim *et al.* (2004)[60] experimentally validated that after 100 000 cycles of fatigue loading, the crack width in the reference concrete link slab specimen increased from 50 μm to over 0.6 mm, while the crack width in the ECC link slab remained close to 50 μm.

7.3.12 Reflective cracking in repairs

In bridge-deck overlay repairs, premature reflective cracking is often the ultimate failure mode that limits repair service life.[61–63] Concrete or hot mix asphalt (HMA) overlays are primary methods for rehabilitating distressed asphalt and Portland cement concrete pavements subjected to moderate and heavy traffic, including highways, bridge decks, and airfields.[64] Before the more expensive option of removal and re-construction of existing pavements are undertaken, overlays can be used to provide smooth surfaces while utilizing the structure of existing concrete pavements.[65,66] Reflective cracking is often the first distress that occurs in overlays.[67] It is characterized as new cracks reflected through the overlay from pre-existing cracks or joints in the substrate under repeated traffic loads. In most cases cracks propagate through the new overlay during the first few years of service.[68] Under severe conditions, a reflective crack can reach the surface of a 150 mm (6 in.) overlay in less than 1 year.[69] Reflective cracking not only reduces the

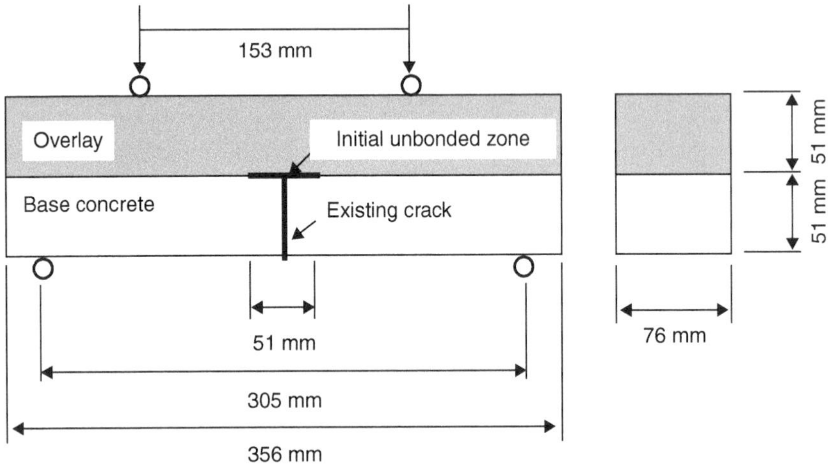

7.14 Layered repair system under four-point monotonic/fatigue bending.[16]

overlay load carrying capacity and service life, but also provides a path for aggressive agents to infiltrate the pavement or bridge-deck structure and cause early deterioration.[70]

Traditionally, efforts undertaken to reduce reflective cracking in overlay repairs have fallen into three broad categories[71]: (a) concrete slab fracturing, which includes rubblization, crack and seat (for plain concrete pavement), and break and seat (for reinforced concrete pavement); (b) stress-relieving interlayer, which is used between the concrete substrate and overlay, and includes paving fabric, modified HMA interlayer, geogrid, and other proprietary interlayers; and (c) modified overlay, which includes increasing overlay strength, increasing overlay thickness, adding fiber reinforcement, introducing saw and seal joints in HMA overlay, etc. If used appropriately, these methods can delay reflective cracking to a certain extent, although they will not fundamentally eliminate the fracture failure of the overlay.

Reflective cracking can be suppressed when ECC is used as the repair material.[72] A layered repair system (Fig. 7.14) was investigated for resistance to reflective cracking and delamination under four-point monotonic and fatigue bending loading.[16,73] This system contained a layer of repair material made of high early strength ECC (HES-ECC) or high early strength concrete cast on a substrate layer of old concrete with an initial vertical crack and interfacial delamination. When subject to monotonic loading, the HES-ECC layered repair system exhibited 100% increased load carrying capacity, and 5–10 times the deformation capacity of a concrete layered repair system. Instead of forming one single crack in the concrete repair layer,

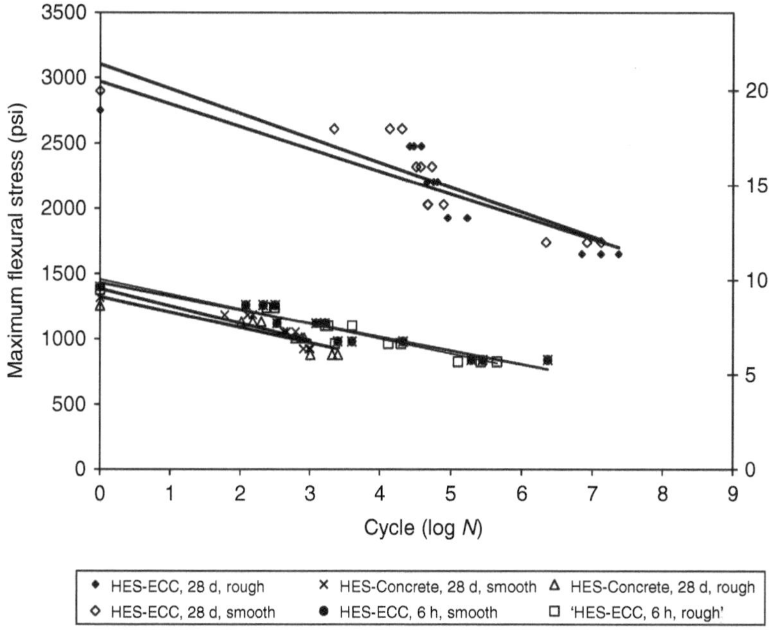

7.15 Fatigue life of HES-ECC and high early strength concrete overlay systems.[16,73]

the initial interfacial crack would kink into the ECC repair layer, and then become trapped there. Upon additional loading, delamination resumed at the interface. The unique kink-trap-cracking behavior repeated itself with increasing loading till the flexural strength of HES-ECC was exhausted. In this way, repair spalling was successfully suppressed by using HES-ECC. Furthermore, when subject to the same level of fatigue loading, the HES-ECC layered repair system showed significantly longer service life than the concrete layered repair system (Fig. 7.15).

7.4 ECC application in bridges

ECC has been applied in bridge construction in the United States and in Japan, including repairs (Sections 7.4.1 and 7.4.2), rehabilitation (Section 7.4.3), and new construction (Section 7.4.4).

7.4.1 ECC patch repair

One of the earliest applications of ECC was a small patch repair on a bridge deck in the US.[26,42] This application offers significant insights into ECC's long-term durability performance. In collaboration with MDOT,

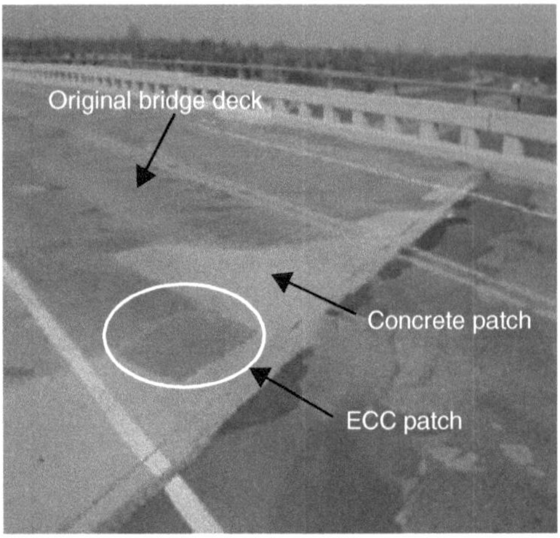

7.16 ECC patch repair on Curtis Road Bridge, Michigan.[26,42]

this ECC patch repair was placed on the bridge deck of Curtis Road over M-14 in Southern Michigan in October 2002. For this project, one section of the deteriorated bridge deck was repaired with ECC, while the remaining portion was repaired with a pre-packaged mixture of Portland cement and plaster of Paris, a commercial concrete patching material (Fig. 7.16). After placement, the ECC and concrete patching repairs were exposed to identical traffic and environmental loads, thereby allowing for a meaningful comparison of their long-term durability performance. Although the bridge has a relatively low average daily traffic of 3000 vehicles per day, it is used frequently by 11-axle trucks heavily loaded with aggregates. Hence, mechanical (traffic) loading on the patch repairs was significant. Potential environmental loading on the patch repairs included shrinkage, temperature change, freezing and thawing, chloride penetration, salt-scaling, and steel corrosion.

The performance of the ECC and adjacent concrete patch repairs was assessed through site visits. Initial visits 2 days after construction showed no visible cracking in the ECC patch repair. However, a clearly visible crack, about 300 μm wide, appeared in the concrete repair. A possible reason for the cracking was restrained shrinkage. After 4-months winter exposure, a number of microcracks with width around 50 μm had formed in the ECC patch repair. In contrast, the initial crack formed in the concrete repair widened to 2 mm and was surrounded by deteriorated and spalling concrete. In 2005, the concrete repair has severely deteriorated with cracks larger than 3.5 mm and had to be re-repaired. In 2007 when the complete bridge deck was reconstructed, the last available data point indicates that the ECC patch had survived the combined mechanical and environmental loading

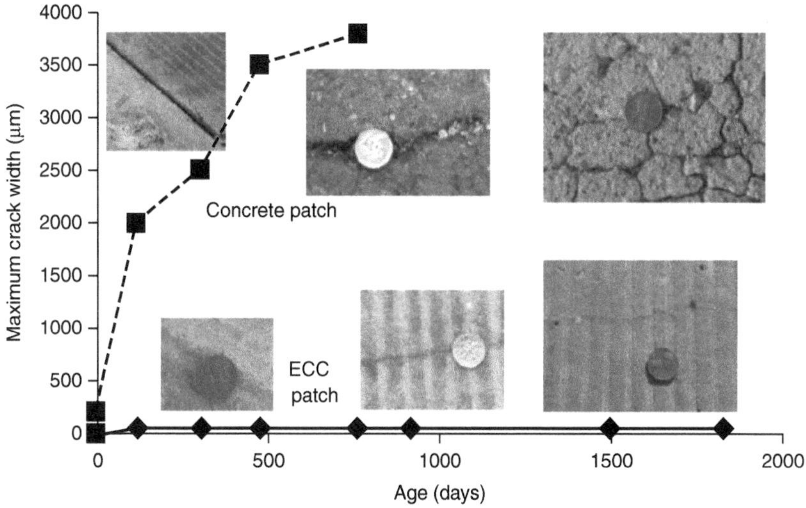

7.17 Deterioration in concrete patch repair compared with ECC patch repair.[26,42]

conditions with minor microcracking limited to less than 50 μm. No spalling or other deteriorations were found. The monitored crack width development is shown in Fig. 7.17. Field performance data revealed that ECC is a durable construction material in the harsh environments of Michigan winters in addition to heavy traffic loads.

7.4.2 High early strength ECC patch repair

In November 2006, a patch repair made of a special version of ECC, HES-ECC, was constructed on the Ellsworth Road Bridge over US-23 (S07 of 81074) in Ann Arbor, Michigan.[16,74] MDOT repair personnel conducted partial depth concrete removal, sandblasted the patch area, replaced damaged reinforcement, and completed patch preparation. Before pouring the patch materials, water fog was sprayed onto the concrete substrate to enhance repair bonding. The average temperature was 55°F (12.8°C) and relative humidity was 59.5% during this construction. 7 cubic feet (0.20 cubic meters) of HES-ECC was mixed using a 12 cubic foot (0.34 cubic meters) capacity concrete gas mixer provided by MDOT. The total mixing time was approximately 30 min. The fresh HES-ECC exhibited desirable viscosity, fiber distribution, and flowability. Flowability testing using a slump cone was conducted immediately after finishing the mixing and before repair placement. With a 25.6 in. (650.2 mm) slump diameter, the HES-ECC achieved a self-compacting state. The HES-ECC material was then poured into

7.18 Placement of self-compacting HES-ECC.[16]

a wheelbarrow, transported to the patch area, and poured into the patch (Fig. 7.18). The self-compacting property of HES-ECC allowed it to easily flow into the corners of the patch area without any vibration. The surface of the HES-ECC patch repair was finalized by using a steel trowel, by hand.

Mixing of Thoroc 10-60 (a patch repair material frequently used by MDOT) was performed using the same mixer. The fresh Thoroc 10-60 was barely flowable, resulting in much more effort and a longer time for construction workers to place and finish (Fig. 7.19). The two patch repairs using HES-ECC and Thoroc 10–60 are shown side by side in Fig. 7.20. Latex-based curing compound was sprayed onto both patches to reduce evaporation and shrinkage.

The two side-by-side patch repairs on the Ellsworth Road Bridge were completed at around 1 p.m. on Tuesday, 28 November 2006, as shown in Fig. 7.20. Taking advantage of the rapid strength gain of the materials, traffic reopened the next morning at 9 a.m.

The performance of the HES-ECC and adjacent Thoroc 10-60 patch repairs has been assessed through a series of 12 site visits. This field performance data for 2 years and 10 months of patch life reveals the promise of HES-ECC as a durable repair material in the harsh environments of Michigan winters (with combined effects of restrained shrinkage, freezing

7.19 Placement of Thoroc 10-60.[16]

7.20 HES-ECC and Thoroc 10-60 patch repairs side by side.[16]

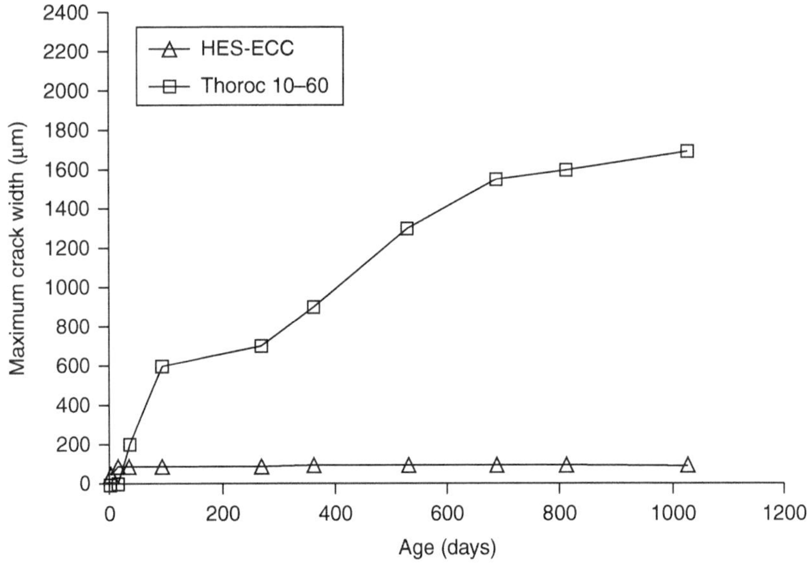

7.21 Maximum surface crack width of HES-ECC and Thoroc 10-60 patch repairs.[16]

and thawing, exposure to deicing salts, temperature change, etc.) in addition to heavy traffic loads, in terms of minimized surface cracking and old/new interfacial delamination (Figs 7.21 and 7.22).

7.4.3 ECC bridge link slab

Many highway bridges are composed of multiple span steel or prestressed concrete girders simply supported at piers or bents. The girders support cast-in-place concrete decks. A mechanical joint is typically used at the end of decks to allow deck deformation imposed by temperature variation, concrete shrinkage, and girder deflection. The bridge-deck joints are expensive to install and maintain. Debris accumulation can often cause deterioration of the expansion joints, which subsequently will lead to severe damage in the bridge deck and substructure. Thermal expansion caused by a high temperature combined with debris filled expansion joints can cause cracks or spalls near the expansion joints. Water leakage through the joints can cause beam-end corrosion, spalling, and deterioration of pier caps.

A possible approach to address this problem is the replacement of expansion joints by a concrete link slab in multi-span bridges. A link slab is a section of the deck that connects the two adjacent simple-span

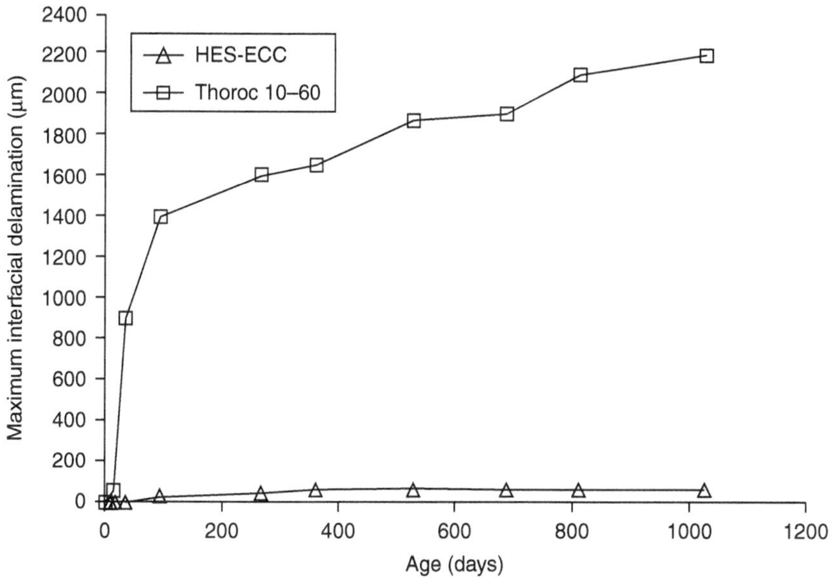

7.22 Maximum interfacial delamination width of HES-ECC and Thoroc 10-60 patch repairs.[16]

girders. A number of concrete link slabs have been constructed in other states, including Michigan and Florida, to address the continuing problem of deteriorating and leaking expansion joints (Fig. 7.23). In general, the link slabs are subjected to negative bending under traffic conditions and environmental loads, and tensile cracks may form at the top of the link slab.

To address these challenges, a high performance and durable link slab made of ECC was constructed in Michigan to replace expansion joints for multi-span simply supported bridges[75] (Fig. 7.23). The ECC link slabs can maintain the simple-span performance of the bridge while accommodating mechanical and environmental (i.e. thermal) loads typically accounted for by the expansion joints. The ductile link slabs allow for a joint-free bridge deck, eliminating leaking problems while creating a smoother riding surface. Notably, the ECC slabs can accommodate large deformation with self-controlled microcrack width to far below the AASHTO limit, even without relying on steel reinforcement. In this way, the ECC link slab eliminates the tensile cracking problem at the top of conventional concrete link slabs. Furthermore, an integrated life cycle assessment and life cycle cost analysis (LCA-LCC) study showed that the bridge system incorporating ECC link slab consumed 40% less total primary energy, and 40% less carbon dioxide and other greenhouse gases, than the conventional bridge-deck system with

7.23 ECC link slab in Michigan to replace bridge expansion joints.[75]

expansion joints.[76] Apparently, the long-term durability of ECC link slab leads to minimized maintenance and repairs, resulting in the significantly reduced economic and environmental impacts.

7.4.4 ECC steel composite bridge deck

A composite ECC/steel bridge deck was constructed in 2004 for a cable-stayed bridge (Mihara Bridge) in Hokkaido, Japan.[77,78] The bridge, opened to traffic in 2005, is 1000 m long with the middle span of 340 m. The full deck area is 20 000 square meters (Fig. 7.24). The deck comprised a 38 mm thin layer of ECC placed on top of steel. This design took advantage of the high durability of ECC and its intrinsic crack control capacity, to protect the underlying steel from the penetration of moisture and chloride. This led to a super light-weight design with a 40% weight reduction, and thus a reduction of initial construction cost by 50%. With an expected service life of 100 years, the life cycle costs associated with maintenance, repair, and rehabilitation are also expected to be significantly lower than conventional bridge-deck systems.

7.24 Mihara Bridge in Hokkaido, Japan, courtesy of Dr Tetsu Kanda.[77]

7.5 Conclusions

ECC, by fundamentally eliminating concrete brittleness, has opened a new world of applications in bridge decks and other structures. This chapter described the field application of ECC in bridge-deck repairs, link slabs, and a new composite deck. The discussion on the properties of ECC under various mechanical and environmental loading conditions has provided insights on why the ductile behavior of ECC can lead to the durable and sustainable (i.e. low environmental impact) performance of ECC repair and new construction systems. ECC technology is still evolving, which includes the design of different versions of ECC materials, the creation of smart and multifunctional ECCs, the understanding of their behavior under combined mechanical and environmental conditions, the establishment of reliable ECC constitutive models under complex loading, and the development of ECC structural design specifications and codes for the wide application of this material in civil infrastructure.

7.6 References

1. Naaman, A. E. and Reinhardt, H. W., 'Characterization of High Performance Fiber Reinforced Cement Composites HPFRCC', *High Performance Fiber Reinforced Cementitious Composites*, RILEM Proceeding 31, Edited by A. E. Naaman and H. W. Reinhardt, 1996, pp. 1–23.
2. Li, V. C., 'Engineered Cementitious Composites (ECC) – Materials, Structural and Durability Performance', *Concrete Construction Engineering Handbook*, Edited by E. G. Nawy, Taylor & Francis Group, 2008.
3. Aveston, J., Cooper, G. and Kelly, A., 'Single and Multiple Fracture', *Properties of Fiber Composites*, Guidford, UK, 1971, pp. 15–24.
4. Krenchel, H. and Stang, H., 'Stable Microcracking in Cementitious Materials'. In *Brittle Matrix Composites 2*. Edited by A. M. Brandt and J. H. Marshall, Elsevier Applied Science, 1989, pp. 20–33.
5. Curbach, M. and Jesse, F., 'High-Performance Textile-Reinforced Concrete', *Structural Engineering International*, Vol. **9**, No. 4, 1999, pp. 289–291.
6. Reinhardt, H. W., Krüger, M. and Große, C. U., 'Concrete Prestressed with Textile Fabric', *Journal of Advanced Concrete Technology*, Vol. **1**, No. 3, 2003, pp. 231–239.
7. Mobasher, B., Peled, A. and Pahilajani, J., 'Distributed Cracking and Stiffness Degradation in Fabric-cement Composites', *Materials and Structures*, Vol. **39**, No. 3, 2006, pp. 317–331.
8. Allen, H. G., 'Stiffness and Strength of Two Glass-fiber Reinforced Cement Laminates', *Journal of Composite Materials*, Vol. **5**, No. 2, 1971, pp. 194–207.
9. Lankard, D. R., 'Preparation, Properties and Applications of Cement Based Composites Containing 5-20 Percent Steel Fiber Reinforcement', *Steel Fiber Concrete*, Edited by S. P. Shah and A. Skarendahl, Elsevier Publishers, 1986.

10. Naaman, A. E., 'SIFCON: Tailored Properties for Structural Performance', *Proceedings of the International Workshop on High Performance Fiber Reinforced Cement Composites*, Edited by H. W. Reinhardt and A. E. Naaman, E & FN Spon, London, 1992, pp. 18–38.
11. Li, V. C., Wu, C., Wang, S., Ogawa, A. and Saito, T., 'Interface Tailoring for Strain-Hardening Polyvinyl Alcohol-Engineered Cementitious Composites (PVA-ECC)', *ACI Materials Journal*, September–October 2002, Vol. **99**, No. 5, pp. 463–472.
12. Li, V. C., 'From Micromechanics to Structural Engineering – the Design of Cementitious Composites for Civil Engineering Applications', *JSCE Journal of Structural Mechanics and Earthquake Engineering*, Vol. **10**, No. 2, 1993, pp. 37–48.
13. Li, V. C., 'Reflections on the Research and Development of Engineered Cementitious Composites (ECC)', *Proceedings of the JCI International Workshop – DFRCC*, Takayama, Japan, 2002, pp. 1–21.
14. Fan, S. and Li, M., 'Three-Dimensional X-Ray Computed Microtomography of Microcracks and Self-Healing in Engineered Cementitious Materials', In review, *Cement and Concrete Composites*, 2013.
15. Li, V. C., 'Integrated Structures and Materials Design', *RILEM Journal of Materials and Structures*, Vol. **40**, No. 4, 2007, pp. 387–396.
16. Li, M., 'Multi-Scale Design for Durable Repair of Concrete Structures', Ph.D. dissertation, University of Michigan, 2009.
17. Fischer, G. and Li, V. C. (eds), *RILEM Proceedings PRO 49: International RILEM Workshop on High Performance Fiber Reinforced Cementitious Composites (HPFRCC) in Structural Applications*, 2006, p 600.
18. Van Ziji, G. P. A. G. and Wittman, F. H. (eds), 'Durability of Strain-Hardening Fibre-Reinforced Cement-Based Composites (SHCC)', *RILEM State-of-the-Art Reports*, 2010.
19. Yokota, H., Rokugo, K. and Sakata, N., 'JSCE Recommendations for Design and Construction of High Performance Fiber Reinforced Cement Composites with Multiple Fine Cracks', *Japanese Society of Civil Engineers*, Tokyo, 2008.
20. Li, V. C. and Leung, C. K. Y., 'Theory of Steady State and Multiple Cracking of Random Discontinuous Fiber Reinforced Brittle Matrix Composites', *Journal of Engineering Mechanics*, ASCE, Vol. **118**, No. 11, 1992, pp. 2246–2264.
21. Li, V. C. and Wu, H. C., 'Conditions for Pseudo Strain-Hardening in Fiber Reinforced Brittle Matrix Composites', *Journal of Applied Mechanics Review*, Vol. **45**, No. 8, August, 1992, pp. 390–398.
22. Marshall, D. B. and Cox, B. N., 'A J-Integral Method for Calculating Steady-State Matrix Cracking Stresses in Composites', *Mechanics of Materials*, Vol. **7**, No. 2, 1988, pp. 127–133.
23. Griffith, A. A., 'The Phenomena of Rupture and Flow in Solids', *Philosophical Transactions of the Royal Society of London*, Vol. **A 221**, 1921, pp. 163–198, http://www.cmse.ed.ac.uk/AdvMat45/Griffith20.pdf.
24. Li, M. and Li, V. C., 'Rheology, Fiber Dispersion, and Robust Properties of Engineered Cementitious Composites', *Materials and Structures*, Vol. **46**, No. 3, 2013, pp. 405–420.

25. Li, M. and Li, V. C., 'Non-Brittle Concrete for Durable Infrastructure in Coastal Regions', *The International Conference on Future Concrete*, Doha, Qatar, pp. 1–3 November 2010.
26. Li, V. C. and Li, M., 'Durability Performance of Ductile Concrete Structures', *Proceedings of the 8th International Conference on Creep, Shrinkage and Durability of Concrete and Concrete Structures*, Ise-Shima, Japan, 30 September–2 October 2008.
27. Li, M. and Li, V. C., 'Behavior of ECC/Concrete Layered Repair System under Drying Shrinkage Conditions', *Journal of Restoration of Buildings and Monuments*, Vol. **12**, No. 2, 2006, pp. 143–160.
28. Bhat, P. and Li, M., 'Effect of Elevated Temperature on Strain-Hardening Engineered Cementitious Composites', in review, *Cement and Concrete Composites*, 2013.
29. ACI Committee 224, 'Control of Cracking in Concrete Structures (ACI 224R-01)', American Concrete Institute, Farmington Hills, MI, 2001, p 45.
30. Tsinker, G. P., *Marine Structures Engineering: Specialized Applications,* Ghapman & Hall, 1995, p 548.
31. Mehta, P. K. and Gerwick, B. C., 'Cracking-Corrosion Interaction in Concrete Exposed to Marine Environment', *Concrete International,* Vol. **4**, No. 10, October 1982, pp. 45–51.
32. Gilbride, P., Morgan, D. R. and Bremner, T. W., 'Deterioration and Rehabilitation of Berth Faces in Tidal Zones at the Port of Saint John', *Shotcrete,* Fall 2002, pp. 32–38.
33. Mehta, P. K., 'Durability of Concrete Exposed to Marine Environment – A Fresh Look', *Concrete in Marine Environment,* SP-109, American Concrete Institute, Farmington Hills, MI, 1988, pp. 1–30.
34. Liu, P. C., 'Damage to Concrete Structures in a Marine Environment', *Materials and Structures,* Vol. **24**, 1991, pp. 302–307.
35. British Standards Institution, 'BS 8110: Part 1', BSI, London, 1997.
36. CEB-FIP Model Code 1990, *CEB Information Report* No. 213/214, Comite Euro-International Du Beton, Lausanne, May 1993.
37. 'Code of Practice for the Structural Use of Concrete – Part 1. Design, Materials and Workmanship', *British Standards Institution Publication CP 110,* London, UK, November 1972 (amended May 1977).
38. 'Standard Specification for Design and Construction of Concrete Structures – 1986, Part 1 (Design)', Japan Society of Civil Engineers, SP-1, Tokyo, Japan, 1986.
39. ACI Committee 222, 'Corrosion of Metals in Concrete (ACI 222R-89)', American Concrete Institute, Farmington Hills, MI, 1989, p 30.
40. Li, M. and Li, V. C., 'High-Early-Strength Engineered Cementitious Composites for Fast, Durable Concrete Repair – Material Properties', *ACI Materials Journal*, Vol. **108**, No. 1, January–February 2011, pp. 3–12.
41. Stang, H. and Li, V. C., 'Classification of Fiber Reinforced Cementitious Materials for Structural Applications', in *Fiber-Reinforced Concretes,* Proc, BEFIB 2004, Edited by M. di Prisco, R. Felicetti and G. A. Plizzari, 2004, pp. 197–218.
42. Li, V. C. and Lepech, M., 'Crack Resistant Concrete Material for Transportation Construction', In *TRB 83rd Annual Meeting*, Washington, D.C., CD ROM, Paper 04-4680.

43. Whiting, D. and Nagi, M., 'Strength and Durability of Rapid Highway Repair Concretes', *Concrete International,* Vol. **16**, No. 9, 1994, pp. 36–41.
44. ASTM C666/C666M – 03 (2008), Standard Test Method for Resistance of Concrete to Rapid Freezing and Thawing, American Society for Testing and Materials, 2008.
45. Li, M. and Li, V. C., 'Influence of Material Ductility on the Performance of Concrete Repair', *ACI Materials Journal*, Vol. **106**, No. 5, September 2009, pp. 419–428.
46. Tsukamoto, M., 'Tightness of Fiber Concrete', *Darmstadt Concrete*, Vol. **5**, 1990, pp. 215–225.
47. Wang, K. F., Jansen, D., Shah, S. and Karr, A., 'Permeability Study of Cracked Concrete', *Cement and Concrete Research*, Vol. **27**, No. 3, 1997, pp. 381–393.
48. Lepech, M. and Li, V. C., 'Water Permeability of Cracked Cementitious Composites', *Proceedings of Eleventh International Conference on Fracture*, Turin, Italy, 20–25 March 2005.
49. Batis, G., Pantazopoulou, P. and Routoulas, A., 'Corrosion Protection Investigation of Reinforcement by Inorganic Coating in the Presence of Alkanolamine-Based Inhibitor', *Cement and Concrete Composites*, Vol. **25**, 2003, pp. 371–377.
50. Elsener, B., 'Macrocell Corrosion of Steel in Concrete – Implications for Corrosion Monitoring', *Cement and Concrete Composites*, Vol. **24**, 2002, pp. 65–72.
51. The State of the Nation's Highway Bridges: Highway Bridge Replacement and Rehabilitation Program and National Bridge Inventory, *Thirteenth Report to the United States Congress*, Federal Highway Administration, Washington, D. C., May 1997.
52. Lewis, D. A., 'Some Aspects of the Corrosion of Steel in Concrete', *Proceeding of the First International Congress on Metallic Corrosion*, London, 1962, pp. 547–555.
53. Scheissl, P., 'Corrosion of Steel in Concrete', *RILEM Report*, Chapman and Hall, London, 1988.
54. Hime, W. G., 'The Corrosion of Steel – Random Thoughts and Wishful Thinking', *Concrete International*, Vol. **15**, No. 10, 1993, pp. 54–57.
55. Sahmaran, M., Li, M. and Li, V. C., 'Transport Properties of Engineered Cementitious Composites Under Chloride Exposure', *ACI Materials Journal*, Vol. **104**, No. 6, November 2007, pp. 604–611.
56. AASHTO T 259-02: Standard Method of Test for Resistance of Concrete to Chloride Ion Penetration, American Association of State Highway and Transportation Officials, Washington, DC, 2002.
57. Sahmaran, M., Li, V. C. and Andrade, C., 'Corrosion Resistance Performance of Steel-Reinforced Engineered Cementitious Composite Beams', *ACI Materials Journal*, Vol. **105**, No. 3, 2008, pp. 243–250.
58. *AASHTO LRFD Bridge Design Specifications*, 3rd ed., AASHTO, Washington, D.C., 2004.
59. Suthiwarapirak, P., Matsumoto, T. and Kanda, T., 'Flexural Fatigue Failure Characteristics of an Engineered Cementitious Composite and Polymer Cement Mortars', *JSCE Journal of Materials, Concrete Structures and Pavements*, Vol. **718**, No. 57, 2002, pp. 121–134.

60. Kim, Y. Y., Fischer, G. and Li, V. C., 'Performance of Bridge Deck Link Slabs Designed with Ductile ECC', *ACI Structural Journal*, Vol. **101**, No. 6, 2004, pp. 792–801.
61. Tayabji, S. D. and Okamoto, P. A., 'Thickness Design of Concrete Resurfacing', Proc., *3rd International Conference on Concrete Pavement Design and Rehabilitation*, 1985, pp. 367–379.
62. ERES Consultants, Inc., 'Evaluation of Unbonded Portland Cement Concrete Overlays', *NCHRP Report 415*, Transportation Research Board, Washington, D. C., 1999.
63. Huang, Y. H., *Pavement Analysis and Design*, 2nd ed., Pearson Education, Upper Saddle River, NJ 07458, 2004.
64. MDOT, *Pavement Design and Selection Manual*, Michigan Department of Transportation, Lansing, Michigan, 2005, p 65.
65. Jones, H. W., 'Asphalt Overlays; Crack & Seat vs. SAMI vs. Rubblizing vs. Removal How to Choose', *2009 Michigan Concrete Repair Seminar*, Michigan's Local Technical Assistant Program, Michigan Concrete Paving Association, Michigan Transportation Asset Management Council, Howell, Michigan, 2009.
66. National Concrete Pavement Technology Center, *Guild to Concrete Overlays: Sustainable Solutions for Resurfacing and Rehabilitating Existing Pavements, Second Edition*, September 2008.
67. Transport Research Board (TRB) of the National Academies, 'Crumb Rubber Modified (CRM) Stress-Absorbing Membrane Interlayers for New Pavement Construction, Maintenance and Repair of Existing Pavements and Bridge Decks', *Research Needs Statements*, http://rns.trb.org/dproject.asp?n=13076, accessed on 31 August 2009.
68. Newman, K. and Shoenberger, J. E., 'Polymer Concrete Micro-Overlay for Fuel and Abrasion Resistant Surfacing: Laboratory Results and Field Demonstrations', US Army Engineer Research and Development Center, Vicksburg, Mississippi, USA, Presented for the *2002 Federal Aviation Administration Airport Technology Transfer Conference*, May 2002.
69. RILEM Technical Committee 157-PRC, 'Third RILEM International Conference Reflective Cracking in Pavements', *Materials and Structures*, Vol. **30**, June 1997, pp. 317–318.
70. National Concrete Pavement Technology Center, *Guild to Concrete Overlay Solutions*, January 2007.
71. Qian, S., 'Influence of Concrete Material Ductility on the Behavior of High Stress Concentration Zones', Ph.D. Dissertation, University of Michigan, 2007.
72. Zhang, J. and Li, V. C., 'Monotonic and Fatigue Performance in Bending of Fiber Reinforced Engineered Cementitious Composite in Overlay System', *Journal of Cement and Concrete Research*, Vol. **32**, No. 3, 2002, pp. 415–423.
73. Li, M. and Li, V. C., 'Durability of HES-ECC Repair Under Mechanical and Environmental Loading Conditions', Proc., HPFRCC5, Edited by H. W. Reinhardt and A. E. Naaman, Mainz, Germany, 2007, pp. 399–408.
74. Li, V. C., Li, M. and Lepech, M., 'High Performance Material for Rapid Durable Repair of Bridges and Structures', Michigan Department of Transportation Research Report RC-1484, December 2006, p 142.

75. Li, V. C., Lepech, M. and Li, M., 'Field Demonstration of Durable Link Slabs for Jointless Bridge Decks Based on Strain-Hardening Cementitious Composites', Michigan Department of Transportation Research Report RC-1471, December 2005, p 265.
76. Keoleian, G. A., Kendall, A., Dettling, J. E., Smith, V. M., Chandler, R. F., Lepech, M. D. and Li, V. C., 'Life Cycle Modeling of Concrete Bridge Design: Comparison of ECC Link Slabs and Conventional Steel Expansion Joints', *Journal of Infrastructure Systems*, March 2005, pp. 51–60.
77. Construction of the Mihara Bridge, (in Japanese), http://nissei-kensetsu.co.jp/bihara.html.
78. Li, V. C., 'Engineered Cementitious Composites', *Proceedings of ConMat'05*, Vancouver, Canada, 22–24 August 2005, CD-documents/1-05/SS-GF-01_FP.pdf.

8

The use of carbon fiber-reinforced polymer (CFRP) composites for cable-stayed bridges

W. XIONG, Southeast University, China, C. S. CAI, Louisiana State University, USA and R. C. XIAO, Tongji University, China

DOI: 10.1533/9780857097019.2.210

Abstract: This chapter introduces fiber-reinforced polymer (FRP) applications for cable-stayed bridges with carbon fiber-reinforced polymer (CFRP) stay cables, hybrid stay cables, and/or a CFRP bridge deck. For each of these CFRP components, key design parameters and design strategies are determined and the appropriate value of each key design parameter is suggested. Using the suggested parameter values, six types of cable-stayed bridges with a main span length of 1400 m are selected and modeled with the finite element method, through which the procedures of designing composite cable-stayed bridges with CFRP components are presented in detail. From a mechanical-behavior viewpoint (static and dynamic) a comparative study of composite cable-stayed bridges with various CFRP components is performed through numerical simulation. The economic aspects of each type of cable-stayed bridge is also comparatively studied with respect to material prices. With their high strength-to-weight ratio, and other advantages of CFRP materials, it is proven in this study that the use of CFRP components in super-long-span cable-stayed bridges is feasible, and that these types of composite cable-stayed bridges could become excellent alternatives to steel cable-stayed bridges where super-long spans are desired.

Key words: cable-stayed bridges, CFRP bridge deck, CFRP stay cables, design theory, hybrid stay cables.

8.1 Introduction

Fiber-reinforced polymers (FRPs) materials are increasingly being used in engineering bridges with long spans (Meier, 1987; Schurter and Meier, 1996; Saadatmanesh and Ehsani, 1998; Rabinovitch and Frostig, 2003; Schilde and Seim, 2007). The use of cable-stayed bridges with FRP composite stay cables (including hybrid stay cables) and/or FRP composite bridge decks has become more popular. With their light weight and superior strength, FRP composite materials can be used to improve load-carrying efficiency and to extend the span length of cable-stayed bridges over 1000 m. The excellent durability and corrosion resistant behavior of FRP composite materials make FRP composite cable-stayed bridges suitable in aggressive environments, such as crossing seawater. These features of FRP materials may make them very attractive and appropriate when designing cable-stayed bridges with super-long spans (Meier, 1987). Due to strength considerations, carbon fiber-reinforced polymers (CFRPs) become the best choice of FRP materials, with the widest range of practical applications, which will be discussed in the following sections (Noisternig and Jungwirth, 1998; Noisternig, 2000).

8.2 Design of carbon fiber-reinforced polymer (CFRP) bridge decks

A number of commercial CFRP bridge decks are available in the market today; however, how to mechanically design a good CFRP bridge deck is still an unsolved issue, especially in making the most of the CFRP's advantages. In this chapter, two design theories of CFRP bridge decks, strength criterion design and stiffness criterion design, are proposed in detail. It can be concluded that the buckling and local stiffness need to be carefully designed when using CFRP materials for bridge decks.

8.2.1 Strength criterion design

The majority of CFRP bridge-deck systems generally use carbon-reinforcing fibers set in a polyester, epoxy, or vinyl ester resin matrix, resulting in a lightweight decking system that can be rapidly erected while providing excellent long-term durability. Typical CFRP bridge-deck systems consist of two principal types:

- pultruded tubes that are bonded together with an adhesive and honeycomb; or
- sandwich core systems that are laid-up by hand or use vacuum-assisted resin transfer molding techniques.

Several examples of CFRP bridge-deck systems are shown in Fig. 8.1. Each deck system is usually factory assembled into panels that are sized appropriately for shipment to site. The panels can then be erected and bonded together *in situ* using high performance adhesives.

A number of commercial CFRP bridge decks are already available in the market. However, the mechanical design of a good CFRP bridge deck is still a matter of debate (Hayes *et al.*, 2000; Cai *et al.*, 2009). According to ACI (2004), the ultimate strength of CFRP materials applied on the bridge deck (2700 MPa for tension and 1440 MPa for compression) is much higher than that of common steel materials (345 MPa for both tension and compression), which is one of the significant advantages of CFRP materials. However,

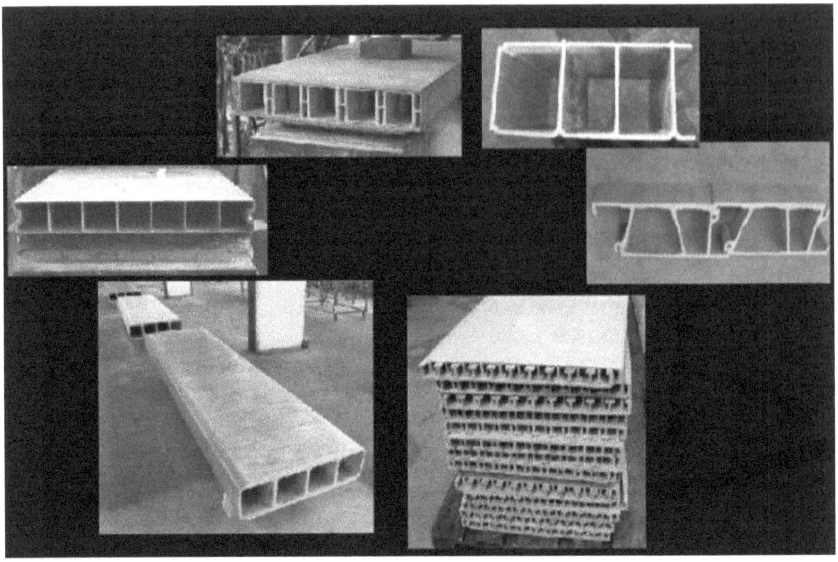

8.1 FRP bridge-deck systems.

usually this high ultimate strength will not be fully utilized, because the bridge deck needs to have a proper safety margin against sudden material damage or failure (or controlled by stiffness). Also, there is no early warning before sudden material damage or failure, due to the linear elastic behavior of CFRP materials and lack of yielding point. For these reasons, an allowable flexural stress of CFRP materials, which is less than the ultimate strength, needs to be specified when designing CFRP bridge decks. This allowable flexural stress can be seen as one of the most significant mechanical parameters of a CFRP bridge deck, determining the whole mechanical performance of the deck during the design process. In view of this, a safety factor for a bridge deck was used to calculate the allowable flexural stress of CFRP materials from their ultimate strength. Hereafter, the conditions under which the allowable flexural stress is reached are designated as 'design conditions,' and the conditions under which the ultimate strength of materials is reached are referred to as 'ultimate conditions' of materials.

A practical and user-friendly approach was used in the present study to determine a proper value for the safety factor of CFRP materials applied in bridge deck (Xiong *et al.*, 2011). Three typical expressions for the safety factor are:

$$k_D = \frac{D_u}{D_d}, \quad k_L = \frac{L_u}{L_d}, \quad k_E = \frac{E_u}{E_d} \qquad [8.1]$$

where k_D, k_L, and k_E = safety factors defined by deformation, load-carrying capacity, and deformation energy, respectively;
D_u, L_u, and E_u = deformation, load-carrying capacity, and deformation energy in the ultimate condition of materials, respectively; and
D_d, L_d, and E_d = deformation, load-carrying capacity, and deformation energy in the design condition of materials, respectively.

It should be pointed out that all the different expressions of the safety factor (k_D, k_L, and k_E) need to be calculated based on the same design and ultimate conditions of materials. For steel materials, the allowable flexural stress for bridge structure/component design (design condition) is typically set as $0.55F_y$ (where F_y is the yield strength, also the theoretical ultimate strength of steel materials in design or ultimate condition) to ensure adequate safety against the catastrophic failure of materials. Then, in the given design and ultimate conditions of steel materials, k_D can be firstly set as 4.00 based on the common ductility ratio in steel structure design (usually 4~6). The other safety factors, k_L and k_E, of steel materials can be easily calculated as 1.82 and 11.24 respectively, based on the pre-set value of k_D (4.00), Equation [8.1], and the idealized bilinear relationship of the stress–strain curve (Xiong *et al.*, 2011).

If both deformation and load-carrying capacity are considered, a new expression for the safety factor can be established as:

$$k_{DL} = k_D^m \cdot k_L^n = \left(\frac{D_u}{D_d}\right)^m \cdot \left(\frac{L_u}{L_d}\right)^n \qquad [8.2]$$

where: k_{DL} = safety factor defined by both deformation and load-carrying capacity; m and n = weight coefficients, both of which are equal to 1.0 in the present study. For steel materials, typically k_{DL} = 7.28 according to Equation [8.2].

As with steel design, the design safety level of CFRP bridge decks can also be developed or adjusted using a proper safety factor value. A safety level of design similar to steel materials is proposed here for CFRP materials before more appropriate strategies are available. This is the basic idea behind the proposed strategy specifying an allowable flexural stress for CFRP materials applied in bridge decks. Assuming one of the four safety factors k_D, k_L, k_E, and k_{DL} of CFRP materials is identical to that of steel materials, the other three values, as shown in Table 8.1, can be determined using Equations [8.1] and [8.2]. Due to the linear elastic behavior of CFRP materials and lack of yielding point, the calculated k_D is the same as k_L, and k_E is the same as k_{DL} for CFRP materials.

Table 8.1 Safety factor of CFRP materials

Safety factor		CFRP	Steel
Case (1)	k_D	4.00	4.00
k_D is identical to that of steel	k_L	4.00	1.82
materials	k_E	16.00	11.24
	k_{DL}	16.00	7.28
Case (2)	k_D	1.82	4.00
k_L is identical to that of steel	k_L	1.82	1.82
materials	k_E	3.31	11.24
	k_{DL}	3.31	7.28
Case (3)	k_D	3.35	4.00
k_E is identical to that of steel	k_L	3.35	1.82
materials	k_E	11.24	11.24
	k_{DL}	11.24	7.28
Case (4)	k_D	2.70	4.00
k_{DL} is identical to that of steel	k_L	2.70	1.82
materials	k_E	7.28	11.24
	k_{DL}	7.28	7.28

As can be seen from case (1) in Table 8.1, if the deformation safety factor of CFRP materials is taken to be the same as that of steel materials, i.e., $k_D = 4.00$, the values of the other three safety factors such as k_L will be much larger than those of steel materials, indicating that the material strength will not be fully used. In contrast, if the safety factor of the load-carrying capacity of CFRP materials is set at the same level as steel materials, i.e., $k_L = 1.82$ as in case (2), the values of the other three safety factors such as k_D will be much lower than those of steel materials, indicating insufficient ductility and possible catastrophic failure of CFRP materials.

If k_E and k_{DL} are given the same values as those of steel materials, i.e., $k_E = 11.24$ in case (3) or $k_{DL} = 7.28$ in case (4), the other three values for the safety factor of CFRP materials will be similar to those of steel materials. In other words, the safety level of designing CFRP materials using $k_E = 11.24$ or $k_{DL} = 7.28$ will be similar to that in designing steel materials. Therefore, they could be used in practical applications before further information on CFRP performance becomes available and design strategies are established. In the present study, the safety factor values resulting from $k_{DL} = 7.28$ in case (4) were selected for the design of a CFRP deck in the following study resulting in a safety factor for load-carrying capacity $k_L = 2.7$. This k_L value of 2.7 is similar to the 2.5 used for steel stay cables. As a result, the allowable flexural stress of CFRP deck materials was calculated as 2700/2.70 =1000 MPa for tension and 1440/2.70 = 533.33 MPa for compression.

It can be clearly seen that the allowable flexural stress of CFRP materials can be set much higher than for steel materials. Therefore, by following the design procedure of traditional steel bridge decks, the strength requirement in designing CFRP bridge decks could be easily satisfied.

8.2.2 Stiffness criterion design

Besides the strength requirement however, the buckling and deflection requirements of CFRP bridge decks, referred to as 'buckling and local stiffness requirement,' should also be satisfied in the design process. Due to the lower elastic modulus of CFRP materials, the buckling of a CFRP bridge deck could be a critical problem in design. Also, the deflection of a CFRP bridge deck subjected to vehicle loads could be larger than that of a steel bridge deck of equal thickness. Excessive deflection will not only accelerate the deterioration of bridge-deck pavement but will also affect ride quality and bridge-deck durability. Therefore, greater attention needs to be paid to buckling and local stiffness when designing CFRP bridge decks.

In the present study, to ensure adequate buckling capacity and local stiffness of CFRP bridge decks, the stiffness was designed to be equal to that of the reference model of traditional steel cable-stayed bridges, since there are no corresponding stiffness design guidelines for CFRP bridge decks. This practice was based on an assumption that the design for the steel bridge deck of traditional cable-stayed bridges completely satisfies the buckling and local stiffness requirements. For a rectangular solid cross-section of a CFRP bridge deck (or equivalently solid, because hollowed sandwich panels are commonly used for CFRP bridge decks), its thickness can be easily calculated based on the required local stiffness of the bridge deck:

$$T_{h\text{CFRP}} = \left(\frac{E_{\text{steel}} T^3_{h\text{steel}}}{E_{\text{CFRP}}} \right)^{1/3} \qquad [8.3]$$

where: $T_{h\text{CFRP}}$ and $T_{h\text{steel}}$ = thicknesses of the CFRP and steel bridge decks, respectively; and E_{CFRP} and E_{steel} = elastic modulus of the CFRP and steel materials, respectively.

For the buckling stiffness case, the thickness of a CFRP bridge deck can also be calculated using the same methodology. From a construction viewpoint, bonding issues between CFRP deck and steel girder are also very important. The materials used to reliably bond CFRP deck and steel girder could be considered the same as when using CFRP laminates to strengthen steel girders. Many related studies have been conducted and reported to ensure adequate adhesive capacity between CFRP laminates and steel girders (Sen *et al.*, 2001; Matta, 2003; Schnerch *et al.*, 2004, 2005; Colombi and Poggi, 2006; Rizkalla *et al.*, 2008; Zhao, 2008). Zhao and Zhang (2007) also gave a state-of-the-art review on the CFRP strengthened steel structures including bond test methods, failure modes in a CFRP bonded steel system, bond strength, and bond-slip relationships. All the studies proved that appropriate bond materials and techniques between the steel and CFRP materials do exist (for example, epoxy resin) and can be reliably applied to strengthening structures. While it is reasonable to believe that a reliable bond between a CFRP deck and a steel girder should also be obtained using the same or similar bond materials and techniques as used for strengthening structures, the bond issue of CFRP bridge-deck design may deserve further research.

Based on the analysis above, the buckling and local stiffness needs to be carefully considered if using CFRP materials for bridge decks. The strength requirement of designing CFRP bridge decks could be easily satisfied, and it is believed that the connection between CFRP deck and steel girder can be also reliably constructed using the same bonding materials and techniques used for strengthening engineering.

8.3 Design of CFRP stay cables

With light weight and superior strength, the applications of CFRP stay cables should highly improve the load-carrying efficiency and extend the span length of cable-stayed bridges to over 1000 m. Herein, the key design parameter for CFRP stay cables and their appropriate span lengths were investigated in order to assure a logical design. A conclusion can be drawn that 1400 m could be a good starting point to apply CFRP stay cables in cable-stayed bridges if considering an obvious advantage in stiffness.

8.3.1 Key design parameter for CFRP stay cables

CFRP materials, being stronger and lighter, can be used as new materials for stay cables, improving their load-carrying efficiency and effectively extending the span length of cable-stayed bridges to over 1000 m. Since the stiffness, weight, and ultimate strength of a stay cable largely depends on its cross-sectional area, in the present study the cross-sectional area of stay cables is determined as the key design parameter for CFRP stay cables. A stay cable design ensures adequate allowable strength for the stay cables if their design forces are given. Therefore, to obtain an appropriate cross-sectional area, the allowable axial stress of CFRP stay cables should firstly be determined using the following expression (Xiong *et al.*, 2012):

$$f_{\text{CFRP}} = \frac{u_{\text{CFRP}}}{k_S} \quad [8.4]$$

where:
f_{CFRP} = allowable axial stress of the CFRP stay cables (working stress);
u_{CFRP} = ultimate strength (stress) of the CFRP stay cables; and
k_S = safety factor.

According to Xiong *et al.* (2011, 2012), in the present study, u_{CFRP} and k_S were set as 2550 MPa and 2.50, respectively, resulting in:

$$f_{\text{CFRP}} = \frac{2550}{2.50} = 1020 \text{ MPa}$$

Then the appropriate cross-sectional areas of CFRP stay cables can be obtained as:

$$A_{\text{CFRP}} = \frac{F}{f_{\text{CFRP}}} \quad [8.5]$$

where:
A_{CFRP} = cross-sectional area of the CFRP stay cable; and
F = unfactored design force of the stay cable.

For a given cable force the cross-sectional areas of CFRP stay cables can be calculated from the steel stay cables in traditional cable-stayed bridges using the equivalent strength criterion:

$$A_{\text{CFRP}} f_{\text{CFRP}} = A_{\text{steel}} f_{\text{steel}}$$

where A_{steel} and f_{steel} are cross-sectional area and allowable axial stress, respectively, of steel stay cables used in traditional cable-stayed bridges.

8.3.2 Appropriate span lengths of CFRP stay cables

As is known, the equivalent stiffness of stay cables determined by cross-sectional area is usually seen as the judging criterion in determining appropriate span lengths. To better understand the effect of equivalent stiffness on the appropriate span lengths of CFRP stay cables, four mechanical parameters (equivalent stiffness and three other related parameters, see Table 8.2) were theoretically investigated through a parametric study by varying the 'cable span lengths' (horizontally projected length of stay cables) from 50

Table 8.2 Mechanical parameters and analytical solutions (Ernst, 1965; Gimsing, 1997)

Mechanical parameters	Analytical solutions
1. Equivalent stiffness	$K_{eq} = E_{eq}A = \frac{1}{\left[(1/E)+(\gamma_c^2 l^2/24)(\sigma_1+\sigma_2/\sigma_1^2\sigma_2^2)\right]A} \approx \frac{EA}{1+(\gamma_c^2 l^2/12\sigma_1^3)E}$
2. Load-carrying efficiency (real gradient of stay cables)	$\frac{V_T}{H} = k\left(-\frac{l}{l_{max}}+1\right)$
3. Cable sag effects	$\frac{\delta}{L_c} = \frac{(\sigma_2-\sigma_1)\gamma_c+(1/L_c)\left[\sigma_2^2 sh(\gamma_c^l/\sigma_2)-\sigma_1^2 sh(\gamma_c^l/\sigma_1)\right] - (4E/L_c)\left[\sigma_2 sh(q_0 l/2\sigma_2 A)-\sigma_1 sh(a_0 l/2\sigma_1 A)\right]}{2\gamma_c E ch(q_0 l/2\sigma_2 A)}$
4. Self-weight stress	$\sigma_g = \frac{\gamma_c l\sqrt{1+k^2}\sqrt{1+k^2(1+l/l_{max})^2}}{2k(1+l/l_{max})}$

Notes: K_{eq} = equivalent stiffness of stay cables; E_{eq} = equivalent elasticity modulus; E = elasticity modulus; A = area of the cross-section of stay cables; $\gamma_c = q_0/A$ = density of stay cables; q_0 = distributed loads (uniform gravity loads) per unit length of stay cables; l = horizontally projected length of stay cables; σ_1 and σ_2 = stress of stay cable in two load conditions (usually dead load and dead + live load conditions), respectively; V_T = vertical component of the cable force in the actual curved cable; H = horizontal component of the cable force in the actual curved cable; h = vertically projected length of stay cables; $k = h/l$; l_{max} = ultimate horizontally projected length of stay cables; δ/L_c = cable sag effect; L_c = chord length of stay cables; *ch*( ) = the function of hyperbolic cosine; and σ_g = self-weight stress of stay cables. The detailed derivations of these equations are not presented here and can be found in the literature (Ernst, 1965; Gimsing, 1997).

to 25 000 m using analytical models (see Table 8.3). These four mechanical parameters enable the equivalent stiffness of stay cables to be comprehensively described, directly or indirectly, and all of them are the key study objectives for span length studies in literature (O'Brien and Francis, 1964; Ernst, 1965; Irvine, 1981; Ahmadi-Kashani and Bell, 1988; Gimsing, 1997; Freire *et al.*, 2006). Based on the results of this parametric study the appropriate span lengths of CFRP stay cables can be recommended.

Figure 8.2a–8.2d shows the performance of CFRP stay cables with regard to the investigated mechanical parameters when varying the cable span lengths from 0 to 30 000 m. The numbers 0.3, 0.4, and 0.5 in the figures denote the nominal gradients of stay cables (k) for the investigated cable models. Firstly, Fig. 8.2a–8.2d shows that the reduction or deterioration of mechanical behavior representing the equivalent stiffness of stay cables becomes more and more obvious when the cable span length increases from 0 to 30 000 m. Furthermore, 5000 m could be a significant cable span length

Table 8.3 Design parameters of analytical models (CFRP stay cables)

Design parameters	Values		
Nominal gradient of stay cables ($k = h/l$)	0.3	0.4	0.5
Cable span length (m)	50~30 000	50~30 000	50~30 000
Vertically projected length of stay cables (m)	15~9000	20~12 000	25~15 000
Area (m^2)	0.00638		
Design stay cable (N)	5.102E + 6		
Elasticity modulus (N/m^2)	1.37E + 11		
Density (kg/m^3)	1.60E + 03		
Breaking strength (MPa)	1.02E + 03		
Safety factor	2.50		

after which all the investigated mechanical behaviors deteriorate more rapidly and suddenly than before, or already have an unacceptable reduction. Indeed, almost a 20% reduction in equivalent stiffness, 18% reduction in load-carrying efficiency, 0.0028 cable sag effects (about 70 m cable sagging), and 10% as the ratio of self-weight stress to breaking strength are observed at the 5000 m point in Fig. 8.2a–8.2d, respectively. By comparison Fig. 8.3 demonstrates that steel stay cables (area = 0.00974 m^2, design stay cable force = 5102 kN, elasticity modulus = 200 GPa, density = 7850 kg/m^3, yield strength = 668 MPa, and safety factor = 2.50 (ACI 2004)) show an obvious advantage in stiffness before the cable span length reaches 1400 m; however, beyond this length CFRP stay cables have the advantage. In other words, 1400 m could be a suitable starting point for CFRP stay cables in cable-stayed bridges. Therefore, in the present study, 1400~5000 m was determined as an appropriate span length of CFRP stay cables from a stiffness viewpoint. Additionally, 30 000 m could be considered as the maximum span length for pure CFRP stay cables as beyond this span length almost all the stiffness has been lost.

8.4 Design of CFRP–steel hybrid stay cables

Directly replacing steel stay cables with CFRP stay cables could result in an increase in displacement under loading, which is actually the main obstacle to the development of such bridges with CFRP stay cables. Hybrid stay cables using CFRP and steel materials are a novel concept in applying CFRP materials to bridge engineering (Xiong *et al.*, 2011, 2012). The basic idea of this hybrid concept is to create a new structure incorporating the advantages of both materials. There are several possible methods to form the hybrid stay cables. One of the most straightforward ways is to simultaneously install both CFRP stay cables and steel stay cables at the same

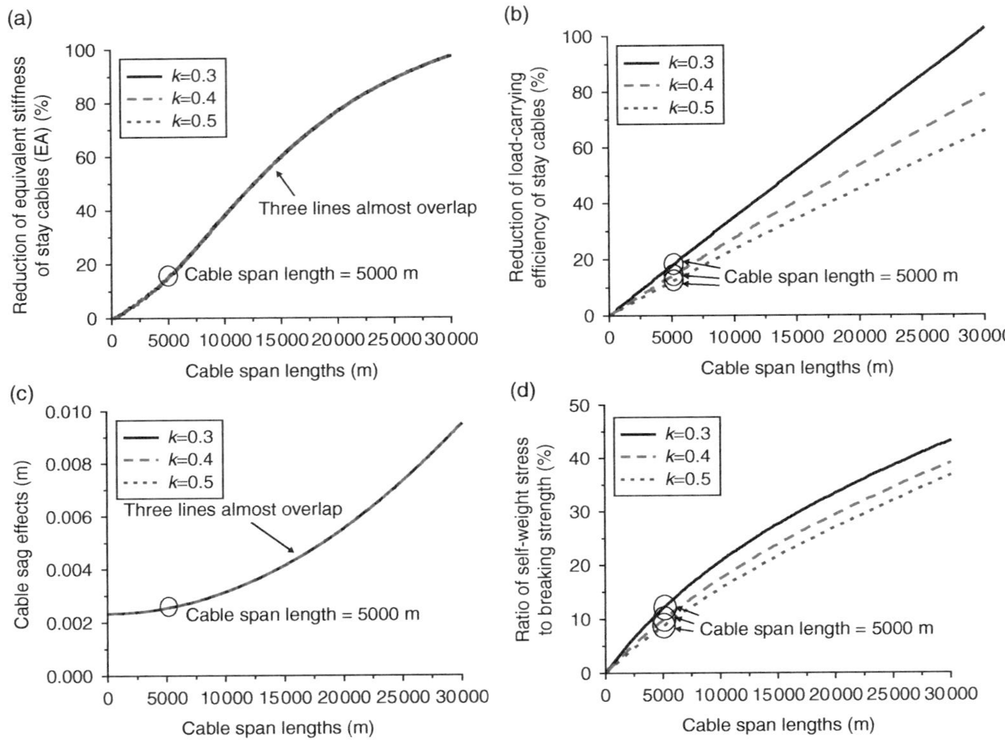

8.2 Performance of CFRP stay cables vs cable span lengths. (a) Reduction of equivalent stiffness vs cable span lengths; (b) reduction of load-carrying efficiency vs cable span lengths; (c) cable sag effects vs cable span lengths; (d) self-weight stress vs cable span lengths.

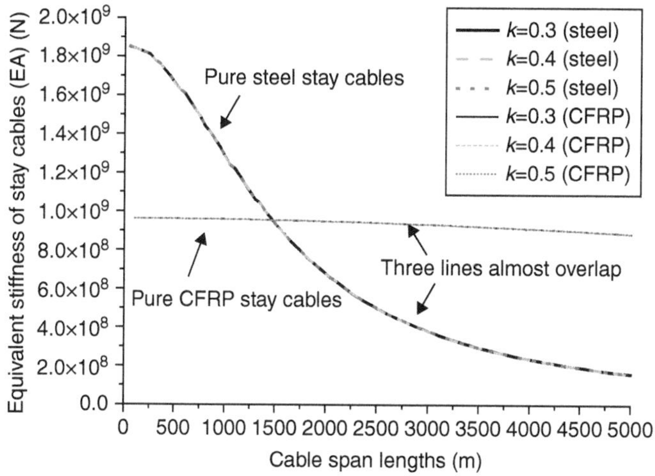

8.3 Comparison for the equivalent stiffness (EA).

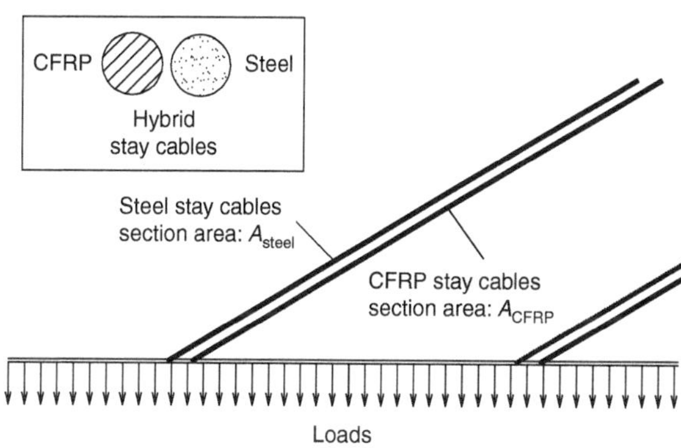

8.4 Hybrid stay cables.

location as shown in Fig. 8.4. By doing this, CFRP stay cables retain all the advantages of CFRP materials, such as their light weight and high strength, while steel stay cables increase the axial stiffness and reduce the total cost of the materials. The steel and CFRP cables can be installed separately in order to control the forces in each cable component.

Due to the different roles played by these two materials, the material proportion of the hybrid stay cables is expected to be a principal factor affecting the behavior and cost of the stay cable. For this reason, in the present study, the ratio of the CFRP section area to the entire section area, $\rho = A_{CFRP}/A_{steel\,+\,CFRP}$, is determined as the key design parameter for the hybrid stay cables. By adjusting the initial cable forces of both the steel and CFRP components, the possible effects of the differential elastic modulus of the two materials on the composite strength calculation can be resolved. It should be pointed out that the present study only aims at the mechanical or economical behavior of the hybrid stay cables without considering construction-related details, which will be the topic of a separate study.

The analysis conducted by the authors indicated the following:

cable span lengths 0~700 m – pure steel stay cables are more appropriate;
cable span lengths 1400~5000 m – pure CFRP stay cables are more appropriate;
cable span lengths 700~1400 m – combined application of steel and CFRP stay cables may be more appropriate.

As a result, a type of hybrid stay cable system was proposed in the present study for cable-stayed bridges with cable span lengths of 700~1400 m. Through a rational design (see the details later), the hybrid stay cable system can be a good alternative to the traditional pure steel stay cables or pure CFRP stay cables in this particular span length range. In other words, this span length range, i.e., 1400~2800 m for main span lengths of cable-stayed bridges, or 700~1400 m for the longest cable span lengths, can be regarded as the appropriate application range for the hybrid stay cable system.

8.4.1 Hybrid stay cable system

Also, for this type of hybrid stay cable system, a further increase in the equivalent stiffness could be obtained with a reasonable increase in their safety factor, i.e., increasing the section area of stay cables. It should be noted that this increasing-safety-factor-method may not be the best option for either pure steel or pure CFRP stay cables because it could result in increased weight or higher cost. When building appropriate connections between the steel stay cables and CFRP stay cables during or after construction, the hybrid stay cable can theoretically be regarded as a single cable with a composite section.

As stated earlier, in the present study, the ratio of the section area of CFRP stay cables to the whole cable area, i.e., $\rho = A_{CFRP}/(A_{steel} + A_{CFRP})$, was determined as the key design parameter for the hybrid stay cable system. It is easy to understand that this parameter can directly determine almost all the mechanical behaviors of the hybrid stay cables, such as strength, density, elastic modulus, and self-weight.

8.4.2 Analytical study for equivalent stiffness of the hybrid stay cable system

Before the introduction of design strategy (design criteria) an analytical study of the equivalent stiffness of the hybrid stay cable system was carried out. In the present study the equivalent stiffness of the hybrid stay cable system is defined as $E_{eq}A$:

$$K_{eq} = E_{eq}A = \frac{EA}{\left[1+\left(\gamma_c^2 l^2/12\sigma_1^3\right)\right]E} \qquad [8.6]$$

By assuming that the steel and CFRP stay cables can be pulled separately to achieve a desired force level for the hybrid stay cable system, each section area can be designed as:

$$A_{\text{steel}} = \frac{F_{\text{steel}}}{\sigma_{a\text{steel}}} = \frac{F_{\text{steel}}}{\left(\sigma_{l\text{steel}}/\eta_{\text{steel}}\right)} \qquad [8.7]$$

$$A_{\text{CFRP}} = \frac{F_{\text{CFRP}}}{\sigma_{a\text{CFRP}}} = \frac{F_{\text{CFRP}}}{\left(\sigma_{l\text{CFRP}}/\eta_{\text{CFRP}}\right)} \qquad [8.8]$$

where: A_{steel} and A_{CFRP} = section areas of the steel and CFRP stay cables, respectively; F_{steel} and F_{CFRP} = stay cable forces of the steel and CFRP stay cables, respectively; $\sigma_{a\text{steel}}$ and $\sigma_{a\text{CFRP}}$ = allowable stresses of the steel and CFRP stay cables, respectively; $\sigma_{l\text{steel}}$ and $\sigma_{l\text{CFRP}}$ = limit stresses of the steel and CFRP stay cables, respectively; and η_{steel} and η_{CFRP} = safety factors of the steel and CFRP stay cables, respectively.

The density of the hybrid stay cable system can be written as:

$$\gamma_c = \frac{\gamma_{\text{steel}}A_{\text{steel}} + \gamma_{\text{CFRP}}A_{\text{CFRP}}}{A_{\text{steel}} + A_{\text{CFRP}}} \qquad [8.9]$$

where γ_{steel} and γ_{CFRP} = densities of the steel and CFRP stay cables, respectively.

By safely assuming the deformation compatibility between the steel and CFRP stay cables after they are installed, the elastic modulus of the hybrid stay cable system can be given as:

$$E = \frac{E_{\text{steel}} A_{\text{steel}} + E_{\text{steel}} A_{\text{CFRP}}}{A_{\text{steel}} + A_{\text{CFRP}}} \qquad [8.10]$$

where E_{steel} and E_{CFRP} = elastic modulus of the steel and CFRP stay cables, respectively.

Substituting Equations [8.9] and [8.10] into Equation [8.6], the equivalent stiffness of the hybrid stay cable system can then be described as:

$$\begin{aligned} K_{\text{eq}} &= E_{\text{eq}} \left(A_{\text{CFRP}} + A_{\text{steel}} \right) \\ &= \frac{\left(A_{\text{CFRP}} + A_{\text{steel}} \right)}{\begin{array}{c}\left(\left(A_{\text{steel}} + A_{\text{CFRP}} \right) / E_{\text{steel}} A_{\text{steel}} + E_{\text{CFRP}} A_{\text{CFRP}} \right) \\ + \left(l^2 / 12\sigma_1^3 \right) \left(\left(\gamma_{\text{steel}} A_{\text{steel}} + \gamma_{\text{CFRP}} A_{\text{CFRP}} \right) / A_{\text{steel}} + A_{\text{CFRP}} \right)^2 \end{array}} \end{aligned} \qquad [8.11]$$

With the definition of $\rho = A_{\text{CFRP}}/(A_{\text{steel}} + A_{\text{CFRP}})$, Equation [8.11] can be further transformed into:

$$K_{\text{eq}} = \frac{\left(A_{\text{CFRP}} + A_{\text{steel}} \right)}{\left(1 / \left(E_{\text{steel}} (1-\rho) + E_{\text{CFRP}} \rho \right) \right) + \left(l^2 / 12\sigma_1^3 \right) \left(\gamma_{\text{steel}} (1-\rho) + \gamma_{\text{CFRP}} \rho \right)^2} \qquad [8.12]$$

If considering material costs, the ratio of equivalent stiffness to material cost of unit cable length can be calculated as:

$$\xi = \frac{\left(A_{\text{CFRP}} + A_{\text{steel}} \right)}{\begin{array}{l}\left(1 / \left(E_{\text{steel}} (1-\rho) + E_{\text{CFRP}} \rho \right) \right) + \\ \left(l^2 / 12\sigma_1^3 \right) \left(\gamma_{\text{steel}} (1-\rho) + \gamma_{\text{CFRP}} \rho \right)^2 \end{array}} \Bigg/ \left(\kappa_{\text{CFRP}} A_{\text{CFRP}} \gamma_{\text{CFRP}} + \kappa_{\text{steel}} A_{\text{steel}} \gamma_{\text{steel}} \right) \qquad [8.13]$$

where: ξ = ratio of equivalent stiffness to the material cost of unit cable length; κ_{steel} and κ_{CFRP} = costs of unit mass for the steel and CFRP stay cables with the consideration of materials and construction, respectively.

By defining $\lambda = \kappa_{\text{CFRP}} / \kappa_{\text{steel}}$, Equation [8.13] can be further written as:

$$\xi = \frac{1}{\begin{array}{l}\left(1 / \left(E_{\text{steel}} (1-\rho) + E_{\text{CFRP}} \rho \right) \right) + \\ \left(l^2 / 12\sigma_1^3 \right) \left(\gamma_{\text{steel}} (1-\rho) + \gamma_{\text{CFRP}} \rho \right) \end{array}} \Bigg/ \left(\lambda \kappa_{\text{steel}} \rho \gamma_{\text{CFRP}} + \kappa_{\text{steel}} (1-\rho) \gamma_{\text{steel}} \right) \qquad [8.14]$$

If the total cost for designing stay cables (including materials and construction) of unit cable length is defined as ψ, the section area of the hybrid stay cable system can be designed as:

$$A = \frac{\psi}{\left(\kappa_{\text{CFRP}}\rho\gamma_{\text{CFRP}} + \kappa_{\text{steel}}(1-\rho)\gamma_{\text{steel}}\right)} \qquad [8.15]$$

$$A_{\text{CFRP}} = A \cdot \rho \qquad [8.16]$$

$$A_{\text{steel}} = A\,(1-\rho) \qquad [8.17]$$

Then, substituting Equations [8.16] and [8.17] into Equation [8.12] the equivalent stiffness of the hybrid stay cable system under a given (pre-set) cost ψ (or 'budget') can be obtained.

For the convenience of discussion the costs per unit cable length of designing pure steel and pure CFRP stay cables with a safety factor of 2.5 are denoted as ψ_{steel} and ψ_{CFRP} respectively and are regarded as the basic costs. Correspondingly, any other possible costs per unit cable length with other safety factors for the hybrid section can be expressed as $\alpha\psi_{\text{steel}}$ or $\alpha\psi_{\text{CFRP}}$.

$$\psi_{\text{steel}} = \frac{F_{\text{steel}}}{\sigma_{a\text{steel}}} \cdot \gamma_{\text{steel}} \cdot \kappa_{\text{steel}} \qquad [8.18]$$

$$\psi_{\text{CFRP}} = \frac{F_{\text{CFRP}}}{\sigma_{a\text{CFRP}}} \cdot \gamma_{\text{CFRP}} \cdot \kappa_{\text{CFRP}} \qquad [8.19]$$

$$\alpha\psi_{\text{steel}} = \alpha \times \psi_{\text{steel}}, \quad \alpha = 1,2,\ldots,n \qquad [8.20]$$

$$\alpha\psi_{\text{CFRP}} = \alpha \times \psi_{\text{CFRP}}, \quad \alpha = 1,2,\ldots,n \qquad [8.21]$$

where $\alpha\psi_{\text{steel}}$ and $\alpha\psi_{\text{CFRP}} = \alpha$ times ψ_{steel} and ψ_{CFRP}, respectively.

All the discussed parameters and expressions that describe the characteristics of the equivalent stiffness of stay cables will be used in the following studies.

8.4.3 Design criteria of the hybrid stay cable system

With the consideration of equivalent stiffness and cost, three design criteria for the proposed hybrid stay cable system are introduced in the following sections. Designers can choose one of these criteria in a design process, or other design criteria can be similarly developed.

Design criterion 1: Best equivalent stiffness with a given safety factor

In this design criterion the best equivalent stiffness of stay cables with a given safety factor can be pre-designed by adjusting the area ratio ρ discussed earlier. To better understand the effects of ρ on equivalent stiffness, the equivalent stiffness of stay cables with different safety factors were theoretically investigated through a parametric study. Based on the results of this study, a design methodology for the hybrid stay cable system using this design criterion was recommended in assisting engineers to perform and optimize their designs.

To broaden the possible design conditions in the parametric study, four analytical models of stay cables with different horizontally projected lengths of 400, 700, 1000, and 1300 m respectively, and a constant designed cable gradient of 0.4, were studied mainly using Equations [8.7]–[8.12]. The designed cable force was set as 5102 kN and retained as constant in all the models for convenience of comparison. Also, to verify the proposed methodology for stiffness improvement by increasing the safety factor, eight safety factor values were considered (2.5, 3, 3.5, 4, 4.5, 5, 5.5, and 6), although some of these safety factors are too high in practice. The other geometric and material properties used in the parametric study were kept the same as those applied previously.

Figure 8.5a–8.5d plots the equivalent stiffness of stay cables vs area ratio (0~1.0) obtained with different safety factors (2.5~6) and cable span lengths (400, 700, 1000, and 1300 m) separately. The effects of increasing safety factors on stiffness improvement are clear from these figures. In each figure, the equivalent stiffness of stay cables at 0.0 and 1.0 area ratios represent the two extremes, i.e., pure steel stay cables and pure CFRP stay cables, respectively. The results corresponding to the area ratios between 0.0 and 1.0 are calculated from the hybrid cases. Taking the results with a safety factor of 5 for example, the superior range of the hybrid stay cable system following design criterion 1 is marked by a dashed line between two circles where the equivalent stiffness is higher than that of stay cables with both pure steel ($\rho = 0$) and pure CFRP ($\rho = 1$) materials. By repeatedly doing this, all the superior ranges of the hybrid stay cable system (the ranges of area ratio) with regard to different safety factors and cable span lengths can be obtained, and are shown in Fig. 8.6 as a design guide.

Based on the design guide shown in Fig. 8.6 the hybrid stay cable system can be easily designed using design criterion 1 as follows:

- Step 1: Determining the span lengths of stay cables;
- Step 2: Determining the safety factor used in the cable design;
- Step 3: From the design guide shown in Fig. 8.6, the appropriate area ratio range for the hybrid stay cable system (dashed area) can be selected based on the cable span length and safety factor used in

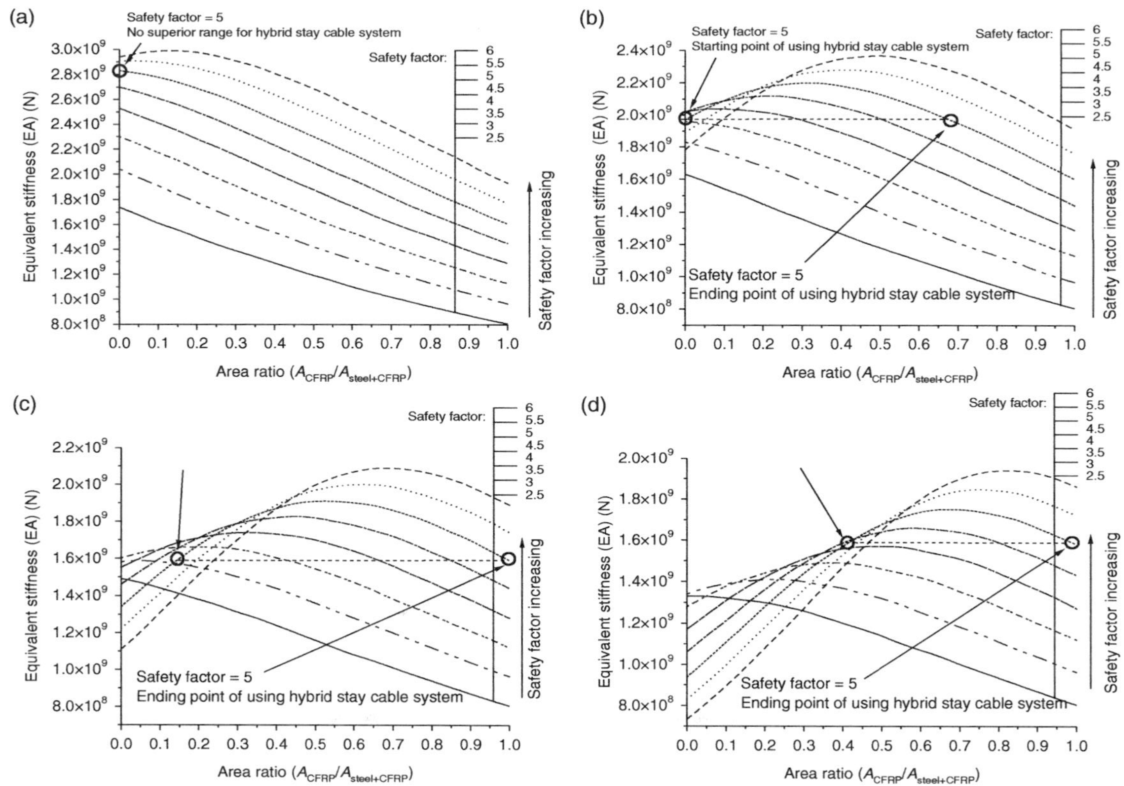

8.5 Equivalent stiffness of stay cables with different cable span lengths. (a) Stay cables with cable span length of 400 m; (b) stay cables with cable span length of 700 m; (c) stay cables with cable span length of 1000 m; (d) stay cables with cable span length of 1300 m.

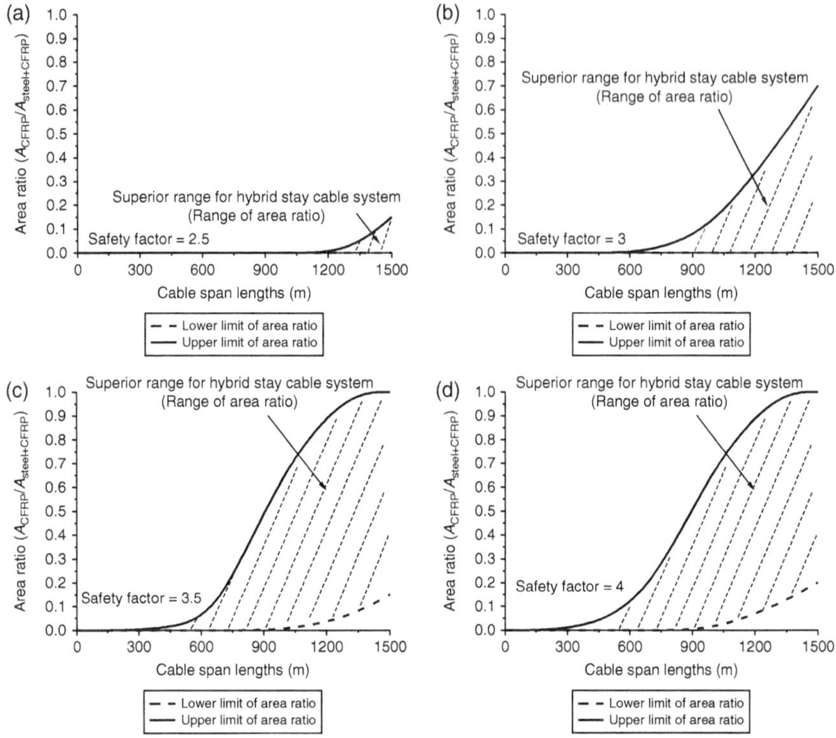

8.6 Design guide for hybrid stay cable system based on design criterion 1. (a) Design guide with safety factor 2.5; (b) design guide with safety factor 3; (c) design guide with safety factor 3.5; (d) design guide with safety factor 4; (e) design guide with safety factor 4.5; (f) design guide with safety factor 5; (g) design guide with safety factor 5.5; (h) design guide with safety factor 6.

(*Continued*)

design; furthermore, by using Fig. 8.5, the optimal area ratio corresponding to the highest equivalent stiffness can be obtained;

- Step 4: If no appropriate area ratio can be found based on the design criterion 1, pure steel stay cables can be selected for short cables near the pylon and pure CFRP stay cables for long cables further away from the pylon.

Design criterion 2: Highest ratio of equivalent stiffness to material cost with a given safety factor

In this design criterion the highest ratio of equivalent stiffness to material cost with a given safety factor can be pre-designed by adjusting the area ratio ρ with a given price ratio of CFRP stay cables to steel stay cables. As with design criterion 1, to better understand the effects of ρ on this

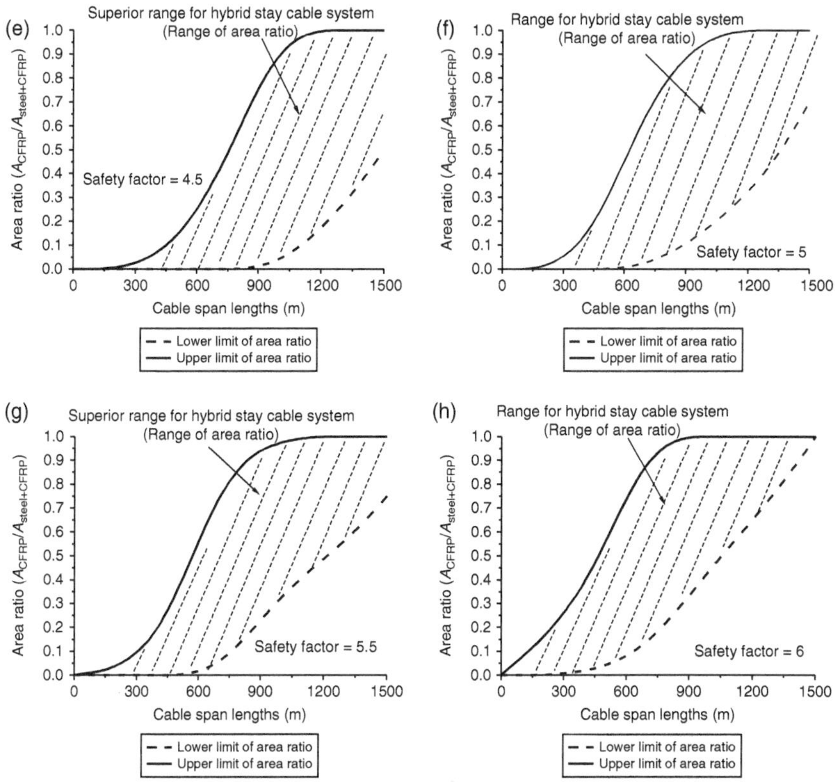

8.6 Continued

stiffness-to-cost ratio, a series of parametric studies were carried out in order to obtain a design methodology using design criterion 2.

The analytical models and investigated safety factors for the parametric study are identical to those used for design criterion 1. However, due to the cost factor in this design criterion, different price ratios of the two materials used in the hybrid stay cable system are studied to consider future cost changes. In this study, present prices of CFRP and steel stay cables are assumed as \$77.1/kg and \$2.9/kg, respectively, resulting in a price ratio of 27:1 (Xiong *et al.*, 2011, 2012) and more price ratios of CFRP stay cables to steel stay cables are also studied, which are 20:1, 15:1, 10:1, 8:1, 5:1, 2:1, and 1:1. For purposes of brevity, only the results of 700 m (Fig. 8.7) and 1000 m (Fig. 8.8) cable span lengths with different price ratios are presented here.

Figures 8.7a–8.7e and 8.8a–8.8e plot the ratio of equivalent stiffness to material cost vs area ratio (0~1.0) obtained with different price ratios (27:1~2:1), safety factors (2.5~6), and cable span lengths (only the 700 and

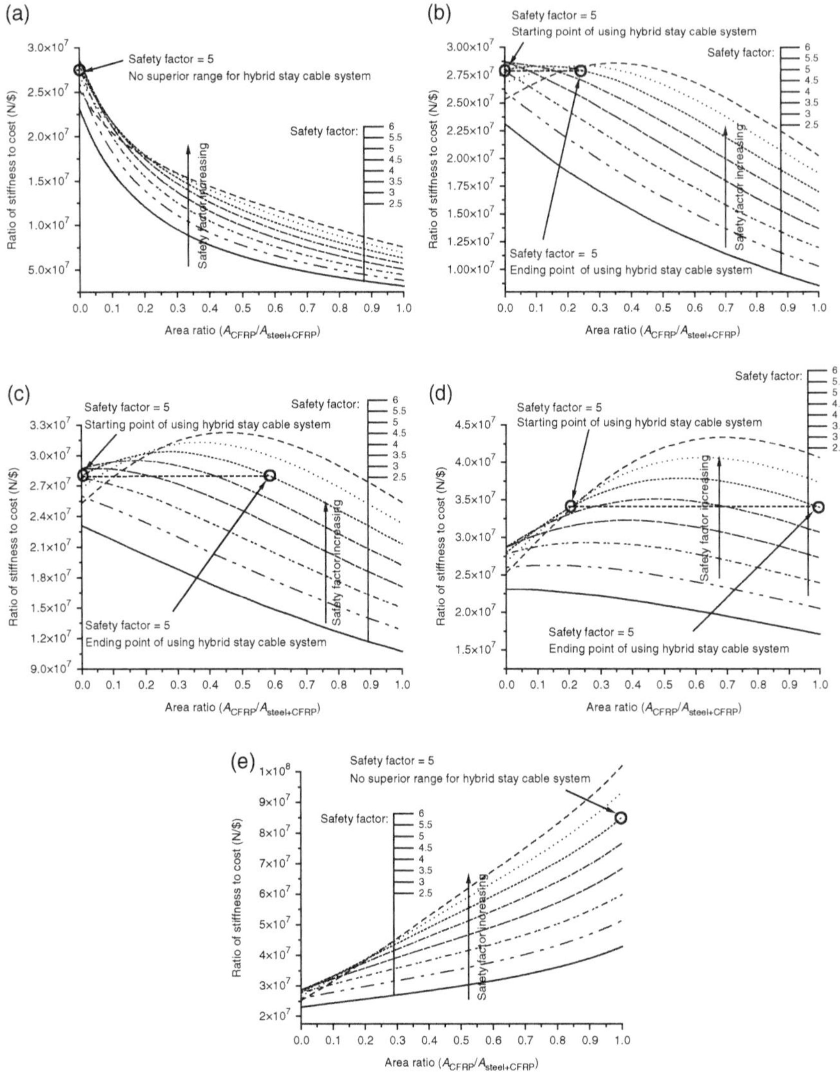

8.7 Ratio of stiffness to cost for stay cables with 700 m span length and different price ratios. (a) Price ratio = 27:1; (b) price ratio = 10:1; (c) price ratio = 8:1; (d) price ratio = 5:1; (e) price ratio = 2:1.

1000 m) separately. The results demonstrate the beneficial effects of increasing safety factors on stiffness improvement. Following the method used in design criterion 1, the superior ranges of the hybrid stay cable system (the ranges of area ratio) with regard to different price ratios, safety factors, and cable span lengths can also be obtained using design criterion 2, which are shown in Figs 8.9 (price ratio = 15:1) and 8.10 (price ratio = 10:1) as a design

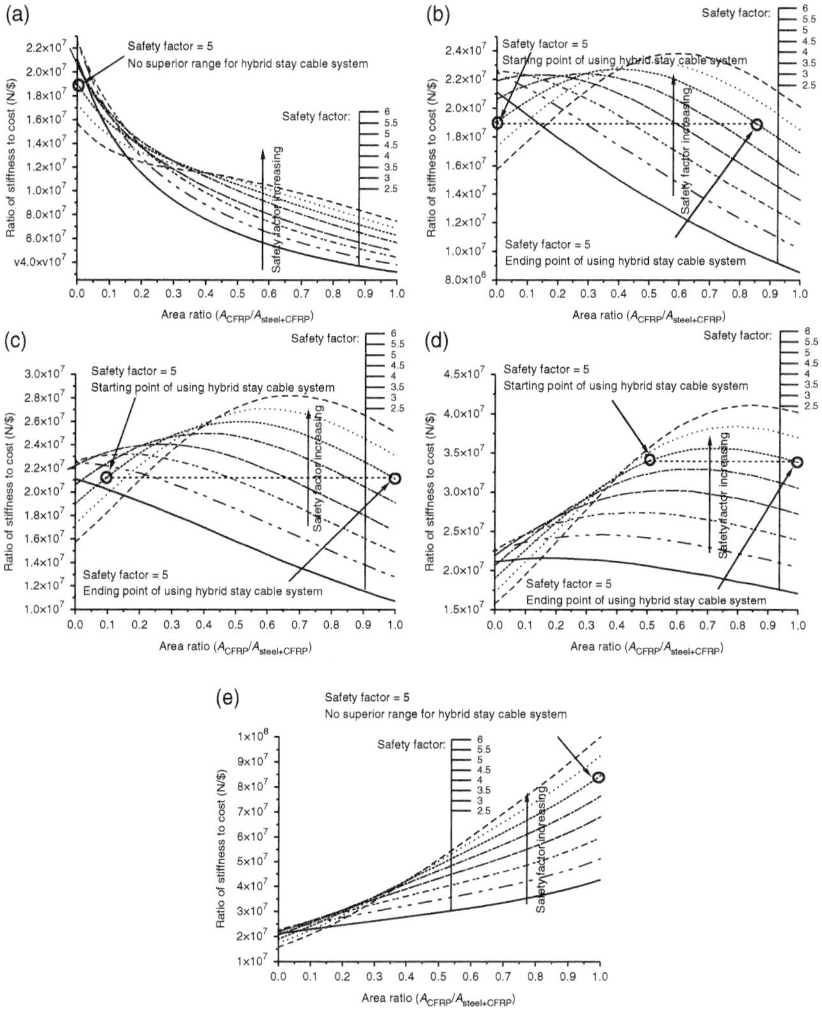

8.8 Ratio of stiffness to cost for stay cables with 1000 m span length and different price ratios. (a) Price ratio = 27:1; (b) price ratio = 10:1; (c) price ratio = 8:1; (d) price ratio = 5:1; (e) price ratio = 2:1.

guide. Again, the results of other price ratios are not shown here, for reasons of brevity.

Based on the design guide shown in Figs 8.9 and 8.10, and other figures with more price ratios, the hybrid stay cable system can be easily designed using design criterion 2 as follows:

- Step 1: Determining the span lengths of stay cables;
- Step 2: Checking the recent price ratio of CFRP stay cables to steel stay cables;

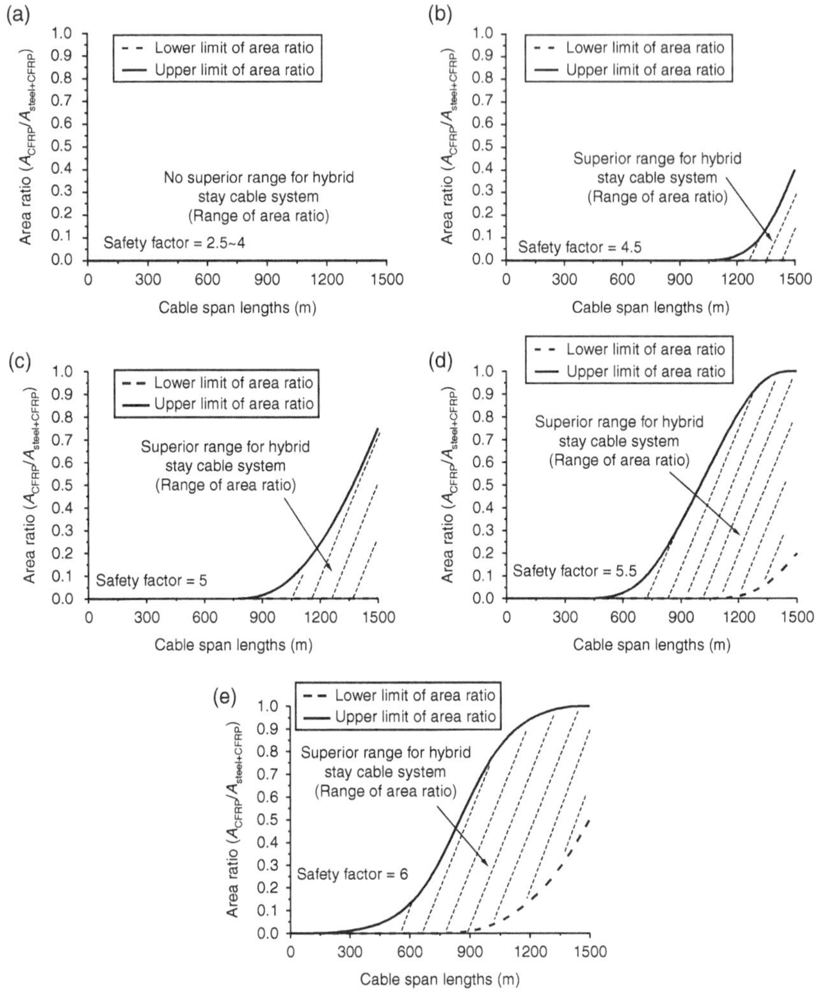

8.9 Design guide for hybrid stay cable system based on design criterion 2 (price ratio = 15:1). (a) Design guide with safety factor 2.5~4; (b) design guide with safety factor 4.5; (c) design guide with safety factor 5; (d) design guide with safety factor 5.5; (e) design guide with safety factor 6.

- Step 3: Determining the safety factor used in the cable design;
- Step 4: From the design guide shown in Figs 8.9, 8.10, and other figures with more price ratios, the appropriate area ratio range for the hybrid stay cable system (dashed area) can be selected based on the recent price ratio, cable span length, and safety factor used in design; furthermore, by using Figs 8.7 and 8.8, the optimal area ratio corresponding to the highest equivalent stiffness-to-cost ratio can be obtained;

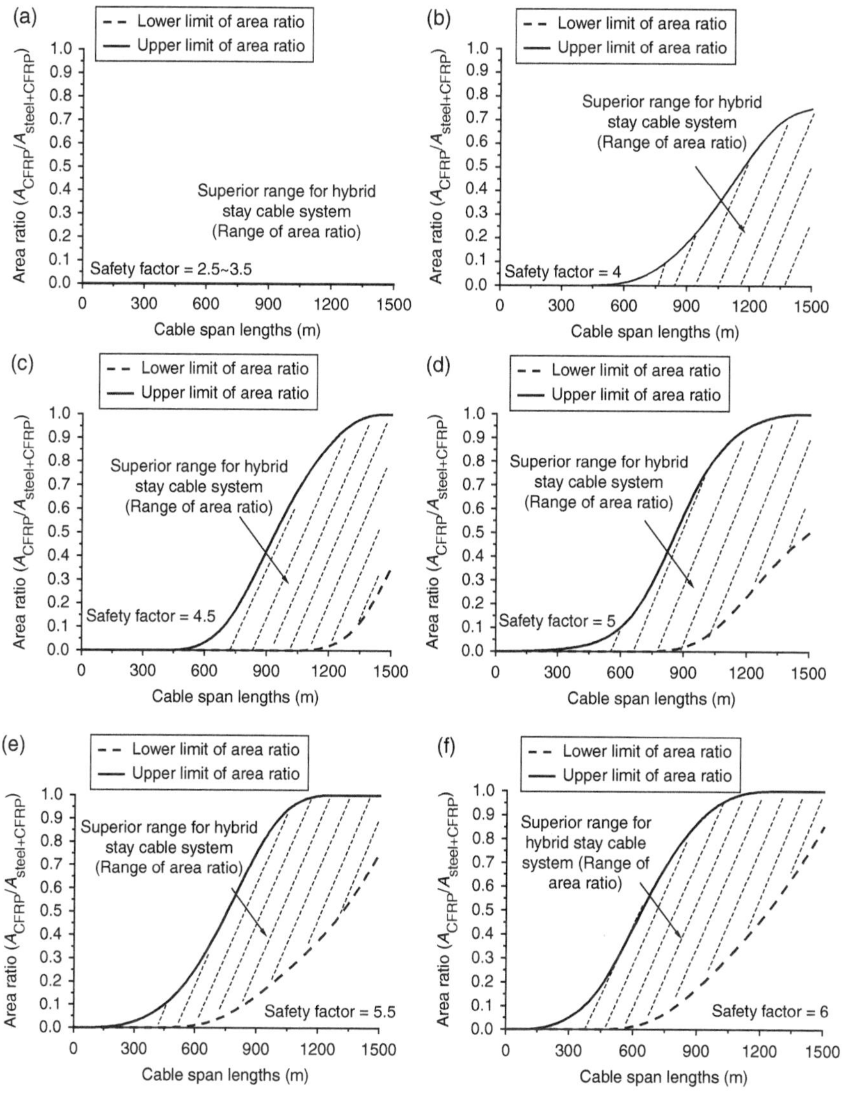

8.10 Design guide for hybrid stay cable system based on design criterion 2 (price ratio = 10:1). (a) Design guide with safety factor 2.5~3.5; (b) design guide with safety factor 4; (c) design guide with safety factor 4.5; (d) design guide with safety factor 5; (e) design guide with safety factor 5.5; (f) design guide with safety factor 6.

- Step 5: As with criterion 1, if no appropriate area ratio can be found based on design criterion 2, pure steel stay cables can be selected for short cables near the pylon and pure CFRP stay cables for long cables further away from the pylon.

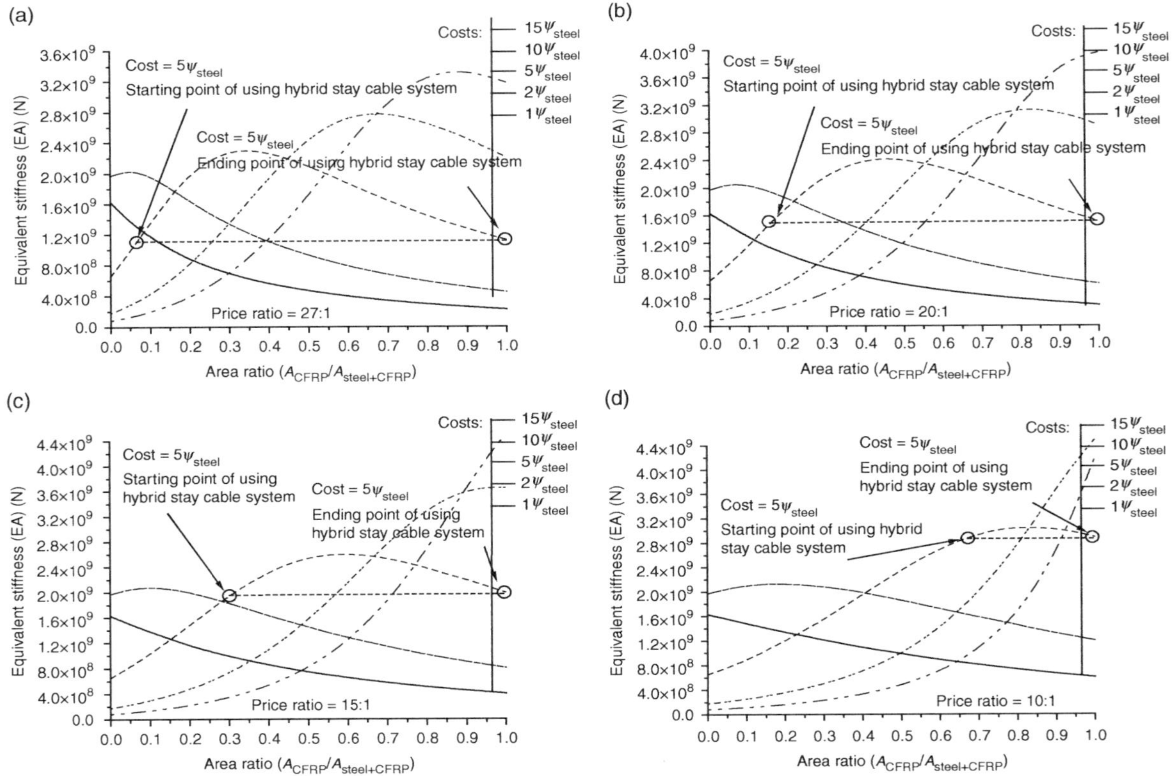

8.11 Equivalent stiffness of stay cables with 700 m span length and different price ratios. (a) Price ratio = 27:1; (b) price ratio = 20:1; (c) price ratio = 15:1; (d) price ratio = 10:1; (e) price ratio = 5:1; (f) price ratio = 2:1.

(*Continued*)

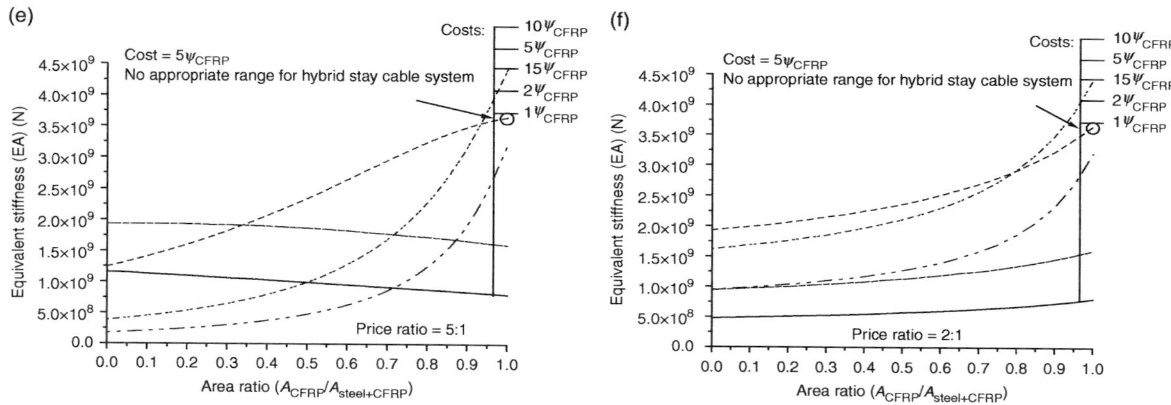

8.11 Continued

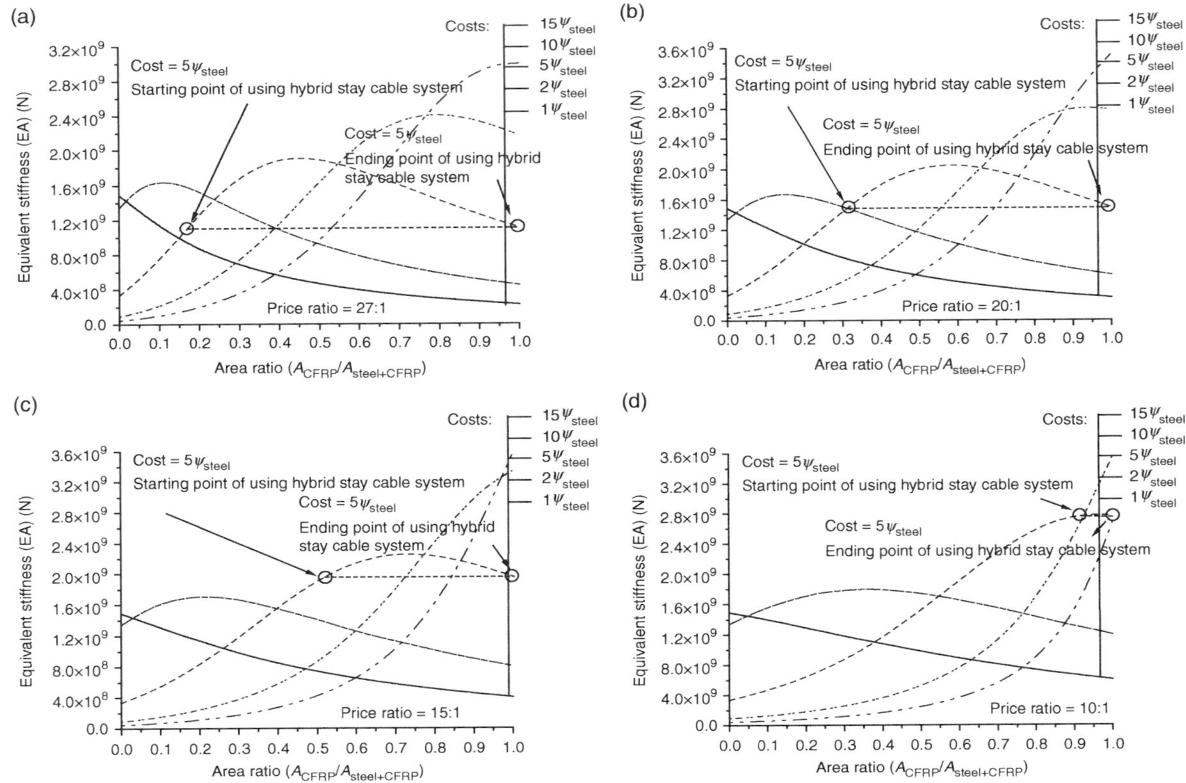

8.12 Equivalent stiffness of stay cables with 1000 m span length and different price ratios. (a) Price ratio = 27:1; (b) price ratio = 20:1; (c) price ratio = 15:1; (d) price ratio = 10:1; (e) price ratio = 5:1; (f) price ratio = 2:1.

(*Continued*)

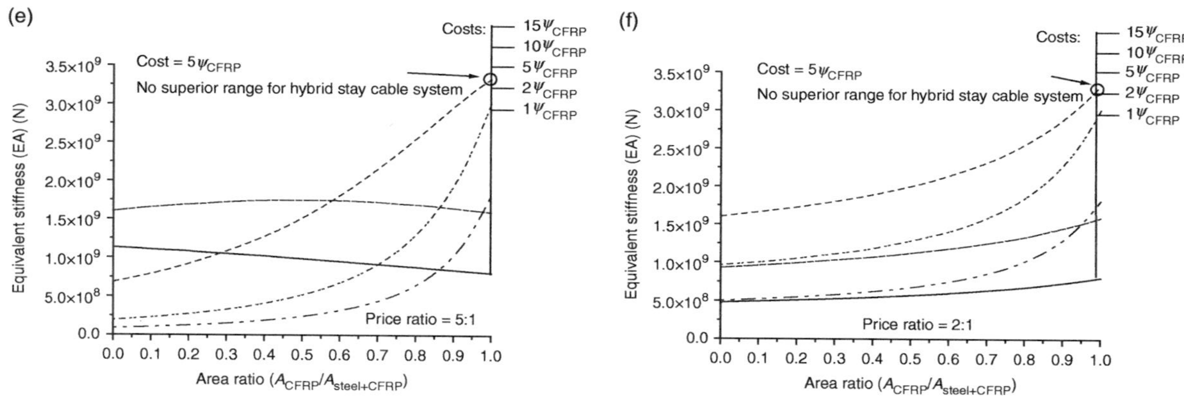

8.12 Continued

Design criterion 3: Best equivalent stiffness under a given cost

In this design criterion, the best equivalent stiffness of stay cables under a given cost can be pre-designed by adjusting the area ratio ρ under a given price ratio of CFRP stay cables to steel stay cables. The safety factor value used for cables will be automatically determined when the sectional area of each stay cable is set using Equations [8.15]–[8.17] as well as satisfying the strength requirement. As before, a series of parametric studies were carried out in order to obtain a design methodology using design criterion 3.

The analytical models and investigated price ratio for the parametric study are identical to those used for design criteria 1 and 2. However, in addition to the developing price ratio, various given costs ($\alpha\psi_{steel}$ and $\alpha\psi_{CFRP}$, using Equations [8.18]–[8.21] to define them) prepared for the hybrid stay cable system should also be considered, which are $1\psi_{steel}$, $2\psi_{steel}$, $5\psi_{steel}$, $10\psi_{steel}$, $15\psi_{steel}$, $1\psi_{CFRP}$, $2\psi_{CFRP}$, $5\psi_{CFRP}$, $10\psi_{CFRP}$ and $15\psi_{CFRP}$. Once again, only the results of 700 m (Fig. 8.11) and 1000 m (Fig. 8.12) cable span lengths with different price ratios are presented here.

Figures 8.11a–8.11f and 8.12a–8.12f plot the equivalent stiffness of stay cables versus area ratio (0~1.0) obtained with different given costs ($1\psi_{steel}$~$15\psi_{steel}$ and $1\psi_{CFRP}$~$15\psi_{CFRP}$), price ratios (27:1~2:1), and cable span lengths (only the 700m and 1000m). The reason for switching ψ_{steel} to ψ_{CFRP} for the cases of 5:1 and 2:1 price ratios is that in these cases the cost of building CFRP stay cables has been similar to or even less than that of steel stay cables. As with the previous criteria, the dashed line between two circles in each figure shows the superior ranges of the hybrid stay cable system with cost $5\psi_{steel}$ or $5\psi_{CFRP}$. The superior ranges of the hybrid stay cable system (the ranges of area ratio) with regard to different costs, price ratios, and cable span lengths can also be obtained using design criterion 3, which are shown in Figs 8.13 (price ratio = 15:1) and 8.14 (price ratio = 10:1) as a design guide. Again, for reasons of brevity, other price ratios are not shown.

Based on the design guide shown in Figs 8.13, 8.14, and other figures with more price ratios, the hybrid stay cable system can be easily designed using design criterion 3 as follows:

- Step 1: Determining the span lengths of stay cables;
- Step 2: Pre-determining the costs (budget) for the cable design including the materials and construction;
- Step 3: Checking the recent price ratio of CFRP stay cables to steel stay cables;
- Step 4: From the design guide shown in Figs 8.13, 8.14, and other figures with more price ratios, the appropriate area ratio for the hybrid stay cable system (dashed area) can be selected based on the pre-determined costs, recent price ratio, and cable span length used in design. Furthermore, by using Figs 8.11 and 8.12, the optimal

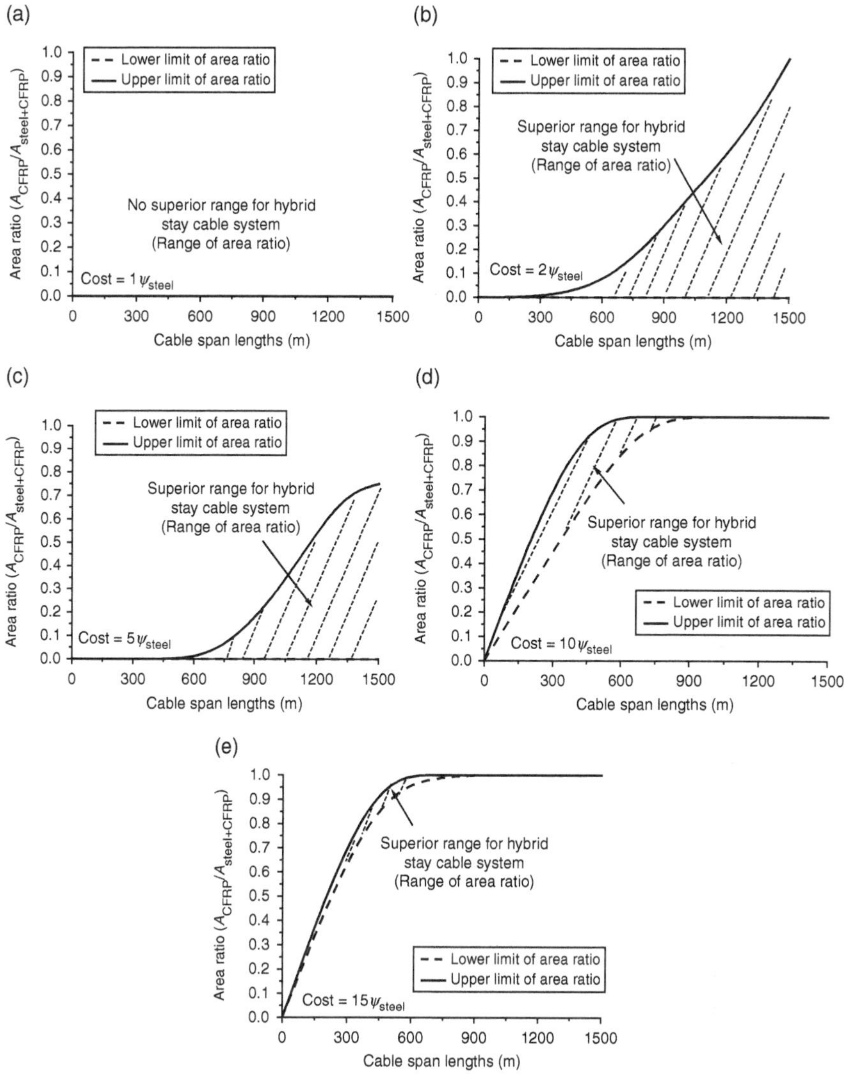

8.13 Design guide for hybrid stay cable system based on design criterion 3 (price ratio =15:1). (a) Design guide with cost $1\psi_{steel}$; (b) design guide with cost $2\psi_{steel}$; (c) design guide with cost $5\psi_{steel}$; (d) design guide with cost $10\psi_{steel}$; (e) design guide with cost $15\psi_{steel}$.

area ratio corresponding to the highest equivalent stiffness can be obtained for a given cost;

- Step 5: As with criteria 1 and 2, if no appropriate area ratio can be found based on design criterion 3, pure steel stay cables can be selected for short cables near the pylon and pure CFRP stay cables for long cables away from the pylon.

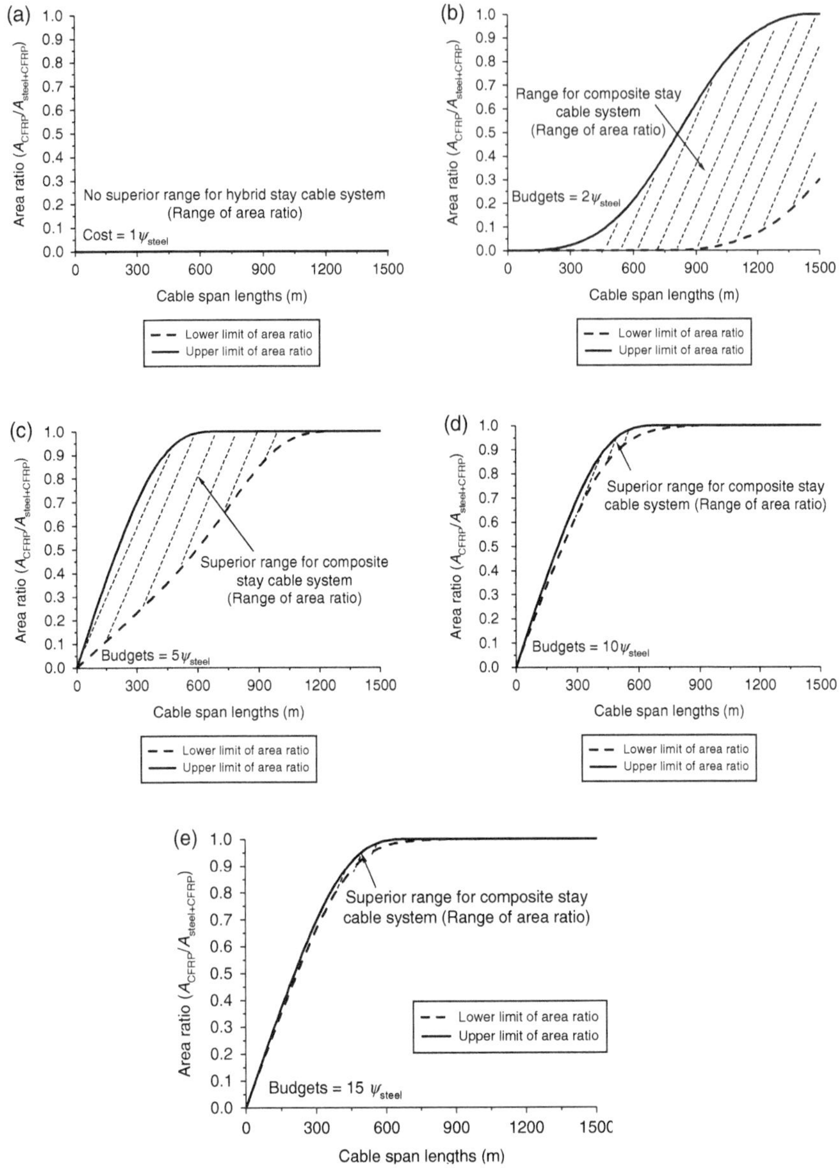

8.14 Design guide for hybrid stay cable system based on design criterion 3 (price ratio =10:1). (a) Design guide with cost 1ψ_{steel}; (b) design guide with cost 2ψ_{steel}; (c) design guide with cost 5ψ_{steel}; (d) design guide with cost 10ψ_{steel}; (e) design guide with cost 15ψ_{steel}.

8.4.4 Cable force distribution of the hybrid stay cable system

Once the value of area ratio is determined for the hybrid stay cable system, the cable force distribution between the steel and CFRP stay cables is the next issue. Appropriate cable force distribution affects the stiffness design of stay cables (Ernst, 1965; Ahmadi-Kashani and Bell, 1988; Freire, 2006). Therefore, this study went on to investigate the effects of cable force distribution (defining $\beta = F_{CFRP}/(F_{CFRP} + F_{steel})$, herein) on the equivalent stiffness of stay cables through a parametric study.

From Figs 8.15 and 8.16 and other figures with more cable span lengths (not shown here for reasons of brevity), the appropriate value of β for each study case usually ranges from 0.4 to 0.7, where the best equivalent stiffness of stay cables can be reached. The variation of the safety factor values has little effect on the β design. It can also be observed that the β value with

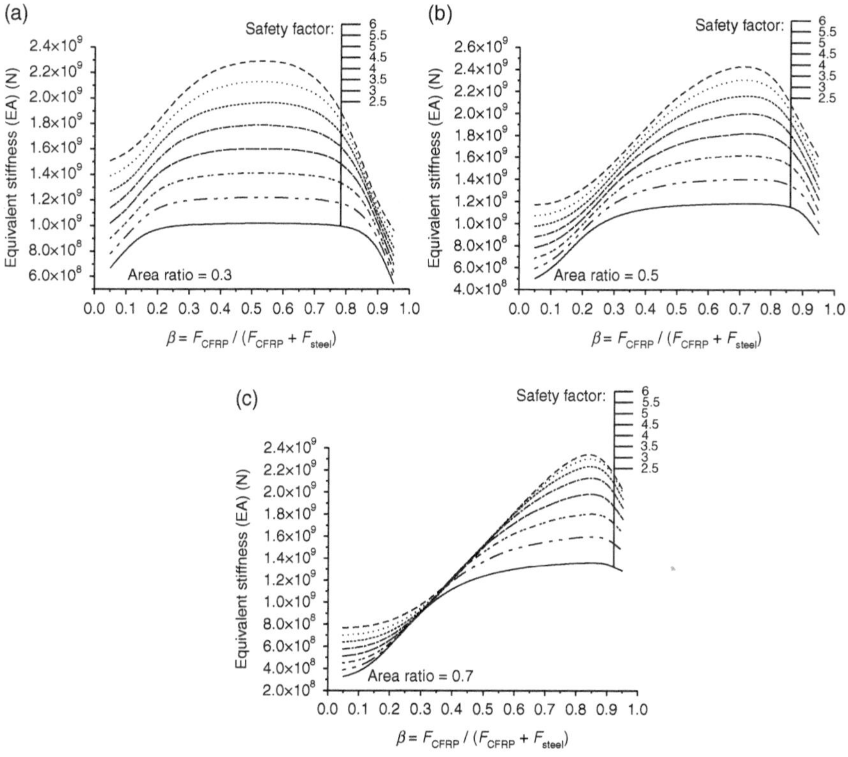

8.15 Results for hybrid stay cable system with cable span length of 700 m. (a) Results with area ratio 0.3; (b) results with area ratio 0.5; (c) results with area ratio 0.7.

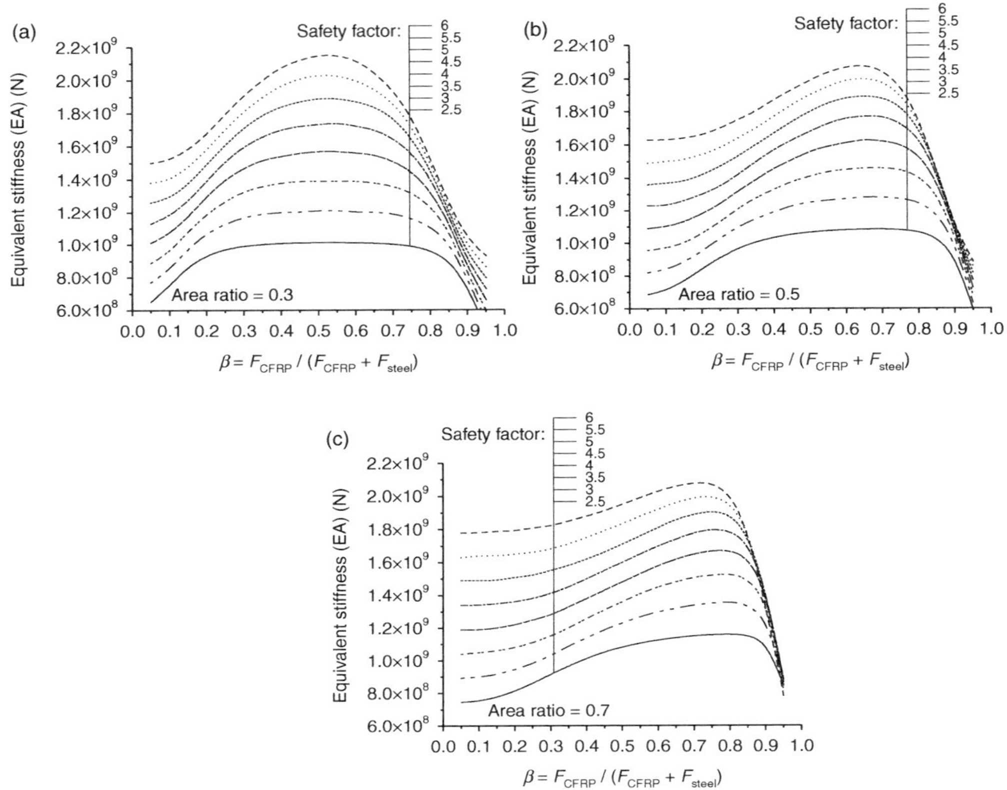

8.16 Results for hybrid stay cable system with cable span length of 1000 m. (a) Results with area ratio 0.3; (b) results with area ratio 0.5; (c) results with area ratio 0.7.

regard to the equivalent stiffness of stay cables increases with the increase of an area ratio. This can be explained in that CFRP stay cables with higher area ratio need greater cable force applied to maintain their high equivalent stiffness. The final decision as to cable force distribution can only be made when the cable stresses also satisfy the strength requirements. In practice, a good choice of the β value can complement the three design criteria to ensure a high equivalent stiffness for the hybrid stay cable system.

8.5 Case study: 1400 m cable-stayed bridges

From a mechanical-behavior viewpoint (static and dynamic), a comparative study of composite cable-stayed bridges with different CFRP components was performed through numerical simulations. The economical behavior of each type of cable-stayed bridges was also comparatively studied considering the varying material price. With the high strength-to-weight ratio and other advantages of CFRP materials, it was proven that the use of CFRP stay cables and CFRP bridge decks in super longspan cable-stayed bridges is feasible and these types of composite cable-stayed bridges can become an excellent alternative to steel cable-stayed bridges where super longspans are desired.

8.5.1 Bridge models

For a comparative study, six selected cable-stayed bridges are shown in Table 8.4 with different combinations of materials. Figure 8.17a shows the span arrangement of the six cable-stayed bridge models with two pylons and double cable planes. The center and side spans were designed as 1400 and 636 m, respectively. For the side span, three intermediate piers were used at a distance of 100 m in order to increase the in-plane flexural rigidity of the bridge. The full height of the A-shape pylon is 357 m, and 287 m of the pylon is above the girder (see Fig. 8.17b). Figure 8.17c shows the cross-section of the girder with a width (B_u) of 32.5 m and a depth (H_w) of 4.5 m. The span-

Table 8.4 Definition of the six types of the proposed cable-stayed bridges

Type	Girder	Bridge deck	Stay cables	Pylons
I	Steel	Steel	Steel	Concrete
II	Steel	Steel	Hybrid	Concrete
III	Steel	Steel	CFRP	Concrete
IV	Steel	CFRP	Steel	Concrete
V	Steel	CFRP	Hybrid	Concrete
VI	Steel	CFRP	CFRP	Concrete

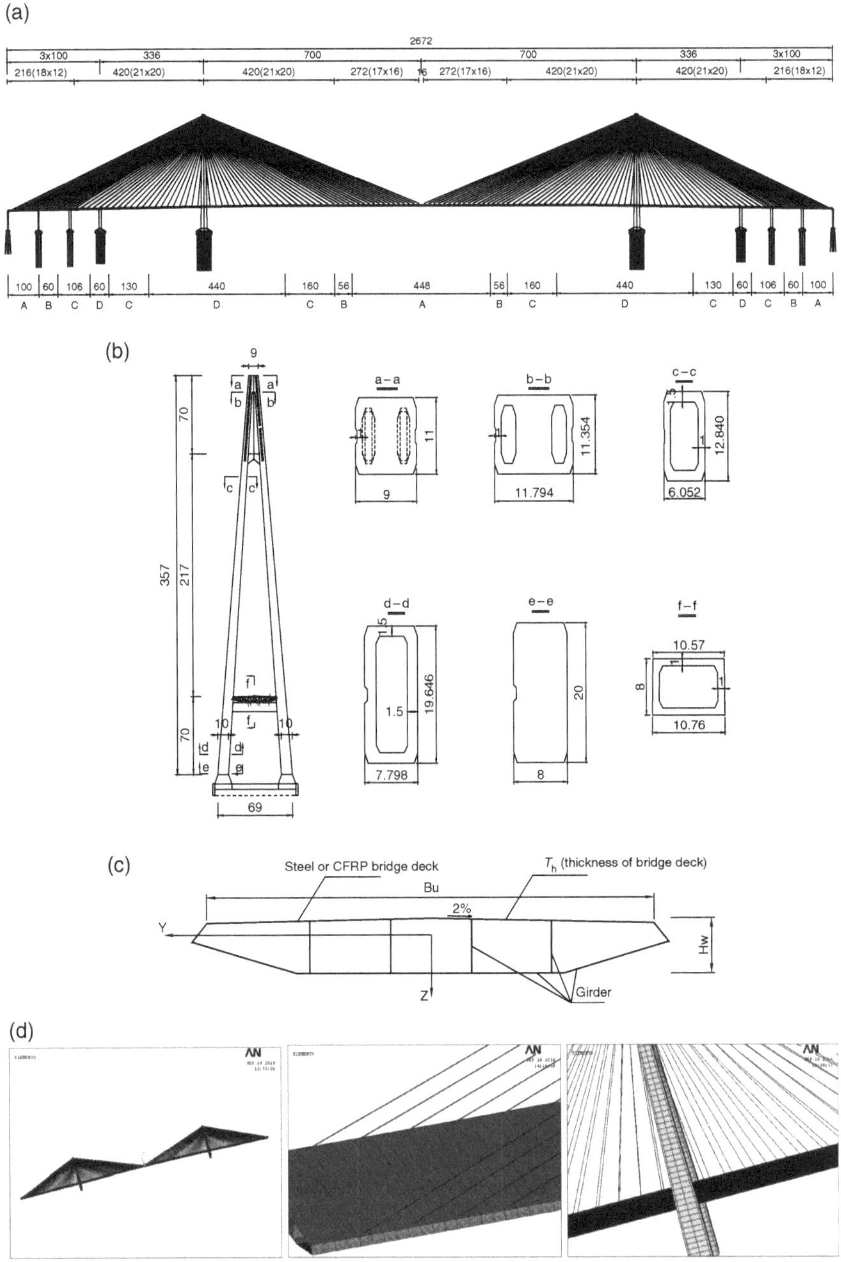

8.17 1400 m span cable-stayed bridge model (unit: m). (a) Span arrangement; (b) pylon; (c) cross-section of girder (A, B, C, and D); (d) finite element model.

Table 8.5 Cross-sectional properties

Cross-section		B_u (m)	H_w (m)	A (m²)	T_h (m)	I_y (m⁴)	I_z (m⁴)	J (m⁴)
Types I, II, and III (Steel bridge deck)	A	32.5	4.5	1.688	0.018	5.861	235.418	20.076
	B	32.5	4.5	2.016	0.022	7.078	281.029	24.361
	C	32.5	4.5	2.360	0.026	8.308	332.812	28.793
	D	32.5	4.5	2.947	0.033	10.433	417.841	36.327
Types IV, V, and VI (CFRP bridge deck*)	A	32.5	4.5	1.560	0.020	5.343	219.586	18.142
	B	32.5	4.5	1.860	0.024	6.447	262.137	22.024
	C	32.5	4.5	2.195	0.029	7.630	317.376	26.123
	D	32.5	4.5	2.712	0.037	9.494	388.970	32.822

*Moment inertia for bending (I_y and I_z) and torsion (J) of the cross sections of the girder with a CFRP bridge deck listed in Table 8.3 were obtained using an equivalent pure steel cross-section. B_u, width of girder; H_w, depth of girder; A, area of girder; T_h, thickness of bridge deck; I_y and I_z, moment inertia for bending along y and z directions, respectively; and J, moment inertia for torsion.

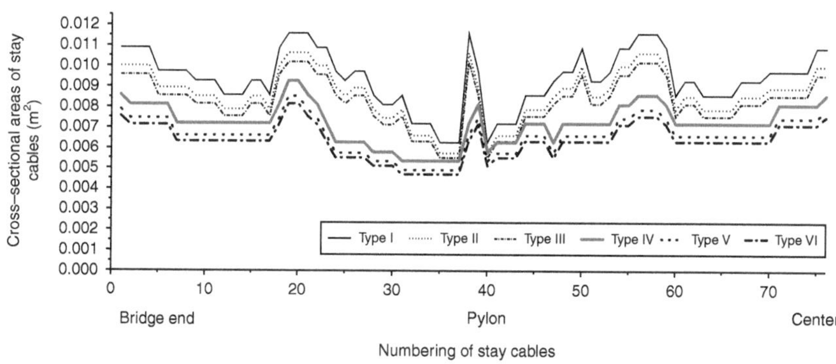

8.18 Cross-sectional areas of stay cables for each bridge type.

to-width ratio of all the six bridges is 43, larger than the commonly adopted minimum value 40 (Leonhardt and Zellner, 1991). The span-to-depth ratio is 311, which is also larger than those used for other long span cable-stayed bridges, e.g. 285 for the Normandy Bridge (France) and 272 for the Sutong Bridge (China). Four typical cross sections (sections A, B, C, and D) with different properties (Table 8.5) were used in different locations along the span (Fig. 8.17a). The girder is suspended by diagonal stay cables anchored to the girder at 12, 16, and 20 m intervals. The cross-sectional areas of stay cables used in each bridge type are shown in Fig. 8.18. Using design criterion 1, a ρ value of 0.60 for hybrid stay cables was determined as an appropriate value balancing mechanical behavior and cost and was therefore employed for the present case studies. The safety factor of hybrid stay cables was kept as 2.5, remaining consistent with other bridge types for a convenient comparison. It can be seen that a Type I bridge requires the maximum and a Type VI the minimum stay cable areas.

There are several commercially produced CFRP stay cables (namely, Leadline, manufactured and distributed by Mitsubishi Chemical Company, and CFCC, developed by Tokyo Rope Mfg. Co. Ltd) with an elastic modulus from 130 to 160 GPa (Zhang *et al.*, 2001; Grace *et al.*, 2002; Jackson *et al.*, 2005). For the CFRP stay cables (including the CFRP part of the hybrid stay cables) in this case study a value of 137 GPa was used for the elastic modulus based on a widely-used type of CFCC-brand CFRP stay cables (Xiong *et al.*, 2011, 2012). For the CFRP bridge deck the elastic modulus was set as 135 GPa (Xiong *et al.*, 2011). Steel with Chinese grades of Q345qD and Q420qD (with yield stress = 345 and 420 MPa; allowable stress = 190 and 200 MPa for normal static load, and 280 and 345 MPa for wind load, respectively) were used for the girder but the latter could only be used locally. C55 and C60 types of concrete (with allowable compression stress = 17.75 and 19.25 MPa, respectively) were used for the pylons. The cable-stayed bridges were designed to carry eight traffic lanes in two directions and the secondary dead load was assumed to be 70 kN/m. The design wind velocity V_d for the bridge was 74.9 m/s using the assumed bridge site information.

The finite element (FE) models of the six bridges were created using ANSYS program, see Fig. 8.17d. The bridge decks, girders, diaphragms, and pylons were all modeled using beam elements with certain cross-section inputs (in total 4674 elements were used) which have three translational degrees of freedom (DOFs) and three rotational DOFs for each node (4659 nodes were used in total). The stay cables were modeled using spar (truss) elements with ten elements for each stay cable and three translational DOFs for each node (Freire *et al.*, 2006). The nodes of girders located at the diaphragms were meshed using structural mass elements with an initial input of torsion moments of inertia (the rotary inertia can be assigned in each coordinate direction). By doing this, the torsion moments of inertia provided by the diaphragms that increase the torsion stiffness of the bridge girder can be independently added to the girders of the beam elements. For each model the girder and pylons at the girder–pylon connections are free to move longitudinally.

8.5.2 Mechanical behavior and a comparative study

To compare the use of different combinations of CFRP and steel materials, the results of all six bridges for one specific parameter are discussed. The Type I bridge using only traditional steel and concrete materials is treated as a reference model in the following comparative study.

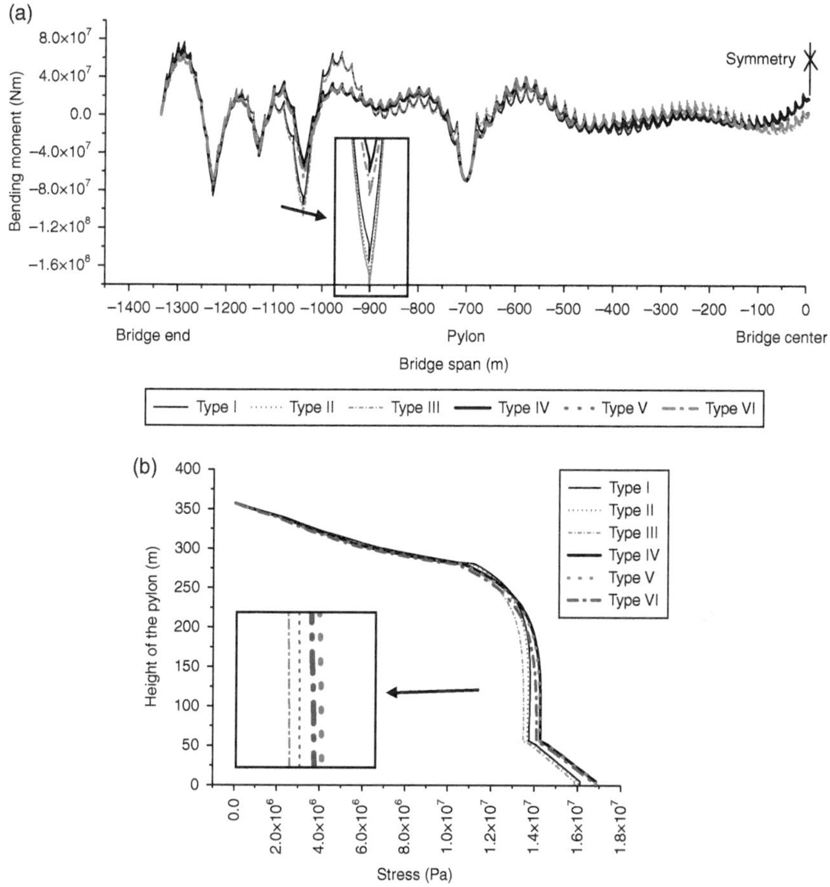

8.19 Reasonable dead load conditions for six bridges (nonlinear analysis). (a) Bending moments of girders; (b) stresses of pylons.

The reasonable dead load conditions for the six bridges (Types I–VI) as shown in Fig. 8.19a and 8.19b were obtained using the proposed design strategies, analytical procedures, and finite element models. As expected, the bending moments of girders (half span) (Fig. 8.19a) and the stresses of pylons (Fig. 8.19b) in the reasonable dead load conditions are within an appropriate range as predicted on the basis of practical engineering experience (Gimsing, 1997; Nagai *et al.*, 2004). The bending moments of girders with a CFRP bridge deck i.e. Types IV, V, and VI, are slightly lower than those with a steel bridge deck i.e. Types I, II, and III. It is noted that the peaks of moments at the locations of piers shown in Fig. 8.19a can be greatly reduced in the construction process if necessary. Based on the results, it can be concluded that following the introduced analytical procedure the reasonable dead load conditions of the cable-stayed bridges with CFRP components i.e. Types II to VI bridges can be designed to be quite similar to those of the reference Type I bridge.

The reasonable stay cable forces are shown in Plate VIII (in the color section between pages 172 and 173) (half span). As expected, due to the light self-weight of CFRP bridge decks, the reasonable stay cable forces of Types IV, V, and VI bridges are much smaller than those of Types I, II, and III. Therefore, using CFRP bridge decks can significantly decrease the amount of material used for stay cables, further reducing the self-weight of bridges and reducing the necessary foundations. However, only negligible differences in the reasonable stay cable forces are found if the stay cable materials are changed from steel to CFRP materials (Types I vs III and IV vs VI).

Figure 8.20a–8.20f show the structural performance of several components of interest under a moving HS20 truck (AASHTO, 2004) vs the vehicle location of the rear axle along the bridge span. It can be seen from Fig. 8.20a that the vertical displacements at the mid span of the bridges with a CFRP bridge deck i.e. Types IV, V, and VI, are clearly larger than those of the bridges with a steel bridge deck i.e. Types I, II, and III, which indicates a decrease in global stiffness by using CFRP bridge decks. This decrease is mainly due to the lower elastic modulus of CFRP materials and lower stay cable forces and stiffness. Furthermore, among all the cases using CFRP bridge decks, the vertical displacement at the mid span of the Type VI bridge with CFRP stay cables is larger than that of the Type V bridge with hybrid stay cables. Both are larger than that of the Type IV bridge with steel stay cables. This indicates that both CFRP bridge decks and CFRP stay cables tend to reduce the global stiffness of bridges, although the decrease is insignificant and acceptable in engineering practice. This conclusion can also be supported by the results of horizontal displacements at the top of the pylons and bending moments at their bottom, as shown in Fig. 8.20b and 8.20c respectively, since both have a direct correlation with the vertical displacements at the mid span.

Figure 8.20d shows that the bending moments at mid span are very close to each other, which means that the low global stiffness of the bridge due to CFRP components has only marginal effect on the bending moments of the girders. Figure 8.20e and 8.20f show an insignificant effect of CFRP components on the stay cable force variations under the moving truck loading. Cable No. 1 is the longest cable in the side span and No. 76 is the longest one in the center span.

The cases using heavier or more trucks show similar results. It should be noted that all these results are only for bridges of 1400 m span length. With the extension of bridge span lengths, the equivalent elasticity modulus of steel stay cables will be greatly reduced in contrast to that of the

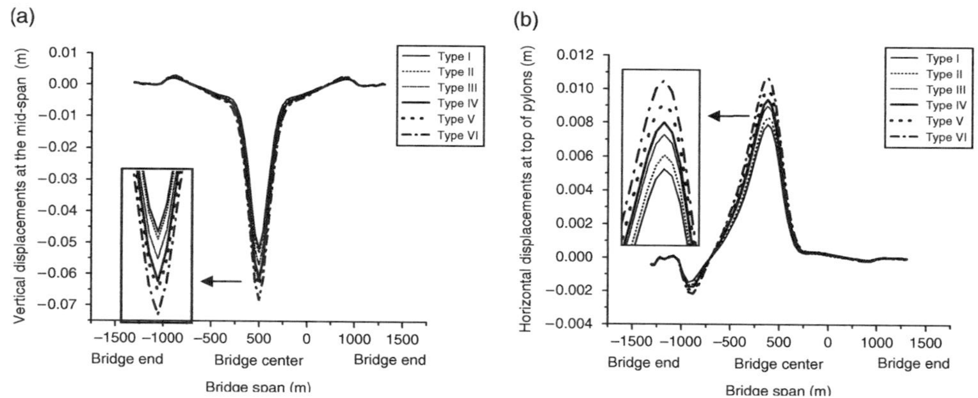

8.20 Structural performance under a moving truck (nonlinear analysis). (a) Vertical displacements at mid span; (b) horizontal displacements at top of pylons; (c) bending moments at bottom of pylons; (d) bending moments at mid span; (e) stay cable forces of side stay cables (No. 1); (f) stay cable forces of center stay cables (No. 76).

(*Continued*)

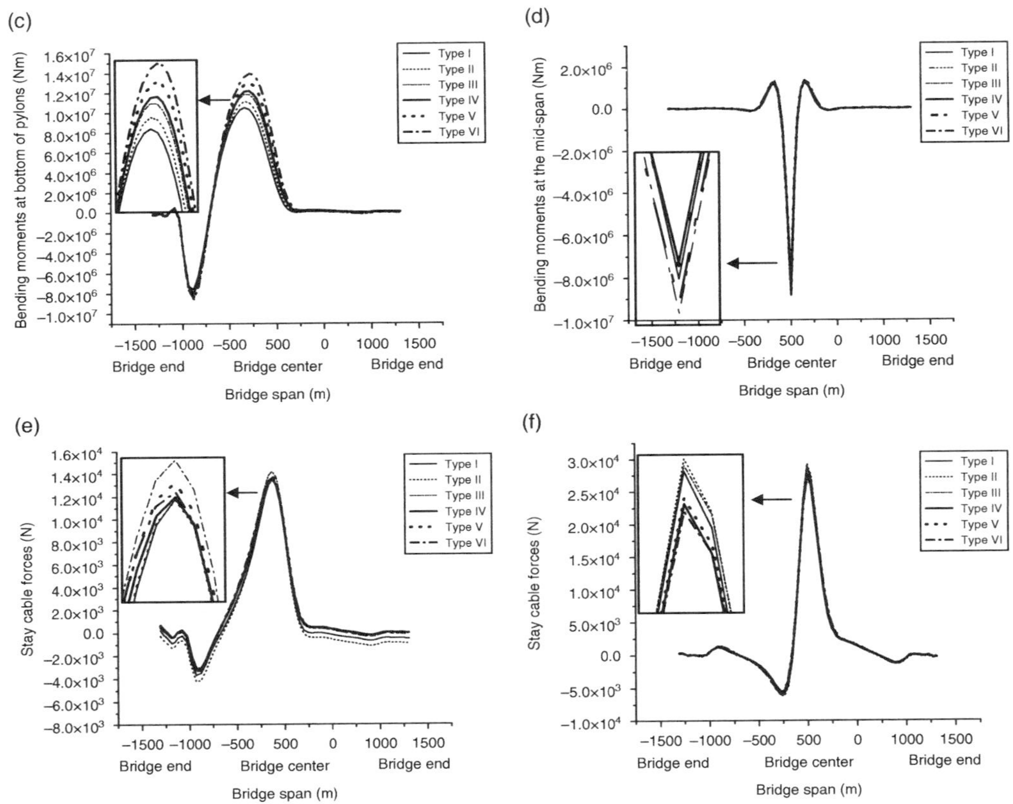

8.20 Continued

lighter CFRP stay cables. Therefore it can be predicted that the global stiffness of bridges with hybrid and CFRP stay cables may eventually be higher than those with steel stay cables. A study in this regard is ongoing by the authors.

The natural frequencies and mode shapes of the six bridges are presented in Table 8.6 and several significant conclusions can be drawn. The reason for including 16 modes for each case is to include a torsion mode shape. By generally comparing the results of Types I, II, and III bridges with Types IV, V, and VI it is clear that CFRP bridge decks can result in a significant increase in natural frequency.

The natural frequencies of the vertical bending modes of the bridges with hybrid and CFRP stay cables i.e. Types II and III are slightly lower than those of the bridge with steel stay cables i.e. Type I. This is mainly due to the lower equivalent elasticity modulus of hybrid and CFRP stay cables, which can decrease the supporting stiffness to the girder by the stay cables. The opposite conclusion could be reached if considering the various effects of stay cable materials on bridge stiffness when the bridge span is extended beyond 1400 m.

On the other hand, the natural frequencies of the lateral bending modes of Types II and III bridges are higher than those of Type I. Since the contribution of stay cables to the lateral stiffness of the girders is small, the light weight of CFRP materials may be the main reason for the higher frequencies. In addition, the natural frequencies of torsion modes in Types II and III bridges are higher than those in Type I, due to the light weight of CFRP materials. The same conclusions can also be obtained for the cases of CFRP bridge decks shown at the bottom part of Table 8.6 (Types IV, V, and VI).

Using the natural frequencies of torsion and vertical bending modes in Table 8.6, the critical wind speed due to a flutter phenomenon (V_{cr}) for each bridge type was calculated and shown in Table 8.7. It can clearly be seen that all the V_{cr} are higher than the design wind velocity V_d, i.e., 74.9 m/s. By comparing these values, high natural frequencies of torsion modes can increase the V_{cr} value; however, a decrease of V_{cr} can also be observed when using light CFRP bridge decks, due to the lower μ values. Referring to $V_{cr} = 51$ m/s (bridge over the Dala, Switzerland) and $V_{cr} = 111$ m/s (bridge at Diepoldsau, Switzerland) (Walther *et al.*, 1999) all the V_{cr} of the six bridges are acceptable in most field applications, although the bridges with a CFRP bridge deck could have slightly lower V_{cr} than those using steel bridge decks. When necessary, the critical wind velocity of bridges with a CFRP deck can be enhanced with many other countermeasures, such as streamlining the deck shape, though it is beyond the scope of the present study.

Again, the safety factor of the hybrid stay cables remained as 2.5, consistent with other types of bridges to provide a convenient comparison. Actually, as

Table 8.6 Natural frequencies and mode shapes of each bridge type

Order of mode	Type I		Type II		Type III	
	Freq. (Hz)	Mode shape	Freq. (Hz)	Mode shape	Freq. (Hz)	Mode shape
1	0.0523	1st sym. L (girder)	0.0533	1st sym. L (girder)	0.0537	1st sym. L (girder)
2	0.0638	LM (girder) and LB (pylon)	0.0650	LM (girder) and LB (pylon)	0.0655	LM (girder) and LB (pylon)
3	0.1239	1st anti. L (girder)	0.1250	1st anti. L (girder)	0.1255	1st anti. L (girder)
4	0.1722	sym. L (pylon)	0.1735	sym. L (pylon)	0.1742	sym. L (pylon)
5	0.1723	anti. L (pylon)	0.1737	anti. L (pylon)	0.1743	anti. L (pylon)
6	0.1986	1st sym. V (girder)	0.1852	1st sym. V (girder)	0.1771	1st sym. V (girder)
7	0.2312	2nd sym. L (girder)	0.2266	1st anti. V (girder)	0.2160	1st anti. V (girder)
8	0.2447	1st anti. V (girder)	0.2353	2nd sym. L (girder)	0.2373	2nd sym. L (girder)
9	0.3098	2nd sym. V (girder)	0.2962	2nd sym. V (girder)	0.2868	2nd sym. V (girder)
10	0.3568	2nd anti. V (girder)	0.3457	2nd anti. V (girder)	0.3373	2nd anti. V (girder)
11	0.3852	2nd anti. L (girder)	0.3824	3rd sym. V (girder)	0.3697	3rd sym. V (girder)
12	0.4041	3rd sym. V (girder)	0.3906	2nd anti. L (girder)	0.3933	2nd anti. L (girder)
13	0.4624	3rd anti. V (girder)	0.4372	3rd anti. V (girder)	0.4228	3rd anti. V (girder)
14	0.4626	sym. T (girder) and L (pylon)	0.4694	sym. T (girder) and L (pylon)	0.4530	4th sym. V (girder)
15	0.4635	anti. T (girder) and L (pylon)	0.4703	anti. T (girder) and L (pylon)	0.4727	sym. T (girder) and L (pylon)
16	0.5029	4th sym. V (girder)	0.4713	4th sym. V (girder)	0.4737	anti. T (girder) and L (pylon)

Notes: L, lateral bending; V, vertical bending; T, torsion; LM, longitudinal moving; LB, longitudinal bending; sym., symmetric; anti., antisymmetric.

discussed earlier, increasing the safety factor could be a good way to enhance the mechanical performance of the hybrid stay cables. Therefore, according to this comparative study, cable-stayed bridges with hybrid stay cables (Types II and IV) could potentially have better mechanical performances by increasing the safety factor under different design criteria.

Type IV		Type V		Type VI	
Freq. (Hz)	Mode shape	Freq. (Hz)	Mode shape	Freq. (Hz)	Mode shape
0.0595	1st sym. L (girder)	0.0608	1st sym. L (girder)	0.0615	1st sym. L (girder)
0.0704	LM (girder) and LB (pylon)	0.0717	LM (girder) and LB (pylon)	0.0722	LM (girder) and LB (pylon)
0.1445	1st anti. L (girder)	0.1461	1st anti. L (girder)	0.1468	1st anti. L (girder)
0.1735	sym. L (pylon)	0.1746	sym. L (pylon)	0.1751	sym. L (pylon)
0.1737	anti. L (pylon)	0.1747	anti. L (pylon)	0.1752	anti. L (pylon)
0.2026	1st sym. V (girder)	0.1892	1st sym. V (girder)	0.1812	1st sym. V (girder)
0.2582	1st anti. V (girder)	0.2393	1st anti. V (girder)	0.2282	1st anti. V (girder)
0.2698	2nd sym. L (girder)	0.2750	2nd sym. L (girder)	0.2776	2nd sym. L (girder)
0.3230	2nd sym. V (girder)	0.3102	2nd sym. V (girder)	0.3012	2nd sym. V (girder)
0.3749	2nd anti. V (girder)	0.3665	2nd anti. V (girder)	0.3596	2nd anti. V (girder)
0.4255	3rd sym. V (girder)	0.4025	3rd sym. V (girder)	0.3889	3rd sym. V (girder)
0.4485	2nd anti. L (girder)	0.4560	2nd anti. L (girder)	0.4446	3rd anti. V (girder)
0.4672	sym. T (girder) and L (pylon)	0.4595	3rd anti. V (girder)	0.4596	2nd anti. L (girder)
0.4682	anti. T (girder) and L (pylon)	0.4727	sym. T (girder) and L (pylon)	0.4753	sym. T (girder) and L (pylon)
0.4861	3rd anti. V (girder)	0.4733	anti. T (girder) and L (pylon)	0.4759	anti. T (girder) and L (pylon)
0.5109	sym. L (pylon)	0.5017	4th sym. V (girder)	0.4830	4th sym. V (girder)

8.5.3 Economic aspects and a comparative study

In addition to the mechanical behavior, economic aspects (material and construction cost) also need to be studied to examine the feasibility of a new cable-stayed bridge type using CFRP components. Using the six selected cable-stayed bridges, the cost for the stay cables and bridge girder and deck

Table 8.7 Critical wind speed due to flutter phenomenon

	η_s	η_α	f_T	f_B	ε	r (m)	m (kg/m)	b (m)	μ	ω_B	V_{cr} (m/s)
Type I	0.9	0.7	0.4626	0.1986	2.3293	11.8096	26588	16.25	24.6539	1.2478	96.7
Type II	0.9	0.7	0.4694	0.1852	2.5346	11.8096	26588	16.25	24.6539	1.1636	99.0
Type III	0.9	0.7	0.4727	0.1771	2.6691	11.8096	26588	16.25	24.6539	1.1128	100.1
Type IV	0.9	0.7	0.4672	0.2026	2.3060	11.8651	20511	16.25	19.0190	1.2730	87.5
Type V	0.9	0.7	0.4727	0.1892	2.4984	11.8651	20511	16.25	19.0190	1.1888	89.1
Type VI	0.9	0.7	0.4753	0.1812	2.6231	11.8651	20511	16.25	19.0190	1.1385	90.0

$\eta_s = V_c(0°)/V_{co}$, correction coefficient of cross-section shape; $\eta_\alpha = V_c(\alpha)/V_c(0°)$, correction coefficient of an angle of incidence σ; V_{co} = theoretical critical wind speed of flat plate due to flutter phenomenon f_T and f_B = natural frequencies of torsion mode and vertical bending mode, respectively; ω_T and ω_B = natural circular frequencies of torsion mode and vertical bending mode, respectively; $\varepsilon = \omega_T/\omega_B = f_T/f_B$, ratio of natural frequency of torsion mode to that of vertical bending mode; r = radius of gyration; m = mass per unit length of girder; b = half width of deck; ρ = density of air (1.3 kg/m^3); $\mu = \pi\rho b^2$; and V_{cr} = critical wind speed due to flutter phenomenon.

can be directly calculated from the previous design. However, the cost of the substructure (including pylons) needs to be estimated following the discussions below.

Analytical model for cost study of substructures

In this section, the pylon component is simplified as a column under an axial compression load, as shown in Fig. 8.21. Some assumptions need to be followed:

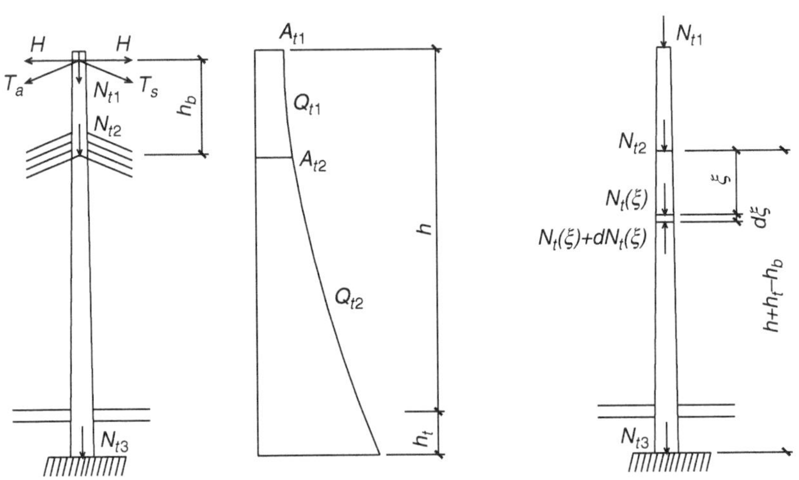

8.21 Calculation sketch for pylons. (a) Material amount (weight) of anchorage zone; (b) material amount (weight) of no-anchorage zone.

(a) cross-sectional area of the pylon in the anchorage zone is varied linearly, based on the varying compression loads (A_{t1} = cross-sectional area at the top pylon; and A_{t2} = cross-sectional area at the bottom of the anchorage zone in the pylon);
(b) a no-anchorage zone is seen as a column under a constant axial compression load.

I. Material amount (weight) of anchorage zone in the pylon Q_{t1}

$$Q_{t1} = \frac{1}{2}\gamma_t (A_{t1} + A_{t2})h_b \quad [8.22]$$

where:
γ_t = density of pylon materials; and
h_b = height of anchorage zone in pylon.

The axial compression loads at the top pylon, N_{t1}, can be calculated as:

$$N_{t1} = H(\tan\alpha + \tan\beta) \quad [8.23]$$

where:
H = horizontal component of the first stay cable forces, assuming the one on the left side equal to the right side (in Fig. 8.21a); and
α and β = incline angle of stay cables on the left and right side of the pylon, respectively.

Then the cross-sectional area at the top pylon can be obtained as:

$$A_{t1} = \frac{N_{t1}}{\sigma_t} = \frac{H(\tan\alpha + \tan\beta)}{\sigma_t} \quad [8.24]$$

where:
σ_t = allowable compressive stress of the pylon materials.

Other parameters can be found in Fig. 8.21a.

The axial compression loads at the bottom of the anchorage zone in the pylon, N_{t2}, can also be calculated using the equilibrium equations shown below:

$$N_{t2} - N_{t1} = (w + p)\left(\frac{l_s + l_m}{2}\right) + Q_s + Q_m + Q_{t1} \quad [8.25]$$

$$Q_{t1} = \gamma_t h_b \left(\frac{N_{t1}}{\sigma_t} + \frac{N_{t2}}{\sigma_t}\right) \Big/ 2 \quad [8.26]$$

where:
w and p = uniform gravity loads and distributed live loads per unit length of the girder, respectively;
l_s and l_m = side span and main span length of the cable-stay bridge, respectively;
Q_s and Q_m = self-weight of stay cables in the side span and main span of the cable-stay bridge, respectively (corresponding to one pylon); and
Q_{t1} = self-weight of the pylon in the anchorage zone.
Then, Equations [8.25] and [8.26] give:

$$N_{t2} = \frac{(w+p)(l_s + l_m/2) + Q_s + Q_m + \gamma_t h_b (N_{t1}/\sigma_t)}{(1-(\gamma_t h_b/2\sigma_t))} + N_{t1} \tag{8.27}$$

Substituting Equations [8.23] and [8.27] into Equation [8.26], the material amount of an anchorage zone in the pylon, Q_{t1}, can be finally estimated.

II. Material amount (weight) of no-anchorage zone in the pylon Qt_2

Figure 8.21b apparently gives:

$$N_t(\xi) + dN_t(\xi) = N_t(\xi) + \frac{\gamma_t}{\sigma_t} N_t(\xi) d\xi \tag{8.28}$$

where:
$N_t(\xi)$ = axial compression loads in the pylon; and
ξ = section distance from where N_{t2} applies.
Therefore,

$$dN_t(\xi) = \frac{\gamma_t}{\sigma_t} N_t(\xi) d\xi \tag{8.29}$$

The differential equation of $N_t(\xi)$ is then calculated as:

$$N_t(\xi) = c_1 \exp\left(\frac{\gamma_t}{\sigma_t}\xi\right) \tag{8.30}$$

where c_1 = any arbitrary constant.
Substituting the boundary conditions that $\xi = 0$ and $N_t(0) = N_{t2}$ into Equation [8.30], the variable axial force $N_t(\xi)$ can be solved:

$$N_t(\xi) = N_{t2} \exp\left(\frac{\gamma_t}{\sigma_t}\xi\right) \tag{8.31}$$

Then the material amount of a none-anchorage zone in the pylon, Q_{t2}, can be easily calculated:

$$Q_{t2} = N_{t3} - N_{t2} = N_t\left(h + h_t - h_b\right) - N_{t2} = N_{t2}\left\{\exp\left[\frac{\gamma_t}{\sigma_t}\left(h + h_t - h_b\right)\right] - 1\right\}$$

[8.32]

where N_{t3} = axial compression loads at the bottom of the pylon.

III. Cost for one pylon

Based on the values of Q_{t1} and Q_{t2}, the cost for one pylon can be expressed as:

$$C_p = Q_t \times \mu_t = \left(Q_{t1} + Q_{t2}\right) \times \mu_t \quad [8.33]$$

where:
Q_t = material amount of one pylon; and
μ_t = cost (price) per unit weight.

IV. Cost for foundation structures

Based on Equation [8.31] the reaction force at the location of one pylon can be calculated as:

$$N_{t3} = N_{t2} \exp\left[\frac{\gamma_t}{\sigma_t} \cdot \left(h + h_t + h_b\right)\right] \quad [8.34]$$

All the reaction force of the half bridge structure, W, is:

$$W = \left(w + p\right)\left(\frac{l_s + l_m}{2}\right) + Q_s + Q_m + Q_t \quad [8.35]$$

The reaction force taken by the side and abutment piers in the half span, N_b, is:

$$N_b = W - N_{t3} \quad [8.36]$$

Then, the material amount of foundation structures at the location of pylon and piers of the first half span bridge can be written as:

$$Q_j = k_j N_{t3} \quad [8.37]$$

$$Q_b = k_b N_b \quad [8.38]$$

where:

Q_j and Q_b = material amount (weight) of the foundation structures at the location of pylon and piers of the first half span bridge, respectively; and

k_j and k_b = weight coefficients for the pylon and piers respectively, both of which are equal to 4.0 based on the Chinese code for budget estimate of bridge engineering in the present study, or are determined by the designers' experiences (JTG B06–2007, 2007).

Based on the values of Q_j and Q_b, the cost for the foundation structures in the half span can be obtained as:

$$C_f = Q_j \mu_j + Q_b \mu_b \quad [8.39]$$

where μ_j and μ_b = cost (price) per unit weight for the pylon and piers, respectively.

V. Cost for substructures

Finally, the substructure cost (including pylons) for the whole bridge structure can be calculated using the expression below:

$$C_{\text{sub}} = 2\times\left[(Q_{t1} + Q_{t2})\mu_t + Q_j \mu_j + Q_b \mu_b\right] \quad [8.40]$$

Discussion of economic aspects of each bridge type

For the purposes of comparison, the current cost of each main component in the six selected bridge cases is assumed in Table 8.8 (JTG B06–2007, 2007).

Based on Equations [8.22]–[8.40], costs listed in Table 8.8, and design parameters already obtained from bridge models, the overall costs required for the six cable-stayed bridges using CFRP components can be directly calculated and are shown in Table 8.9. The area ratio of the hybrid stay cables (Type II and Type V) is 0.6 based on the results in Section 8.4.3.

Table 8.8 Current cost for each main component

Components	Cost	Components	Cost
Steel stay cables	\$3.0/kg	Steel girder	\$3.3/kg
CFRP stay cable	\$80.0/kg	Concrete pylon	\$0.3/kg
Steel bridge deck	\$3.3/kg	Foundation	\$0.2/kg
CFRP bridge deck	\$85.0/kg	Piers	\$0.2/kg

Table 8.9 Costs for the six types of cable-stayed bridges

Components/bridge types		Type I	Type II	Type III	Type IV	Type V	Type VI
Superstructure (Million $)	Steel stay cables	75.20	26.39	–	54.89	19.79	–
	CFRP stay cables	–	145.21	220.22	–	108.9	162.97
	Total for cables (percentage of total)	75.19 (11.5%)	171.60 (24.1%)	220.23 (31.6%)	54.89 (6.6%)	128.71 (14.6%)	162.97 (18.8%)
	Girder and deck (percentage of total)	234.58 (35.9%)	234.58 (33.0%)	234.58 (33.6%)	524.84 (63.2%)	524.84 (59.5%)	524.84 (60.5%)
	Subtotal (percentage of total)	309.77 (47.4%)	406.19 (57.1%)	454.80 (65.2%)	579.73 (69.8%)	653.54 (74.1%)	687.81 (79.3%)
Pylon (Million $) (percentage of total)		74.41 (11.4%)	66.16 (9.3%)	52.60 (7.5%)	54.32 (6.5%)	49.62 (5.6%)	38.92 (4.5%)
Pier (Million $) (percentage of total)		268.84 (41.2%)	239.05 (33.6%)	190.02 (27.2%)	196.26 (23.6%)	179.29 (20.3%)	140.61 (16.2%)
Total cost (Million $)		653.02	711.40	697.42	830.30	882.45	867.34
Total cost per span length (Million $/m)		0.3207	0.3494	0.3425	0.4078	0.4334	0.426

According to Table 8.9, the cost of both CFRP stay cables and bridge deck is still higher than that of steel design even though their light self-weight can significantly reduce the cost of the substructures. The designs using hybrid stay cables (Types II and V) are a little more expensive than others using pure steel or CFRP materials as stay cables. It can be concluded that the cable-stayed bridges using CFRP materials do not reflect any initial cost benefit from the current price of CFRP materials. However, CFRP materials may have life-cycle cost benefits that are not discussed here.

Also when it comes to costs, the factor of varying price for each material needs to be further considered. The current price ratio of CFRP to steel materials is roughly determined as 27/1. The CFRP material price has been significantly reduced in the last decade and will be further reduced in the near future, which will lower the price ratio used in Table 8.9 and lead to different conclusions.

Figure 8.22 shows the varying cost of the six cable-stayed bridges or their components based on a cost ratio variation of CFRP to steel materials from 27/1 to 2/1. All the costs for components or the entire bridge using CFRP materials could be reduced significantly with a decrease in the price ratio from 27/1 to 2/1, especially for the cases with both CFRP stay cables and CFRP bridge decks (Types III and VI). The cost percentage of each component is also seen in Fig. 8.22b and 8.22d. Both reductions on CFRP stay cables and bridge deck can result in an increasing percentage of costs for substructure (piers and pylons), which is shown in Fig. 8.22e.

In summary, Fig. 8.22 shows that using CFRP materials for the components or entire bridge could be more cost-effective than the traditional design when the price ratio is over a certain value. For example, in Fig. 8.22a for stay cable cost, when the price ratio is less than 8/1 the stay cables using CFRP materials (Types II, III, V, and VI) are more economical than steel ones; while in Fig. 8.22f looking at the total cost, this dividing ratio value is 16/1 for those using CFRP components (Types II–VI). Due to the large amount of material in the bridge deck, the cases using CFRP bridge deck (Types IV–VI) could show greater economic benefit from the reduced price ratio. It is logical to predict that the advantages of using CFRP components will be further enhanced with the rapid decrease in the cost of CFRP materials. In the near future, the proposed cable-stayed bridges using CFRP stay cables and/or CFRP bridge deck could be an excellent alternative to the traditional steel cable-stayed bridges, not only from a mechanical viewpoint but also from an economical viewpoint.

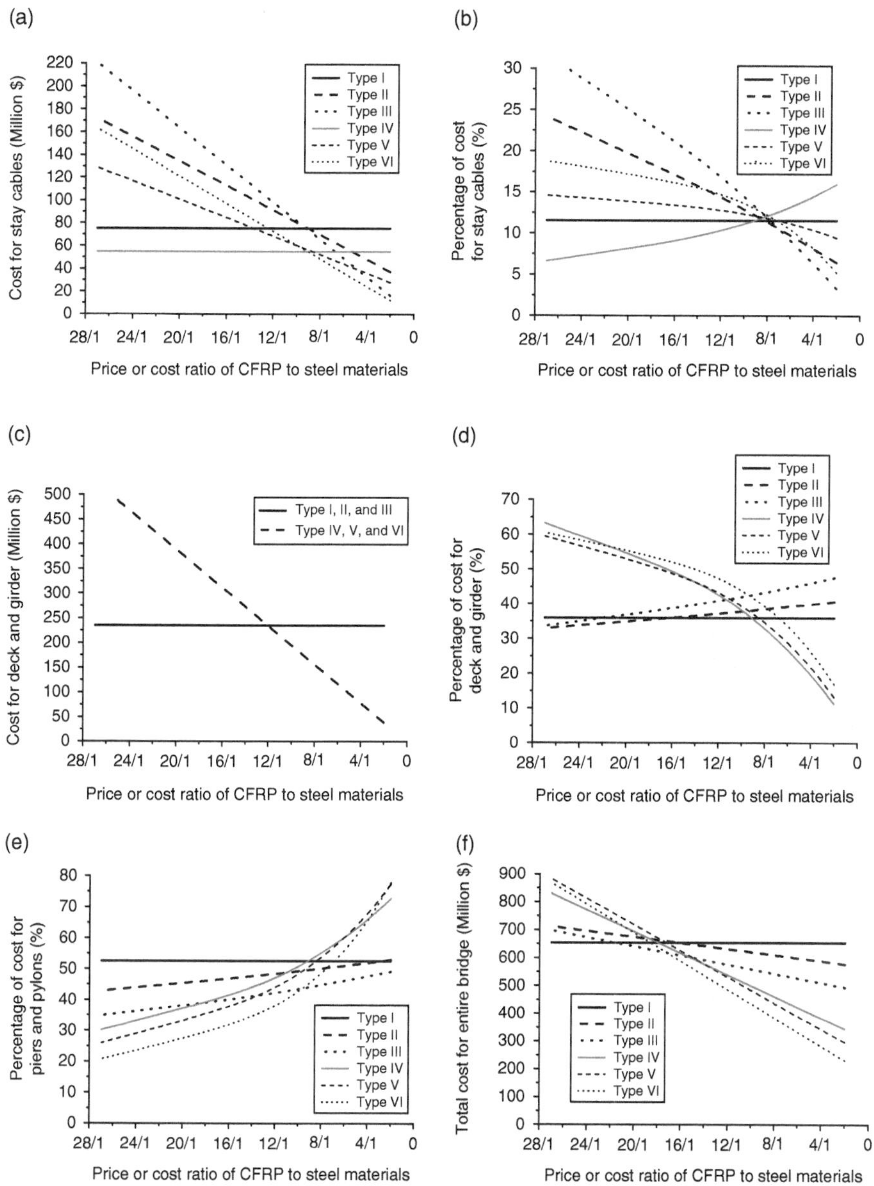

8.22 Economic behavior vs price/cost ratio. (a) Cost for stay cables; (b) percentage of cost for stay cables; (c) cost for deck and girder; (d) percentage of cost for deck and girder; (e) percentage of cost for piers and pylons; (f) total cost for entire bridge.

8.6 Conclusions and future trends

Based on the numerical study, cable-stayed bridges with CFRP components (CFRP bridge deck, CFRP stay cables, and hybrid stay cables) can be an excellent alternative to traditional steel cable-stayed bridges, especially for bridges with super-long spans. Although in the present study only the cable-stayed bridges with a main span length of 1400 m were studied, it can still be predicted that the advantages of using CFRP components will be further enhanced with the continuous extension of bridge span lengths and decreasing cost of CFRP materials. For future work, a study of the span extension/limitation and construction design for the proposed cable-stayed bridges with CFRP components are also needed.

8.7 Acknowledgments

This work was partially supported by the grant from National Natural Science Foundation of China (NSFC) (Project No. 51229801), Specialized Research Fund for the Doctoral Program of Higher Education (Project No. 20120092120058), and Natural Science Foundation of Jiangsu Province of China (Project No. BK2012343). The contents presented reflect only the views of the writers who are responsible for the facts and the accuracy of the data presented herein.

8.8 References

ACI. (2004). *Prestressing Concrete Structures with FRP Tendons*. American Concrete Institute (ACI), Farmington Hills, MI, U.S.A.

Ahmadi-Kashani, K. and Bell, A. J. (1988). 'The analysis of cables subject to uniformly distributed loads'. *Engineering Structures*, **10**(3), 174–184.

AASHTO LRFD Bridge Design Specifications (2004). Association of State Highway and Transportation Officials. Washington, DC, U.S.A.

Cai, C. S., Oghumu, S. and Megers, D. (2009). 'Finite element modeling and development of equivalent properties for FRP bridge panels'. *Journal of Bridge Engineering*, ASCE, **14**(2), 112–121.

Colombi, P. and Poggi, C. (2006). 'Strengthening of tensile steel members and bolted joints using adhesively bonded CFRP plates'. *Construction and Building Materials*, **20**(1–2), 22–33.

Ernst, H. J. (1965). 'Der e-modul von seilen unter beruecksichtigung des durchhanges'. *Der Bauingenieur*, **40**(2), 52–55.

Freire, A. M. S., Negrao, J. H. O. and Lopes, A. V. (2006). 'Geometrical nonlinearities on the static analysis of highly flexible steel cable-stayed bridges'. *Computers and Structures*, **84**(31–32), 2128–2140.

Gimsing, N. J. (1997). *Cable-supported Bridges*, 2nd ed. John Wiley & Sons, New York, U.S.A.

Grace, N. F., Enomoto, T. and Yagi, K. (2002). 'Behavior of CFCC and CFRP leadline prestressing system in bridge construction'. *PCI Journal*, **47**(3), 90–103.

Hayes, M. D., Lesko, J. J., Haramis, J., Cousins, T. E., Gomez, J. and Masarelli, P. (2000). 'Laboratory and field testing of composite bridge superstructure'. *Journal of Composites for Construction,* **4**(3), 120–128.

Irvine, H. M. (1981). *Cable Structures*. The MIT Press, Cambridge, MA, U.S.A.

Jackson, D., Shepheard, B., Kebadze, E., Teles, R., Rossi, R. and Gonçalves, R. C. (2005). 'CFRP mooring lines for modu applications'. *Offshore Technology Conference*, Houston, Texas, U.S.A.

JTG B06–2007 (2007). *Code for Budget Estimate of Highway and Bridge Construction*. China Communications Press, Beijing, China.

Leonhardt, F. and Zellner, W. (1991). 'Past, present and future of cable-stayed bridges'. In *Cable-stayed Bridges: Recent Developments and Their Future,* Ito, M., Fujino, Y., Myiata, T. and Narita, N. (eds.), Elsevier, New York, U.S.A., 1–33.

Matta, F. (2003). 'Bond between steel and CFRP laminates for rehabilitation of metallic bridges'. Ph.D. Thesis, Faculty of Engineering, University of Padua, Padua, Italy.

Meier, U. (1987). 'Proposal for a carbon fiber reinforced composite bridge across the strait of Gibraltar at its narrowest site'. *Proceedings of the Institution of Mechanical Engineers,* **201**(B2), 73–78.

Nagai, M., Fujino, Y., Yamaguchi, H. and Iwasaki, E. (2004). 'Feasibility of a 1,400m span steel cable-stayed bridge'. *Journal of Bridge Engineering,* **9**(5), 444–452.

Noisternig, J. F. (2000). 'Carbon fibre composites as stay cables for bridges'. *Applied Composite Materials,* **7**(2–3), 139–150.

Noisternig, J. F. and Jungwirth, D. (1998). 'CFRP-tendons for structural application – requirements and developments'. In *2nd International Conference ICCI,* 5–7 January 1998, Tucson, Arizona, U.S.A., 115–127.

O'Brien, T. W. and Francis, A. J. (1964). 'Cable movements under two-dimensional loading'. *ASCE Journal of Structural Engineering,* **90**(ST3), 89–123.

Rabinovitch, O. and Frostig, Y. (2003). 'Experiments and analytical comparison of RC beams strengthened with CFRP composites'. *Composites Part B: Engineering,* **34**(8), 663–677.

Rizkalla, S., Dawood, M. and Schnerch, D. (2008). 'Development of a carbon fiber reinforced polymer system for strengthening steel structures'. *Composites Part A: Applied Science and Manufacturing,* **39**(2), 388–397.

Saadatmanesh, H. and Ehsani, M. R. (1998). 'Fiber composites in infrastructure'. In *2nd International Conference ICCI,* 5–7 January 1998, Tucson, Arizona, U.S.A.

Schilde, K. and Seim, W. (2007). 'Experimental and numerical investigations of bond between CFRP and concrete'. *Construction and Building Materials,* **21**(4), 709–726.

Schnerch, D., Stanford, K., Sumner, E. A. and Rizkalla, S. (2004). 'Strengthening steel structures and bridges with high modulus carbon fiber reinforced polymers: resin selection and scaled monopole behavior'. *TRB 2004 Annual Meeting – Transportation Research Record,* Washington, DC, U.S.A.

Schnerch, D., Stanford, K., Sumner, E. A. and Rizkalla, S. (2005). 'Bond behavior of CFRP strengthened steel bridges and structures'. *Proceedings of the International Symposium on Bond Behaviour of FRP in Structures,* Hong Kong, China.

Schurter, U. and Meier, B. (1996). 'Stork bridge winthertur'. *Schweizer Ingenieur und Architekt,* **44**, Oktober 1996, 976–979.

Sen, R., Liby, L. and Mullins, G. (2001). 'Strengthening steel bridge sections using CFRP laminates'. *Composites Part B: Engineering,* **32**(2001), 309–322.

Walther, R., Houriet, B., Isler, W., Moïa, P. and Klien, J. F. (1999). 'Cable stayed bridges'. In R. Walther (ed.), *Dynamic Analysis,* 2nd ed., Thomas Telford, London.

Xiong, W., Cai, C. S., Zhang, Y. and Xiao, R. C. (2011). 'Study of super long span cable-stayed bridges with CFRP components'. *Engineering Structures,* **33**(2), 330–343.

Xiong, W., Cai, C. S., Xiao, R. C. and Deng, L. (2012). 'Concept and analysis of stay cables with a CFRP and steel composite section'. *Journal of Civil Engineering,* **16**(1), 107–117.

Zhang, B. R., Benmokrane, B., Chennouf, A., Mukhopadhyaya, P. and El-Safty, A. (2001). 'Tensile behavior of FRP tendons for prestressed ground anchors'. *Journal of Composites for Construction,* **5**(2), 85–93.

Zhao, X. L. (2008). 'FRP strengthening of thin-walled structures'. *CIMS2008, Fifth International Conference on Coupled Instabilities in Metal Structures, Gregory J Hancock Symposium,* June 2008, Sydney, Australia, 23–25.

Zhao, X. L. and Zhang, L. (2007). 'State-of-the-art review on FRP strengthened steel structures'. *Engineering Structures,* **29**(8), 1808–1823.

9

Repair of deteriorated bridge substructures using carbon fiber-reinforced polymer (CFRP) composites

M. E. WILLIAMS, Walter P. Moore and Associates, Inc., USA

DOI: 10.1533/9780857097019.2.265

Abstract: This chapter presents a case study in the evaluation and repair of a 10-year-old highway bridge with substructures undergoing moderate to severe premature concrete deterioration and expansion due to alkali-silica reaction (ASR). Detailed visual observations of distressed conditions, structural analysis, and results of laboratory materials testing are presented as contributors to the development of a repair course of action. Carbon fiber-reinforced polymer (CFRP) composite applications for confinement and anchorage concepts are discussed as they relate to the structural repair methodology presented. CFRP testing procedures are discussed to ensure quality control in the repair implementation.

Key words: alkali-silica reaction (ASR), concrete bridges, premature concrete deterioration, CFRP.

9.1 Introduction

The condition of an aging highway bridge infrastructure continues to be a subject of concern. The vast majority of bridges built to last 50 years in the USA now have an average age of 43 years. According to the National Bridge Inventory (NBI), one out of every four bridges is structurally deficient and in need of repair or functionally obsolete. As the cost of new bridge construction continues to soar, bridge repair, where possible, has generally become more economical than bridge replacement. Increasing transportation demands, and increasing economic impacts of bridge traffic closures, have been an impetus for pursuing durable and long-lasting repair solutions with minimal impact on bridge operations.

Concrete bridge deterioration can be caused by many factors. Bridges can experience distress due to growing traffic volumes, higher loads, and harsh environments. Bridges can also experience premature concrete deterioration due to material defects or adverse reactions of the concrete

9.1 Overview of bridge and two substructure straddle bents.

constituents. In certain instances, premature concrete deterioration can be more severe than progressive deterioration from corrosion or other related material damage mechanisms. For concrete bridges, alkali-aggregate reactions such as the alkali-silica reaction (ASR) can be detrimental, leading to the potential loss in structural capacity and uncertain bridge life expectancy. This chapter presents a case study in the evaluation and repair of bridge substructures experiencing premature concrete deterioration, using carbon fiber-reinforced polymer (CFRP) composites.

The subject of the case study is a concrete bridge that serves as a major connector for a high traffic volume toll road and has been in service for approximately 10 years. The multi-span bridge consists of precast concrete beams supported by cast-in-place reinforced concrete straddle bents, as shown in Fig. 9.1. Two concrete straddle bents in this case study span over an easement designated for future light rail expansion. The straddle bents are approximately 12 m in height, with inverted T-beams spanning approximately 17 m between two concrete 2 m diameter columns (Fig. 9.2). The two straddle bents are similar in design, and are separated by approximately 30 m. Both straddle bents are supported on drilled pier deep foundations. The straddle bents and bridge were constructed in accordance with the original design drawings and specifications.

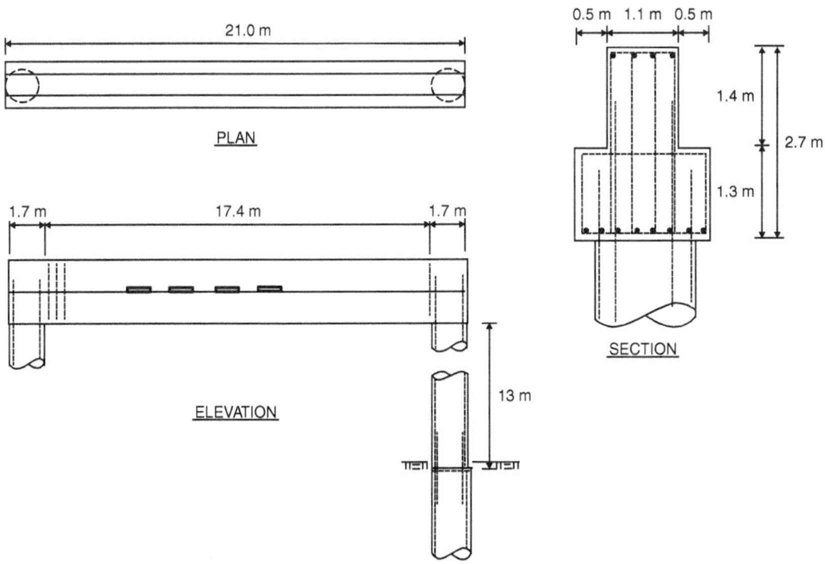

9.2 Schematic drawings of straddle bent. *Note*: Additional reinforcing steel omitted on drawing for clarity.

9.2 Investigating deterioration of concrete in bridges

A prior routine bridge inspection noted cracking distress in the concrete bent substructure, particularly at the ends of the bent beams. Based on a review of bridge maintenance documents, cracking was believed to have initiated approximately 5 years after construction. An in-depth inspection was conducted to study the nature and cause of the cracking, in order to develop a future course of action for repair.

A close-up evaluation of the bridge substructures in this case study was performed in order to understand the nature and extent of the concrete deterioration. The investigation proceeded with a visual investigation of the substructures, followed by non-destructive evaluation, and concrete materials testing in a laboratory. The findings of these three phases of the investigation established the nature and causation of the observed concrete deterioration.

9.2.1 Visual investigation

The primary means of visual evaluation was by up-close observations using aerial lift access. Concrete distress conditions were documented by way of field notes and crack mapping. Predominantly, the two straddle bent beams

9.3 Partial elevation view of straddle bent with visible white staining exuding from cracks.

exhibited map cracking on all surfaces, with visible white staining exuding from cracks, as shown in Figs 9.3 and 9.4. Crack widths ranged from hairline to 3 mm on the surface of the bent beams. Severe concrete deterioration, in the form of wide cracking and concrete spalling, was observed at the ends of the bent beams (Fig. 9.5). At these locations, concrete crack widths up to 12 mm were recorded. Less notable map cracking was also observed near the top of the reinforced concrete columns supporting the substructure bent beams.

9.2.2 Non-destructive evaluation

Non-destructive evaluation techniques were deployed to study the concrete deterioration in the two straddle bents. The substructure elements were scanned using ground (surface) penetrating radar (GPR) to identify the presence of reinforcing steel and to verify as-built conditions. Where possible, the bent beams were scanned in a single horizontal pass for side-by-side comparison with the as-built construction drawings. As shown for a particular bent beam end in Fig. 9.6, the horizontal GPR scan of the T-beam revealed conformance of reinforcing steel stirrups and longitudinal column steel placement with the as-built drawings. This finding was typical for the placement of reinforcing steel at both straddle bent beam end locations. The

9.4 Map cracking with visible white staining exuding from cracks.

reinforcing steel placement was a key factor in determining the type and extents of the CFRP repair solution discussed later in this chapter.

Acoustic sounding techniques were deployed to obtain measurements of crack depth. Impact-echo testing was conducted at cracked concrete elements at selected locations. Impact-echo testing showed typical crack depths of 50–150 mm at locations of map cracking. Most notably, these crack depths exceeded the concrete clear cover at the reinforcing steel. Further testing of the bent beam ends showed full depth cracking, through a bent beam width of approximately 1.1 m. Crack depth measurements were further corroborated with crack length measurements along cores excised through cracks.

9.2.3 Concrete materials testing

Although there were some visual cues to possible mechanisms causing premature concrete deterioration, laboratory testing of excised concrete specimens was utilized to identify the specific deterioration mechanism. Exploratory coring was conducted to evaluate the concrete material properties and deteriorated condition of the concrete. Core samples were excised to obtain representative samples of deteriorated (cracked) and

9.5 Severe cracking at end of straddle bent beam.

non-deteriorated concrete areas for subsequent laboratory testing and comparative study.

Concrete petrography was conducted on excised specimens at a testing laboratory in accordance with ASTM C 856 (standard practice for petrographic examination of hardened concrete) to study the cause of the concrete deterioration. Concrete samples were cut and examined under a stereomicroscope at 45 times magnification, and then as thin sections under polarized light up to 400 times magnification. The petrographic analysis indicated that the general quality of the original concrete was good, based largely on paste properties, water–cement ratio, and consolidation. The excised concrete cores, however, showed signs of randomly-oriented microcracks under microscopic evaluation, with evidence of ASR gel formation in voids and around quartzite aggregate particles. The ASR gel reaction product in the presence of moisture was believed to be the primary contributor to the expansion within the concrete and subsequent premature concrete deterioration. If left unchecked, further expansion of the ASR gel would lead to continued and more severe concrete deterioration of the bridge substructure.

Concrete deterioration due to ASR gel expansion can be influenced by concrete confinement, degree of restraint, and placement of reinforcing steel to control concrete cracking. In focusing on the substructure beam ends experiencing the most severe cracking, a review of the as-built construction drawings revealed the placement of limited reinforcing steel stirrups at the substructure beam ends. While additional reinforcing steel was not required for structural design purposes, the lack of confining steel stirrups provided ripe opportunity for ASR gel expansion in the presence of moisture. In addition, column longitudinal reinforcing steel extending up into the substructure bent beams for continuity was curtailed just below where the most severe cracking developed at the beam ends (Fig. 9.6).

9.3 Analysis of concrete deterioration in bridge substructures

The evaluation of structures undergoing premature concrete deterioration continues to be the object of research work seeking to incorporate long-term effects from deleterious reactions such as ASR. Finite element models of structural behavior, including expansive behavior of concrete gravity dams due to alkali-aggregate reactions, have been developed by Léger *et al.* (1996), Malla and Wieland (1999), and Sellier *et al.* (2009). While the finite element models reported in these research works have been validated with experimental test data, additional work is needed for general use on concrete bridge structures. Research has also been conducted to predict the swelling pressure of concrete undergoing expansion due to ASR. Work by Pike (1967) reported expansive pressures for mortars confined in a special metal container ranged from 2.5 to 14 MPa. Subsequent work by Lenzner and Ludwig (1978) measured expansive pressures of 2.5 to 4 MPa in mortars constrained with reinforcing bars. More recent work by Ferraris *et al.* (1997) measured expansion pressures ranging from 2 to 9 MPa depending on the level of reactivity in the concrete test specimens.

While considerable research has been conducted on ASR in concrete test specimens, there is limited published literature on the structural analysis of concrete bridge elements undergoing ASR expansion and subsequent deterioration. Accordingly, the bridge analysis presented in this case study relies on fundamental principles of structural mechanics with material behavior assumptions for expansive pressures exerted within the concrete bridge elements.

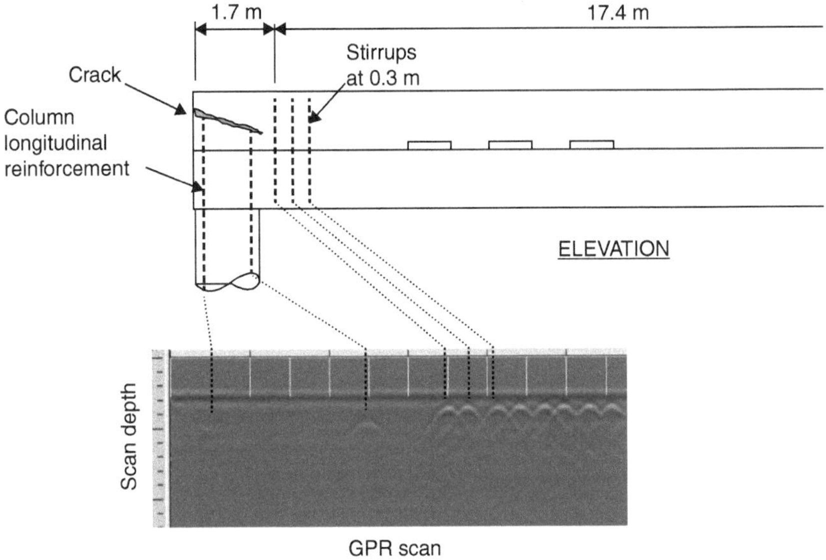

9.6 GPR scan showing as-built reinforcement locations. *Note*: Additional reinforcing steel omitted on drawing for clarity.

As previously shown, premature concrete deterioration was on-going in two concrete substructures. The straddle bent beams were undergoing map cracking throughout and severe cracking and splitting at the bent beam ends. Given the apparent cracking at the surface and confirmation of crack depths beyond the reinforcement cover, the substructure beams were suspected of undergoing expansive stresses exceeding the tensile strength of concrete (modulus of rupture). For the bridge substructures in this case study, the modulus of rupture was calculated to be approximately 4 MPa based on a concrete compressive strength of 42 MPa (AASHTO, 2012). Computer modeling of expansive behavior of rectangular or inverted T-shaped sections to replicate the level of concrete surface cracking would have been a complex endeavor involving nonlinear material behavior.

The extents of concrete cracking were considered in assessing the structural capacity of the substructures. Despite extensive map cracking and severe concrete deterioration at the bent beam ends, the concrete structure appeared to be generally intact, based on acoustic (hammer) sounding, and devoid of corrosion-related activity. Structural analysis of the substructure bent beams showed adequate structural capacity based on the original design parameters. However, knowing the substructures were undergoing irreversible concrete deterioration and the indeterminacy of the substructure capacity due to extensive cracking, structural remediation was the prudent course of action. Due to the expansive nature of ASR, structural remediation by confinement of the expanding concrete elements was required. This confinement was achieved by application of CFRP composites to the straddle bent beams.

9.4 Repair of bridges using carbon fiber-reinforced polymer (CFRP) composites

Repair of bridge substructures undergoing premature concrete deterioration can be achieved using a number of methods. Given the overwhelming need to keep the bridge open to service, partial demolition and repair or section enlargements were dismissed as repair options due to their potential functional and esthetic impact, respectively. Steel jacketing was also considered, and dismissed due to potential long-term maintenance issues. Repair using CFRP composites was preferred, due to its small and light footprint and ease of application while the bridge was in service. Among composite strengthening materials available for concrete, CFRP was preferred to glass-fiber or aramid reinforcement for exterior environments exposed to moisture. The high tensile strength of CFRP composites was also a compelling factor when selecting the material for the repair solution. The end goal of the repair was to counteract expansive pressures exceeding 4 MPa and subsequent surface cracking induced from ASR.

9.4.1 Design basis

In the repair methodology, CFRP composites were designed in a manner to provide external confinement of the bent beams. Research and experimentation in the confinement of concrete elements using CFRP has produced varied results. CFRP composites are more effective in confining circular concrete elements (i.e. columns). Confinement of non-circular concrete sections using CFRP can be achieved to a lesser extent using an equivalent diameter and shape factors (ACI, 2008). An important distinction to make in this case study is that CFRP was provided to confine further concrete expansion rather than to provide additional concrete compressive strength or joint confinement, as is the typical application for CFRP confinement in axially loaded members. In addition, some confinement is already provided by way of the reinforcing stirrups in the bent beams at locations of map cracking. Ideally, confinement and arrestment of future cracking would be achieved by a combination of existing reinforcing stirrups and external CFRP application. One drawback of this CFRP repair methodology is that the concrete element wrapped in CFRP must first expand in order to engage the CFRP and initiate restraint through confinement. This is somewhat counterintuitive, in that the CFRP composites are being provided to prevent further expansion due to ASR. However, applications of CFRP for confinement can be justified considering that some residual expansion is to be expected given the latent moisture conditions contributing to ASR within the bent beams.

CFRP composites were also designed to control expansion at the beam ends where there were minimal steel reinforcing stirrups. Calculation of the amount of CFRP composite strengthening was achieved by determining the design strength of the reinforcing stirrups at the typical spacing within the bent beam, had they been provided at the bent ends. Computation of the CFRP material requirements was similar to that for shear strengthening.

9.4.2 CFRP design for substructure beam

CFRP composites were designed for placement in locations to control concrete expansion along the straddle bent beams. A single ply of CFRP was designed for application along the beam where there was extensive map cracking. Given the angular geometry of the inverted T-beam, a single wrap was not possible, due to re-entrant corners. CFRP wrapping was therefore designed in two pieces. A single ply CFRP was designed to be applied as a three-sided U-wrap to the top leg of the inverted T-beam and a three-sided U-wrap on the flange of the inverted T-beam (Fig. 9.7 – Section 'B'). Since the confinement procedure was designed in two pieces, CFRP end anchorage was provided to create pseudo-continuity between top and bottom of the beam in CFRP applications. End anchorage was provided by saw cutting a slot in the beam and embedding saturated plies in resin in the slot (Fig. 9.7). A single U-wrap application to the bottom flange of the beam was provided where access to the top leg of the inverted T-beam was not possible due to interference with the precast bridge beams and bearing seats (Fig. 9.8).

9.4.3 CFRP design for substructure beam ends

CFRP composites were designed for placement in locations to control concrete expansion at the straddle bent beam ends. In the case of the beam ends, a horizontal CFRP application was designed to control lateral expansion of the outside face of the bent beam end. Additionally, calculations showed that three vertical plies of CFRP would be required to supplement the existing minimal reinforcing steel stirrups, over a distance of approximately 2 m. These three plies were designed to be applied as a three-sided U-wrap to the top leg of the inverted T-beam. The CFRP installation drawings provided work phasing such that the horizontal CFRP application at the outside face of the beam was applied first (to the top leg and then bottom flange of the

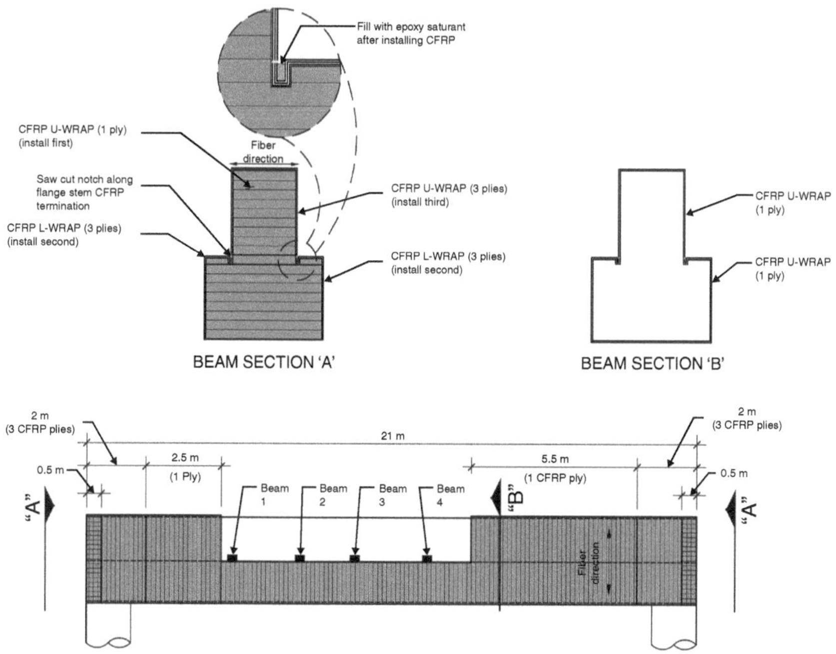

9.7 CFRP design for wrapping straddle bent. The bottom figure shows the bridge straddle bent in elevation. The top two figures show a section cut through that elevation. Section 'A' shows the end of the bridge straddle bent. Section 'B' shows a cut through the bridge straddle bent.

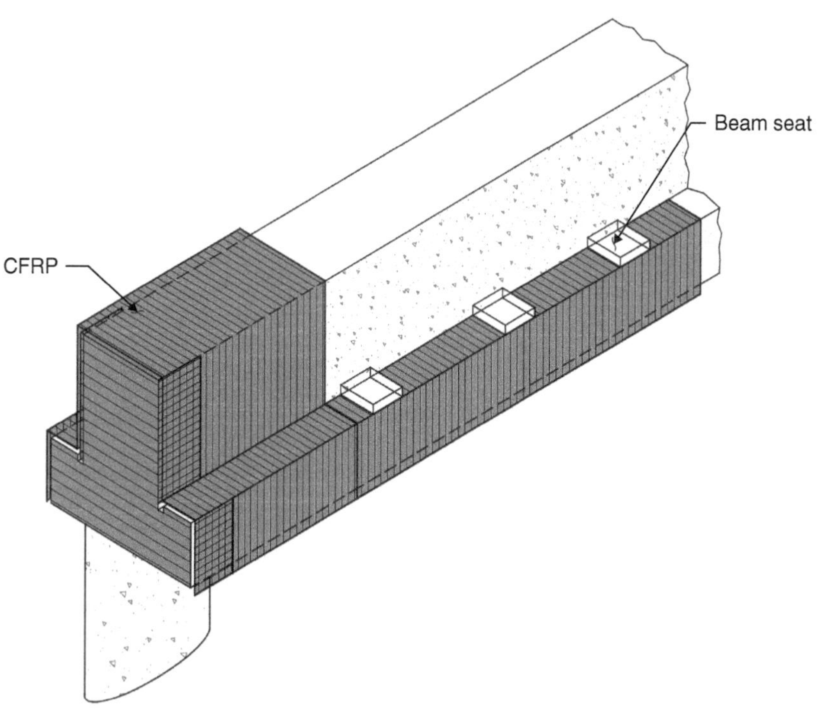

9.8 CFRP design for wrapping straddle bent at beam seats (isometric).

inverted T-beam) followed by the vertical U-wrap application (see Fig. 9.7 – Section 'A').

9.4.4 CFRP design specifications

Drawings and technical specifications were developed prior to the application of CFRP composites. Documentation for concrete surface preparation was provided to include the removal of existing paint finishes on the concrete surfaces as well as surface roughness criteria using contained abrasive blasting. Although most CFRP confinement designs are generally not bond critical, surface roughness was specified to enhance bond to control concrete surface expansion. Procedures for crack repair were also provided in this construction documentation. Cracks with widths less than 6 mm were filled by injection of epoxy. Cracks with widths exceeding 6 mm, at the bent beam ends for example, required additional concrete removal and patching with concrete repair materials (Figs 9.9 and 9.10).

Several approved CFRP systems were specified for possible use on the project. Each system provided a layered application approach consisting of a primer, epoxy putty for surface irregularities, saturant, CFRP fabric, and protective top coat. The CFRP composite could be applied using either a wet or dry layup, depending on the contractor's repair approach. A protective UV stable coating was specified for application over the finished CFRP composite. The coating was color-matched to other coated concrete surfaces in order to minimize the visibility of the CFRP application.

9.5 Review of CFRP repair of bridge substructure

Installation of the CFRP composite strengthening for the substructure elements was overseen for conformance with the repair design specifications. A mock-up was first installed to verify the proper CFRP application steps. Preparation of the concrete substrate to receive the CFRP was reviewed throughout the project, as proper surface preparation is critical to developing the necessary bond between the CFRP and concrete surface. The concrete surface was also observed to be sound and free of delaminations prior to the CFRP application. Temperature and relative humidity readings were recorded during mixing and application of the CFRP components. Finally, material batch numbers were logged for each application area in the event of a location-specific defect in the CFRP installation.

The CFRP composite application occurred in multiple phases. CFRP wrapping began at the end of each straddle bent beam (Fig. 9.11). CFRP

9.9 Cracking and unsound concrete at the termination of longitudinal column reinforcement, prior to repair.

9.10 Severe cracking at end of straddle bent beam, prior to repair.

9.11 Application of CFRP composite U-wrap to the straddle bent beam end.

wrapping then proceeded along the top of the bent beam (Fig. 9.12). CFRP was then applied to the bottom of the inverted T-beam and anchored into the slots previously installed along the beam (Fig. 9.13). Throughout this process, a work platform for the CFRP installation was provided by system scaffolding (Fig. 9.14).

CFRP installations were visually reviewed during and after application phases. Where possible, voids and delaminations were corrected prior to initial set of the CFRP composite. Such defects can occur due to various factors including sagging of CFRP material, outgassing due to temperature variations, improper saturation, and improper rolling and finishing. Defects noted after initial set of the CFRP composites were marked for repair. As discussed later in this chapter, excessive delaminations necessitated removal and reinstallation of CFRP composites (Fig. 9.15).

9.6 Site testing of CFRP repair and laboratory testing of materials

In addition to visual reviews of the CFRP installation, various site testing methods were employed to check the CFRP installation for conformance

9.12 Application of CFRP composite U-wrap along straddle bent beam.

9.13 Application of CFRP composite anchor in slot (prior to final anchor installation).

9.14 Overview of CFRP composite application to straddle bent beam.

9.15 Areas of delamination marked after visual inspection of CFRP.

with the design specifications. Site testing methods include bond testing, infrared thermography, and acoustic (impulse response) testing.

9.6.1 CFRP bond testing

In this case study, CFRP test patches were installed for each application day and batch of resin used. Patches were tested for adhesion after curing to ensure conformance with ASTM D7522/D7522M–09 (standard test method for pull-off strength for FRP bonded to concrete substrate). A minimum pull-off tensile strength of 1.4 MPa was specified for bond.

9.6.2 Acoustic sounding

Acoustic sounding techniques were utilized to rapidly detect CFRP bond defects that were not visible during initial inspection. Delaminations detected by acoustic sounds, typically light tapping, were noted to have a dull, hollow sound upon testing. Although the testing can be performed quickly, there can be a high degree of inaccuracy, often leaving voids under the CFRP systems undetected. Those defects that are detected can be marked and subsequently repaired after acoustic sounding.

9.6.3 Infrared thermography

Infrared (IR) thermography is the science of acquisition and analysis of thermal information from non-contact thermal imaging devices. IR thermography detects emitted radiation in the infrared range of the electromagnetic spectrum. This corresponds to wavelengths longer than the visible light portion of the spectrum. Thermal imaging can therefore be utilized to detect defects in the CFRP installation that may not be visible. Voids or delaminations in the CFRP installation are shown with a unique thermal signature due to entrapped air between the CFRP composite and the concrete substrate. As shown in Fig. 9.16, CFRP delaminations exhibit a dark thermal signature in comparison to well bonded CFRP applications. As such, defects were marked and subsequently repaired after thermal inspection.

9.6.4 Impulse response testing

Impulse response is a non-destructive testing technique used to evaluate the integrity of concrete elements. In this test, a low-strain impact is applied to excite the structure and the response of the structure is measured. The concrete

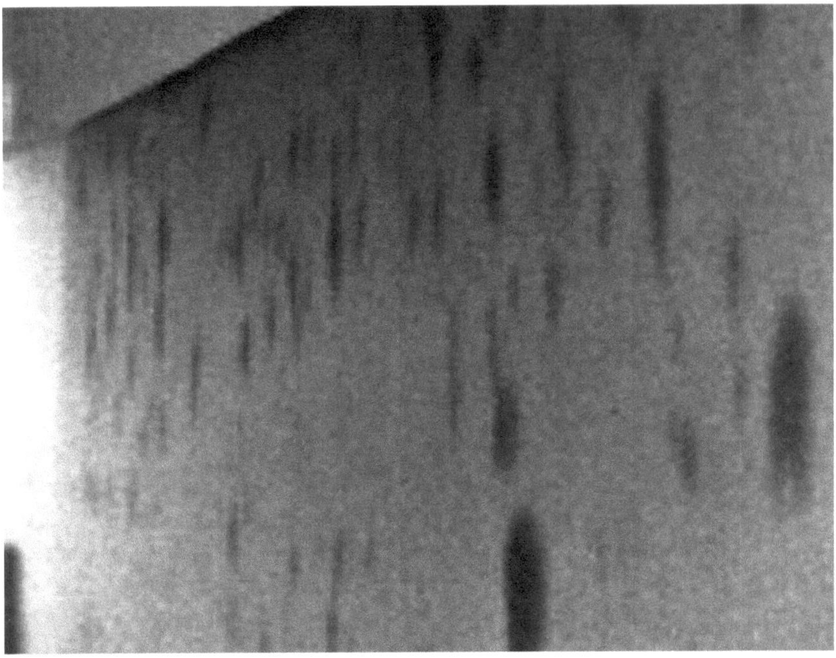

9.16 Areas of CFRP delamination identified by infrared thermography (black spots).

element is struck by using a small impactor, which has a built-in load cell that measures the impulse imparted. The response (vibration) of the concrete element is monitored by a velocity transducer (geophone) placed adjacent to the impact location. The geophone is connected to a data acquisition and processing system, which calculates the dynamic mobility as a function of the frequency of the excitation. The dynamic mobility was analyzed as a secondary means to characterize the condition of the CFRP composite and concrete substrate conditions, and the probability of delaminations in the CFRP application that may not be visible at the surface. Defects identified by impulse response testing were marked and subsequently repaired after inspection (Fig. 9.17).

9.6.5 Laboratory testing of CFRP materials

Laboratory testing is typically conducted to verify the mechanical properties of the CFRP composite. In this case study, witness panels were constructed by the repair contractor for each application day and batch of resin used in order to ensure quality control. Three-ply CFRP witness panels were created to match the installation condition on the straddle bents (Fig. 9.18). A portion of these fully cured witness panels were sent to a qualified testing laboratory with experienced personnel to validate the CFRP material properties with those used in the design. Witness panel testing was

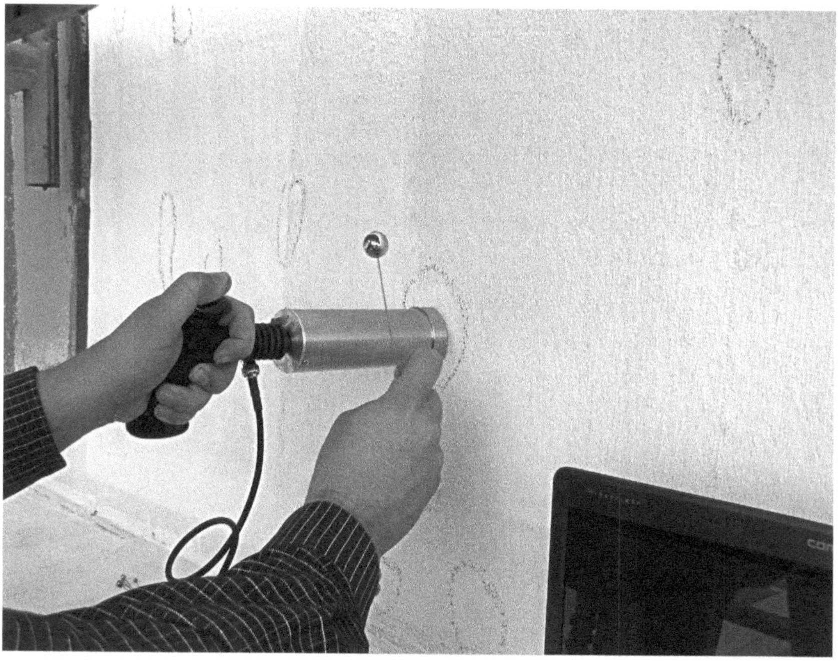

9.17 Impulse response testing to identify extent of CFRP delaminations.

conducted to determine the CFRP composite thickness, tensile strength, modulus of elasticity, and percent elongation. The test results were then compared to the published mechanical properties from the CFRP material manufacturer.

9.7 Dealing with defects in CFRP repairs

Some defects are likely to occur given the labor intensive process of mixing multi-component materials and applying CFRP composites to existing concrete elements at a bridge site. ICC-AC 178 (2008) provides the following guidelines for judging CFRP defects and possible recourses for repair:

- Small entrapped air pockets and voids (approximately 1.6–3.2 mm in diameter) naturally occur in mixed resin systems and generally do not require repair.
- Except at the edges or boundaries, delaminations of less than 130 square mm are generally permissible.
- No more than ten delaminations of this size shall be allowed per 1 m square of the laminated area.
- Any delaminations spanning more that 5% of the surface area shall be repaired by any method approved by the engineer.

9.18 Preparation of CFRP witness panels.

In this case study, several CFRP U-wrap applications to the bent beams had delaminations well in excess of the 5% judging guideline, thereby requiring removal and replacement. The CFRP wraps in question experienced sudden outgassing during initial set due to rapid temperature variations between the application faces. Under direct sun exposure, the outgassing resulted in bubble formations through the CFRP application strips. The contractor removed the CFRP wraps in this location and successfully reinstalled the material at a later date.

9.8 Conclusions

This case study has provided a methodology for assessing and repairing concrete bridge substructures undergoing premature concrete deterioration. Once the deterioration was diagnosed and attributed to ASR expansion, a repair scheme using CFRP composites was implemented for two straddle bent substructures. It is worth noting that this major connector bridge was evaluated and repaired without removing the bridge from service. CFRP applications in lieu of straddle beam replacement or section enlargement provided a cost-effective solution to mitigating ASR expansion and further concrete deterioration. The CFRP repair also proved to be suitable from an

9.19 Overview of straddle bents after repair.

esthetic standpoint, as it generally goes unnoticed by the casual observer (Fig. 9.19). The CFRP repair has been in place for several years and appears to be performing as intended, based on periodic visual reviews. Long-term monitoring of these concrete members will help to validate the success of the remedial repairs in mitigating future concrete deterioration.

9.9 References

American Association of State, Highway, and Transportation Officials (AASHTO). (2012) *LRFD Bridge Design Specifications*.

American Concrete Institute (ACI). (2008) ACI-440.2-R08 *Guide for the Design and Construction of Externally Bonded FRP Systems for Strengthening Concrete Structures*, 35–36.

Ferraris C. F., Clifton J. R., Garboczi E. J. and Davis F. L. In: Scrivener K. L., Young J. F., editors. (1997) 'Stress due to alkali–silica reactions in mortars, mechanisms of chemical degradation of cement-based systems', E & FN Spon; Mechanisms of Chemical Degradation of Cement-Based Systems. *Proceedings of the Materials Research Society's Symposium on Mechanisms of Chemical Degradation of Cement-Based Systems*. November 27–30, 1995, Boston, MA, 75–82.

International Code Council. (2008) *AC178 Acceptance Criteria for Inspection and Verification of Concrete and Reinforced and Unreinforced Masonry Strengthening Using Fiber-Reinforced Polymer (FRP) Composite Systems*.

Léger P., Cote P. and Tinawi R. (1996) 'Finite element analysis of concrete swelling due to alkali-aggregate reaction in dams', *Computer and Structures*, **60**, 601–611.

Lenzner D. and Ludwig V. (1978) 'The alkali aggregate reaction with opaline sand stone from Schleswig Holstein', *Proceedings of the Fourth International Conference on Effects of Alkalies in Cement and Concrete*, Purdue University, 11–34.

Malla S. and Wieland M. (1999) 'Analysis of an arch-gravity dam with a horizontal crack', *Computers and Structures*, **72**, 267–278.

Pike R. G. (1967) 'Pressures developed in cement pastes and mortars by the alkali-aggregate reaction', *HRB Bulletin*, **172**, 34–36.

Sellier A., Bourdarot E., Multon S., Cyr M. and Grimal E. (2009) 'Combination of structural monitoring and laboratory tests for the assessment of AAR-swelling – Application to a gate structure dam', *ACI Materials Journal*, **106** (3), 281–290.

10
Sustainable replacement of aging bridge superstructures using fiber-reinforced polymer (FRP) composites

Y. KITANE, Nagoya University, Japan and A. J. AREF, University at Buffalo – The State University of New York, USA

DOI: 10.1533/9780857097019.2.287

Abstract: This chapter first reviews current structural applications of fiber-reinforced polymer (FRP) composites in bridge structures, and describes advantages of FRP in bridge applications. The chapter then introduces the design of a hybrid FRP–concrete bridge superstructure, which has been developed at the University at Buffalo for the past ten years, and discusses structural performance of the superstructure based on extensive experimental and analytical studies.

Key words: fiber-reinforced polymer (FRP), bridge, superstructure, hybrid structure.

Note: This chapter is an adapted version of Chapter 14 'Fiber-reinforced polymer (FRP) composites for bridge superstructures' by Y. Kitane and A. J. Aref, originally published in *Developments in fiber-reinforced polymer (FRP) composites for civil engineering*, ed. Nasim Uddin, Woodhead Publishing Limited, 2013, ISBN: 978-0-85709-234-2.

10.1 Introduction

In this section, bridge conditions in the United States and advantages of fiber-reinforced polymer (FRP) composites over conventional materials are reviewed.

10.1.1 Bridge conditions in the USA

Civil infrastructure systems play a crucial role in social and economic activities in any society. Considering their importance, it is not surprising that the total investment in civil infrastructure is massive. The Federal Highway Administration (FHWA) estimates that the federal government alone has invested over $1 trillion in the US highway system (Wu, 2005). However, the US is now facing a major challenge to keep the nation's infrastructure systems in usable condition.

In 1967, the collapse of the Silver Bridge turned the bridge engineering community's attentions toward safety of bridges. State Departments of Transportation and FHWA collaborated to establish a systematic evaluation of structural safety, which resulted in the National Bridge Inspection Standards (NBIS) issued in 1971. These standards provide uniform procedures for the collection and maintenance of inventory and inspection data, minimum qualifications for bridge inspection personnel, and standardized methods for evaluating bridge conditions. The data collected by each state are submitted annually to FHWA, and FHWA maintains the data in the National Bridge Inventory (NBI) database. This NBI database contains only data for structures with a span of more than 6.1 m (Dunker and Rabbat, 1995; Small and Cooper, 1998).

A bridge is classified as deficient or not deficient by the inspection. There are two categories for deficient bridges: structurally deficient and functionally obsolete. The former indicates a deficiency in the health of a structure; the latter indicates a deficiency in the performance. Their definitions are as follows: bridges are structurally deficient if they have significant deterioration and have been restricted to light vehicles, and require immediate rehabilitation to remain open, or are closed; and bridges are functionally obsolete if they have deck geometry, load-carrying capacity, clearance, or approach roadway alignment that no longer meet the criteria for the system of which the bridge is an integral part.

The number of deficient bridges is the most common indicator of the overall condition of bridges in the US, and the data for the last 20 years are plotted in Fig. 10.1. As of 2011, nearly 24% of 600 000 public bridges were either structurally deficient or functionally obsolete. Figure 10.2 shows a histogram of ages of US public bridges. The average age is about 40 years and, as can be seen in the figure, a ratio of structurally deficient bridges to the number of bridges increases with age, implying that there will be an increasing need for maintenance work in coming years. According to a study by Brailsford *et al.* (1995), bridge decks are ranked the No. 1 bridge maintenance item by State Department of Transportation (DOT) agencies. Tables 10.1 and 10.2 show the leading bridge maintenance priorities and the typical sources of bridge deterioration, respectively.

Although a quarter of US bridges are classified as deficient, the number of deficient bridges has been decreasing for the past few years, as can be seen in Fig. 10.1. There are a few reasons for this decreasing trend. First of all, the funding for preservation of the highway system has steadily increased to maintain deteriorating bridges. Secondly, the periodic and uniform inspection of bridges has played an important role in this decrease of deficient bridges. Not only does the inspection help State DOTs to locate deficient bridges, but the NBI database helps bridge experts and engineers point out

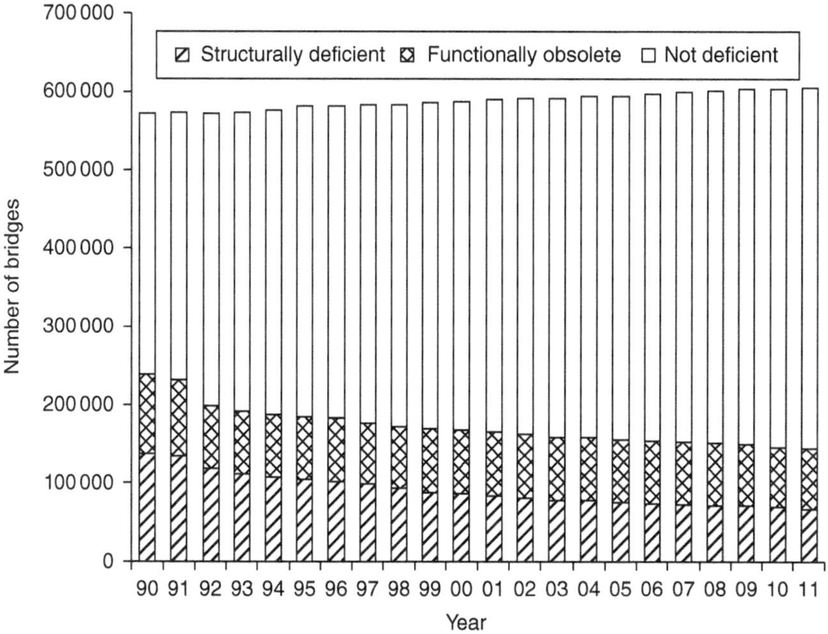

10.1 Number of deficient bridges in the US. (*Source*: National Bridge Inventory.)

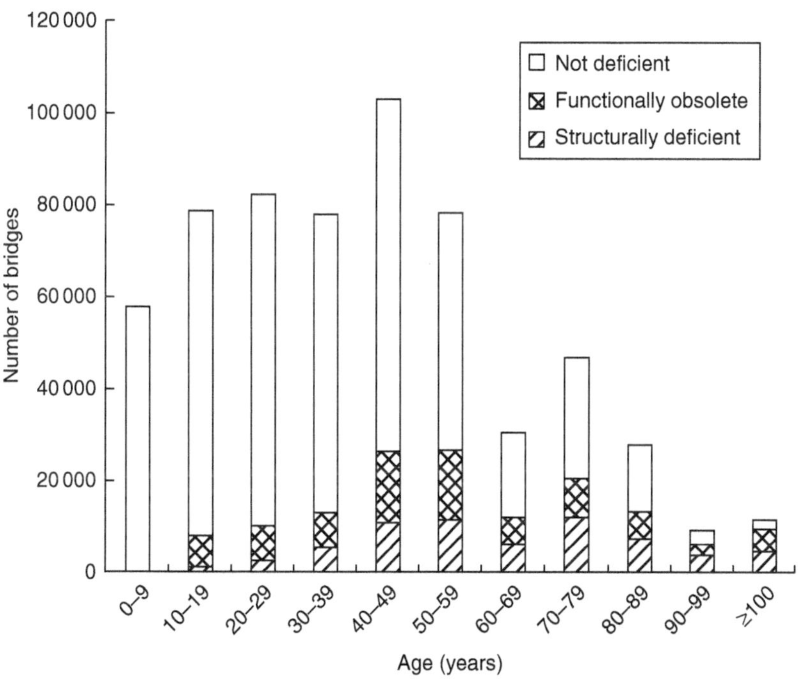

10.2 Age of US bridges.

Table 10.1 Bridge maintenance priorities at State DOTs

Rank	Maintenance items
1	Bridge decking
2	Expansion joints
3	Steel trusses/connections
4	Painting
5	Concrete beams/columns
6	Steel bearings
7	Bridge railings
8	Timber piling
9	Drainage systems
10	Abutments/pier caps
11	Channel protection
12	Impacts
13	Electrical/mechanical

Source: Brailsford *et al.* (1995).

Table 10.2 Typical sources of bridge deterioration

Rank	Deterioration mechanism
1	Road deicing salts
2	Salt-water environment
3	Impact
4	Fatigue
5	Other:
	General aging
	Calcium chloride
	Freeze–thaw
	Alkali–silica reaction of concrete
	Timber decay

Source: Brailsford *et al.* (1995).

critical problems with current bridge design practices. Thirdly, new materials and structural designs that require less maintenance have been developed and implemented in recent years.

However, this decreasing trend in the number of deficient bridges may not continue without better maintenance strategies. A large volume of the bridges were built in the 1960s (the Interstate era), and they will need more maintenance, major rehabilitation, or replacement in the near future. The US DOT also reported that it would require a significant increase in bridge funding in the coming years to fix all current and new deficiencies caused by further deterioration, since the investment that has been committed is only sufficient to preserve the bridge conditions and keep them from getting worse (US Department of Transportation, 2000).

10.1.2 Advantages of FRP composites

It is imperative to build bridge systems that have long-term durability and low maintenance requirements. As mentioned in the previous section, about 11% of US bridges are structurally deficient as of the year 2011, and billions of dollars will have to be spent in maintaining and improving conditions of the deficient bridges. A solution to this challenge may be to use new materials or to implement new structural systems. Among new structural materials, FRP composites have recently gained much attention in the civil engineering community due to their superior material properties, such as high specific stiffness, high specific strength, and high corrosion resistance.

FRP composites were applied in industry for the first time about 80 years ago. Since then, FRP composites have been developed mostly in the defense industry, particularly aerospace and naval applications. In the last two decades, considerable efforts have been made to apply FRP composites in the construction industry, and recently structural applications of FRP composites have started to appear in civil infrastructure systems. State-of-the-art reviews on FRP composites for construction can be found in papers by Karbhari and Zhao (2000) and Bakis *et al.* (2002). Some of the advantages of FRP composites over conventional materials in civil infrastructure applications are: high specific strength and stiffness, corrosion resistance, enhanced fatigue life, tailored properties, ease of installation, lower life-cycle costs, etc.

With these advantages, FRP composites have great potential to be successfully applied in bridge structures. The high corrosion resistance of FRP composites makes them ideal alternative materials to resolve a number of persistent problems that the US highway system is now facing. Moreover, FRP composite structures can be much lighter than those built with conventional structural materials, which leads to reduction of dead load and ease of erection. Some of the research efforts and applications of FRP composites in bridge structures can be found in papers by Zureick *et al.* (1995), Seible *et al.* (1998), Tang and Hooks (2001), and Mufti *et al.* (2002).

10.2 Fiber-reinforced polymer (FRP) applications in bridge structures

Structural applications of FRP composites can be categorized into two large groups: rehabilitation and new construction. In a rehabilitation program, deficiencies of a structure are eliminated so that the structure can meet requirements of design codes as well as social and economic demands, and depending on its objective, the structural rehabilitation can be further divided into three groups: repair, strengthening, and retrofitting. Repairing a structure is to fix deficiencies by using FRP composites, so that the structure can regain its originally designed performance level. Strengthening a structure by applying FRP composites is to enhance the designed performance level. The retrofitting of a structure is to upgrade the seismic capacity of a structure by the use of FRP composites. For new construction, FRP composites can be used as reinforcing bars and tendons in concrete bridges, structural members of pedestrian bridges, deck and superstructure of vehicular bridges, bridge accessories such as inspection decks and manhole covers, etc. FRP applications in bridge structures are reviewed briefly in this section.

10.2.1 Rehabilitation

Repairing or strengthening beams, slabs, or bridge decks with FRP composite laminates is one of the most popular applications of FRP composites in bridge systems. There are many bridges where strengthening of the soffit of beams, slabs, or bridge decks is required. In some cases, concrete beams, slabs, and decks have deteriorated due to corrosion of steel reinforcement and freeze–thaw action. In other cases, structures have to be upgraded to bear higher load levels. Conventionally, either the external post-tensioning or the addition of epoxy-bonded steel plates to the soffit is used for the repair and strengthening of concrete beams, slabs, and decks. Recently, externally-bonded FRP laminates have gained more popularity in this repair and strengthening procedure than the use of steel plates, because FRP is corrosion resistant and very easy to handle.

A number of steel truss bridges in the US need either rehabilitation or replacement. As shown in Table 10.1, steel trusses are ranked No. 3 in the leading bridge maintenance priorities. Because a bridge replacement project can be very costly, a rehabilitation project is usually preferred if it is a viable option. In 2000, rehabilitation of a steel truss bridge was successfully completed by replacing the reinforced concrete deck with an FRP composite deck in the village of Wellsburg, NY (Alampalli and Kunin, 2001). This is the first project using this type of rehabilitation method. In this project, the application of an FRP composite deck reduced dead load of the bridge by 240 metric tons (the dead load became about one-fifth of the original dead load), and resulting in doubling the load ratings. The total cost of the project was $0.8 million, while it would have been $2.2 million to replace the entire bridge. In addition, a significant time reduction for the entire rehabilitation project was demonstrated.

In addition, FRP sheet bonding repair can be applied to repair corrosion-damaged steel bridges. In 2007, corrosion-damaged lower chord members of Asari Bridge, a three-span continuous steel truss bridge in Japan, were successfully repaired by bonding carbon fiber-reinforced polymer (CFRP) strand sheets.

For seismic retrofit, the most popular application of FRP composites is FRP composite jacketing of concrete columns. Recent earthquakes have shown the vulnerability of existing older concrete columns in bridges and buildings. Particularly, reinforced concrete bridge piers, designed before the 1971 San Fernando earthquake, are vulnerable because the transverse reinforcement is inadequate. Lessons learned from the 1971 San Fernando earthquake and the 1989 Loma Prieta earthquake led to the development of a steel jacket system for retrofitting those reinforced concrete (RC) columns with substandard details. Although steel jackets work very well, they are costly because the installation is time-consuming, and they require constant maintenance. To speed up the installation of column jackets and reduce maintenance work, various column-jacketing techniques using FRP composites have been developed and successfully used in seismic retrofit projects.

10.2.2 Reinforcing bars and tendons

FRP reinforcing bars and tendons have been used for new concrete structures. Feasibility studies of FRP reinforcing bars began in the 1950s, and their applications gained much attention to prevent corrosion of steel reinforcing bars in the late 1980s (Bank, 2006). Currently, there is a design guide published by the American Concrete Institute (2006).

FRP composite tendons were implemented in a highway bridge deck in Germany in 1986, for the first time in the world. It has been over two decades since the first application, and the FRP tendons have widely been accepted in bridge construction. Pre- or post-tension tendons and stay cables made of FRP composites have an advantage over steel tendons and cables in their corrosion resistance. For these applications, aramid fiber-reinforced polymer (AFRP) and CFRP composites are often used due to their superior creep resistance. Moreover, FRP tendons and cables are lightweight and much easier to handle than those of steel.

10.2.3 Pedestrian bridges

The first pedestrian FRP composite bridge was built by the Israelis in 1975 (Tang, 1997). The first all-composite pedestrian bridge was installed in 1992 in Aberfeldy, Scotland (American Composites Manufacturers Association, 2004). It is a cable-stayed bridge, with a total length of 113 m. Many others have been built, in Asia, Europe, and North America, since then. The development of pultruded structural shapes has contributed to this increase in the number of FRP composite pedestrian bridges. Advantages of FRP composite pedestrian bridges over those made of conventional materials are light weight, ease in installation, and low maintenance. These advantages have made it possible to easily install pedestrian bridges in areas that are inaccessible by heavy construction equipment and environmentally restrictive. For this reason, many FRP composite pedestrian bridges can be found in State and National Parks in the US.

10.2.4 Decks

As shown in Table 10.1, bridge decks require the most maintenance of all elements in a bridge superstructure because the wearing surface deteriorates or wears out, and the deck system itself deteriorates due to deicing salt and other causes. There has been a need for bridge decks made of new materials that can offer long-term durability and low life-cycle costs. To resolve these issues, FRP composite bridge decks have been developed and installed in many bridges. FRP composite bridge decks have several advantages over conventional bridge decks. First of all, because FRP composites are corrosion resistant, bridge decks made of FRP composites have lower life-cycle costs. Secondly, they are very easy to install due to their light weight. Thirdly, they can be prefabricated. This makes the products uniform in quality and the installation time much shorter. Lastly, the reduction of deck weight is a major benefit to the bridge system. By reducing dead load, the bridge system with an FRP deck can either be used for higher live-load levels or longer spans than a bridge system with a conventional bridge deck system. The use of an FRP bridge deck is reported to reduce the weight of conventional construction by 70–80% (Tang, 1997).

Many deck systems have been developed and tested since the early 1990s. The first US all-composite vehicular-bridge deck was built in 1996 in Russell, Kansas (Tang and Podolny, 1998), where the FRP deck panel is a sandwich construction of composite honeycomb and two composite face sheets. It took only one day to install the bridge deck in Russell County, showing how easy it is to install this type of modular FRP composite bridge deck.

One of the most popular types of FRP composite bridge deck is the one that has a sandwich structure. It consists of a core and top and bottom face sheets, and the core can be foam, honeycomb cells, or made of hexagon and double-trapezoid profiles. Other types of FRP composite decks have also been developed by making use of pultruded structural shapes (see, for example, Hayes *et al.*, 2000).

Based on experiences with different types of FRP decks, maintenance issues peculiar to each type have been found. To have standard guidelines to inspect and evaluate the condition of existing FRP bridge decks, National Cooperative Highway Research Program (NCHRP) conducted a research project and published an inspection manual as NCHRP Report 564 (Telang *et al.*, 2006).

10.2.5 Accessories (such as inspection walkways and manhole covers)

Recently, accessories such as inspection walkways and manhole covers made of FRP composites have begun to be used. Corrosion of conventional steel inspection walkways in the bridges located in corrosive environments is always a problem. The use of FRP inspection walkways requires much less maintenance in such severely corrosive environments than conventional ones. In addition, the light weight of FRP composites makes installation work of FRP inspection walkways much simpler, without any heavy equipment, and FRP manhole covers much easier to handle.

10.2.6 Superstructure

With the knowledge obtained from constructions of pedestrian bridges and bridge decks, many researchers and engineers have been trying to develop cost-effective all-FRP composite vehicular bridges. Advantages of FRP composite vehicular bridges over bridges of conventional materials include corrosion resistance, ease of installation, reduction of construction period, less maintenance requirements, less dead load, a capability to have a longer span, and lower life-cycle costs.

Early examples of all-FRP vehicular bridges include Laurel Lick Bridge, Tom's Creek Bridge, Tech 21 Bridge, Kings Stormwater Channel Bridge.

- The Laurel Lick Bridge, with a span of 6.1 m, was built in 1997. The bridge has modular glass fiber-reinforced polymer (GFRP) decks on pultruded GFRP wide flange beams (Creative Pultrusions, Inc., 2002).
- The Tom's Creek Bridge, with a span of 5.33 m, was built in 1997. The bridge utilized pultruded hybrid FRP double-web beams as superstructure, and the FRP beam was composed of E-glass and carbon fibers in a vinylester matrix (Neely *et al.*, 2004).
- The Tech 21 Bridge, with a length of 10.1 m, was built in 1997. The bridge has GFRP decks of sandwich construction on three GFRP beams with trapezoidal cross-section (Farheyl, 2005). The GFRP beams were composed of E-glass fibers and polyester resin.
- The Kings Stormwater Channel Bridge, with a length of 20.1 m, was built in 2000. The bridge has modular GFRP decks supported by concrete-filled filament-wound CFRP tubes (Seible *et al.*, 1999).

In spite of various advantages, there are several developments required to realize a successful design of a cost-effective all-FRP composite vehicular bridge. Some of the areas needing more developments are:

- Although FRP composites have high specific strength, the stiffness has controlled the design rather than strength in almost all the demonstration FRP vehicular-bridge projects where GFRP is used. It has been reported that the maximum stress level rarely exceeds 10% of the ultimate strength of the materials in these bridges (Karbhari and Zhao, 2000). To use the strength of the materials efficiently, more improvements and developments are needed on FRP composite bridge systems. Examples are structural optimization of the cross-sectional geometry, hybrid use of glass and carbon fiber, and hybrid use of FRP and conventional materials.
- FRP composites are susceptible to stress concentration due to their anisotropic material properties. Because FRP composites are not as ductile as steel, catastrophic failure may occur once the stress reaches the ultimate stress level. Therefore, better ways to connect FRP members have to be developed in order to avoid these high stress concentrations. Some connection details of FRP members developed in the aerospace industry should be examined and transferred for bridge applications.
- Pultruded FRP composite structural shapes should be standardized. This will lead to achieving the uniform quality of structural components and reducing the initial costs. In addition, the standardization of structural shapes is necessary to facilitate the development of design guidelines for FRP composite bridges.
- Environmental durability of FRP composite materials has to be thoroughly investigated and well understood. The experimental and field data on the long-tem durability of FRP composites will facilitate the development of a design philosophy based on life-cycle costs. In addition, long-term performance data of FRP bridges should be accumulated so that proper maintenance procedures can be standardized.
- Initial costs of FRP composite bridges are still too expensive to compete with bridges of other conventional materials. Standardization of structural shapes and design with efficient use of materials are required to reduce the initial costs.

10.3 Hybrid fiber-reinforced polymer (FRP)–concrete bridge superstructures

This section introduces the design concept of a hybrid FRP–concrete bridge superstructure and discusses its structural performance based on experimental and analytical studies.

10.3.1 Hybrid FRP–concrete structural system

To resolve some issues that all-FRP composite bridges have, combinations of FRP and conventional materials have recently been investigated by a number of researchers. A good review on the combined or hybrid construction can be found in the paper by Mirmiran (2001). The advantages of the hybrid structural systems include the cost effectiveness and the ability to optimize the cross-section based on material properties of each component. According to Mirmiran (2001), the most effective use of FRP composites is in the form of hybrid construction with concrete, where FRP acts as a load-carrying constituent and a protective measure for concrete.

The innovative idea of a hybrid FRP–concrete structural system for flexural members was first proposed by Hillman and Murray (1990). They proposed the combination of pultruded FRP sections and concrete to form lightweight decks in order to reduce the dead load in steel frame buildings. The FRP section was designed to serve as both reinforcement and permanent formwork. Most of the concrete was located above the neutral axis of the hybrid section. The study concluded that the weight reduction would be more than 50% when compared with a common type of concrete slab system.

Bakeri and Sunder (1990) investigated hybrid FRP–concrete bridge deck systems. They proposed a deck system of a simply curved membrane of FRP composites filled with concrete that was intended to resist the compressive force. A finite element analysis was performed to evaluate the mechanical performance of the deck system under an HS20-44 truck loading. They concluded that the hybrid FRP–concrete system was promising, particularly from the viewpoint of cost.

Saiidi *et al.* (1994) conducted experimental and analytical studies on composite beams that consist of CFRP structural sections and reinforced concrete slabs. Although in their study an epoxy resin was used to provide the bond between the concrete deck and the CFRP sections, they concluded that the use of epoxy resin to bond concrete to CFRP was only partially efficient, and that mechanical connectors or other reliable means would be needed to develop a good composite action between CFRP elements and a concrete slab.

Deskovic *et al.* (1995a) presented an innovative design of a hybrid FRP–concrete beam. The proposed beam consists of a filament-wound GFRP box section combined with a layer of concrete in the compression zone and a thin CFRP laminate in the tension zone. Three-point bending test results confirmed that the proposed design was feasible for producing an efficient and cost-effective hybrid system. They also studied long-term behavior (creep and fatigue) of the proposed hybrid section experimentally and analytically

(Deskovic *et al.*, 1995b). Their analytical model agreed closely with experimental results. They concluded that the proposed hybrid section had very good time-dependent response characteristics.

Seible *et al.* (1998) designed a two-span, 210 m long, two-lane highway bridge with lightweight concrete-filled circular CFRP composite tubes. Their study showed that the design is stiffness-driven for this modular beam and slab bridge system, and that significant strength reserve remains in the carbon shell. Their preliminary estimates indicate that two different bridge systems, the concrete-filled CFRP beams with RC deck and the concrete-filled CFRP beams with pultruded modular E-glass deck, are 20% and 100% more expensive, respectively, when compared to a conventional RC slab bridge.

Ribeiro *et al.* (2001) investigated the flexural performance of hybrid FRP–concrete beams. In their design, GFRP pultruded channel profiles were assembled with a layer of concrete. Four types of hybrid rectangular beams were tested in the four-point bending. Two of them had all the section filled with concrete, while the other two had concrete only in the upper part of the profile. To compare the flexural performance of each assembly, they used two indices: flexural specific rigidity and synergistic effect. Flexural specific rigidity is a ratio between the flexural rigidity and the specific weight, while synergistic effect is a ratio of the ultimate load of the assembly to the sum of the ultimate loads of its two elements (GFRP channel and concrete). Test results showed that the flexural specific rigidity the inserted for the specimen that has more GFRP in the tension side and concrete only in the compression side. The synergistic effect was the highest for the specimen that did not exhibit a bond failure at the FRP–concrete interface. No special measure to provide the bond at the interface was taken in any specimen.

Fam and Rizkalla (2002) studied flexural behavior of concrete-filled FRP circular tubes experimentally. This hybrid system was developed as an excellent alternative for structural components subjected to aggressive corrosive environments. The FRP tube provides a lightweight permanent formwork for fresh concrete and acts as a non-corrosive reinforcement. The concrete core has two main functions: (1) to provide internal support to the tube and, consequently, prevent the local buckling of the FRP tube; and (2) to provide the internal resistance force in the compression zone and increase the strength and stiffness of the member. A total of 20 beams including steel tubes were tested in four-point bending. GFRP tubes were either filament-wound or pultruded. Some of their conclusions were: (1) the higher the stiffness of the hollow tube, the lower the gain in flexural strength and stiffness resulting from concrete filling; (2)

although concrete-filled pultruded GFRP tubes showed higher stiffness than concrete-filled filament-wound GFRP tubes of the same thickness, they failed prematurely by horizontal shear due to the lack of fibers in the hoop direction; (3) concrete-filled FRP tubes with thicker walls or a higher percentage of fibers in the axial direction tended to fail in compression; (4) a higher flexural strength-to-weight ratio was achieved by providing a central hole in the core; (5) a shear transfer mechanism is necessary at the interface between the concrete core and the GFRP tube in flexural members; and (6) experimental results did not show a significant effect of the confinement of concrete on flexural strength, while ductility was improved.

Van Erp and his group (2002, 2005) have been developing hybrid FRP–concrete beams for a bridge application. The proposed beam is very similar to the one proposed by Deskovic *et al.* (1995a). The beam consists of two parts: concrete and FRP box section. The FRP box section is made of GFRP, and additional CFRP is added to the tensile flange of the box section to increase the stiffness of the beam. A layer of concrete is bonded on the GFRP box section by a high-quality epoxy adhesive. The weight of the hybrid beam is claimed to be about one-third that of a reinforced concrete beam. The top GFRP flange was designed to sustain more compression force after concrete is crushed. Concrete crushing is a warning of failure in this design. A full-scale bridge superstructure with a span of 10 m and a width of 5 m was designed and built with these hybrid beams. The beams were assembled using a high-quality epoxy adhesive to make up the bridge. The bridge was installed in Toowoomba, Australia, in January 2002. Field test results demonstrated high load-carrying capacity, excellent fatigue behavior, outstanding durability, and ability to carry high concentrated loads.

10.3.2 Design concept of the hybrid FRP–concrete bridge superstructure

By utilizing the concept of the hybrid FRP–concrete structural system, an FRP bridge superstructure has been developed at the University at Buffalo for the past ten years. Although many different configurations of hybrid FRP–concrete bridge superstructure are possible, the structural configuration shown in Fig. 10.3a, where supports are not shown, was chosen based on numerous finite element analyses leading to the particular geometrical parameters. In combination, glass fiber reinforcement, and vinyl ester matrix were selected as constituents for the GFRP composites, because glass fiber reinforcement is much cheaper than carbon or

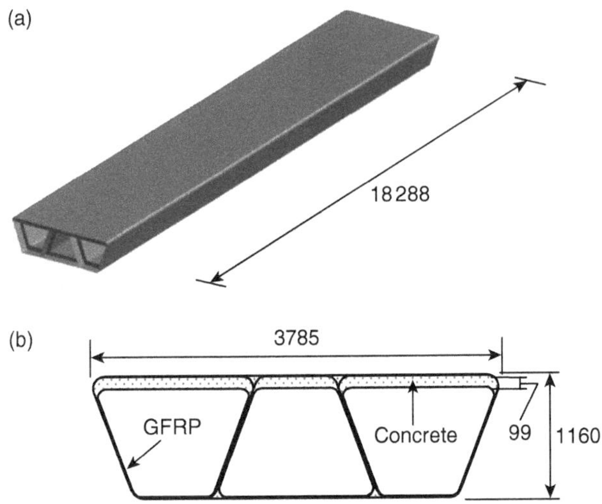

10.3 Hybrid FRP–concrete bridge superstructure (dimensions in mm). (a) Overall configuration; (b) cross-section.

aramid fiber reinforcement, and vinyl ester has high resistance to corrosion. Figure 10.3b shows a cross-section of the proposed bridge superstructure. In this figure, parts in black are made of GFRP laminates, while the shaded portion is filled with concrete. In this design, three trapezoidal box sections are bonded together to form a one-lane superstructure. A trapezoidal shape was chosen so that forces could transfer between box sections efficiently. After being assembled, these trapezoidal sections are wrapped with a GFRP laminate, which ensures that three sections act together to produce an integral bridge superstructure. Advantages of this bridge superstructure are summarized as follows:

- GFRP is corrosion resistant, and concrete is not exposed to environmental conditions; therefore, the system is resistant to corrosion, and it requires less maintenance than conventional bridges.
- Concrete has high strength- and stiffness-to-cost ratio; thus, by using concrete efficiently, the total amount of FRP can be reduced, which leads to the reduction of initial costs.
- Concrete is designed to be always under compression in the longitudinal direction. The fact that concrete is not used in the tension side leads to significant weight reduction when compared to a concrete-filled FRP tube design.
- It has been reported that the local deformation under a loading point may become large for all-composite bridge decks (Bakeri and Sunder,

1990; Aref, 1997). A layer of concrete can reduce this local deformation of the top flange.

- Most parts can be fabricated in the factory; therefore, good quality control can be assured and the construction period can be shortened.

10.3.3 Design features

As the first trial design, a prototype bridge was designed as a simply supported single-span, one-lane bridge with a span of 18.3 m. Design parameters were determined to meet a set of design criteria by applying AASHTO design loads.

Design philosophy and assumptions

The design philosophy and assumptions used in the design process are as follows:

- In flexure, concrete should fail in compression first before GFRP laminates fail in either compression or tension.
- Under the service limit state condition, concrete does not crack due to the bending in the transverse direction.
- Under the strength limit condition, the top flange of the box section should resist the tire pressure with GFRP laminates only for the bending in the transverse direction if the concrete is already cracked.
- A perfect bonding between concrete and GFRP laminates is assumed.
- Confinement effect on the concrete strength is neglected.
- A knock-down factor or strength reduction factor for the GFRP is taken as 0.4. Seible *et al.* (1995) used 0.5 and Dumlao *et al.* (1996) used 0.25 in their designs. The value of 0.4 was chosen simply as a value between 0.25 and 0.5.
- Long-term degradation factor for FRP composites is not considered because the data are not available for the particular material used in this study. Typical values of degradation factors can be found in various sources. For example, see *ICE Design and Practice Guides* (Moy, 2001).

Box section

Box sections were chosen to be a basic structural form of the proposed design. According to Ashby (1991), thin-walled box sections are the most efficient structural forms for beams. However, the thin-walled box sections made of GFRP laminates have some disadvantages (Deskovic *et al.*, 1995a), which include:

- the compressive flange is considerably weaker than the tensile flange due to the local buckling failure;
- the failure of GFRP usually occurs in a catastrophic manner without giving much warning because the stress–strain curve of GFRP in the fiber direction does not show plastic deformation as much as conventional materials such as steel and concrete do;
- the design of a GFRP beam is usually governed by stiffness instead of strength; this is often resolved by introducing more material in the beam; therefore, the design tends to be uneconomical and the strength of the material is not used efficiently.

To overcome these disadvantages, a thin layer of concrete was decided to be placed in the compression zone of the section as shown in Fig. 10.3b. The section was designed in such a way that concrete is surrounded with GFRP, and concrete will be protected from environmental exposure.

Web inclination

Another feature of the design is to have trapezoidal box sections, although it is more common to have a square or rectangular shape for a box section. When many rectangular box sections are put together to comprise a superstructure, the force will be transferred between the adjacent sections mainly through shear and bending. However, FRP composites usually have very low shear stiffness when compared to their normal stiffness in the fiber directions. By having the webs inclined, the structural shape of the cross-section becomes more of a truss, and a greater portion of the applied load will be carried by axial force in FRP laminates within the cross-section.

The inclination angle was determined based on finite element analyses of hybrid bridge superstructures with several different inclination angles subjected to live loads. The inclination of 3/8 (=21°) was chosen because that yielded the minimum deformation at the riding surface.

Stacking sequences

In the design process, it was assumed that the GFRP parts would be fabricated by a hand layup process. The inner trapezoidal tube is fabricated first, and the outer tube laminate is laid up onto the inner tube. After a certain number of the trapezoidal tubes are put together, they are wrapped with the outermost laminate. These three different laminates are shown in Fig. 10.4. Considering that the hand layup process is time-consuming, it is best to make the stacking sequences of these three laminates as simple as

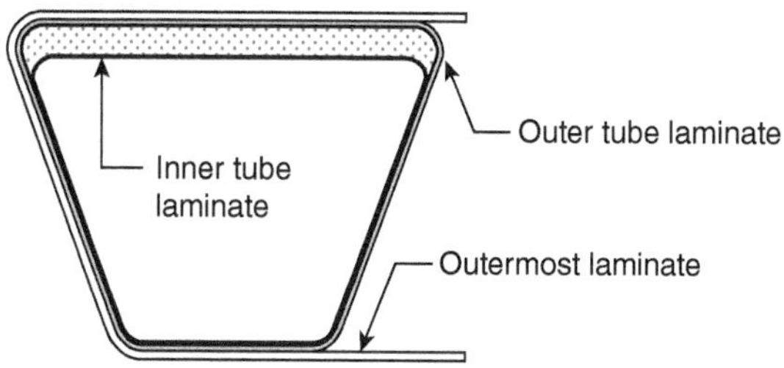

10.4 Three different laminates in the superstructure.

possible. A woven fabric was chosen as reinforcement; therefore, a layer of fabric has reinforcement fibers in the two orthogonal directions. The stacking sequence chosen for the outer tube and inner tube laminates was all 0°, while for the outer tube laminate, ±45° and 0° laminae were used for 67% and 33% of the thickness, respectively. The direction is measured from the longitudinal direction of the bridge. Because the reinforcement type is of a woven fabric, a 0° lamina has fibers in the 0° and 90° directions. By introducing ±45° laminae in the outer tube laminate, the shear stiffness of the hybrid bridge can be increased as the outer tube laminates make a large portion of the interior webs.

Thickness of the concrete layer

The thickness of the concrete layer is a key design parameter to optimize the hybrid FRP–concrete structural system because the increase in the thickness of concrete leads not only to a reduction in initial costs by reducing GFRP composites but also to an increase in the total weight. By examining flexural rigidity, it was found that the flexural rigidity rapidly increases with the concrete thickness until the concrete thickness is about 10% of the superstructure depth. After that point, additional concrete may not be used efficiently to increase the flexural rigidity, although the cracking and ultimate moments will still get benefits from the increase in the concrete thickness. Therefore, the maximum thickness of concrete that will increase flexural rigidity efficiently without adding too much weight was determined as 10% of the bridge depth for the proposed hybrid bridge. In the final design, the thickness was chosen as 99 mm, which is 8.5% of the total depth of the bridge. Thus, the ratio of the area of concrete to the total area in the cross-section comes to 0.3.

Shear keys

To have good composite action between GFRP laminates and concrete, GFRP shear keys, as shown in Fig. 10.5, were designed. They are to be

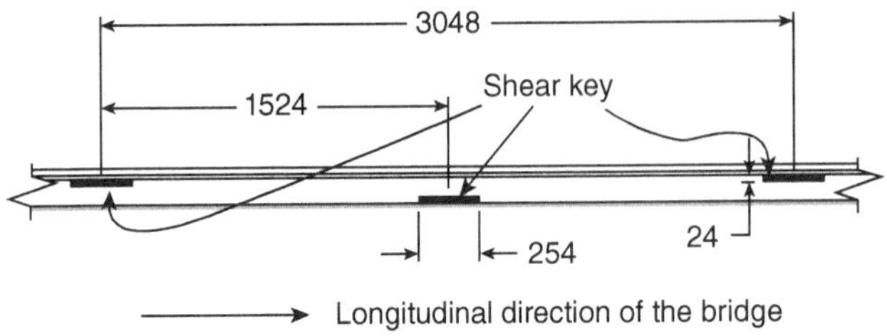

10.5 Shear keys (dimensions in mm).

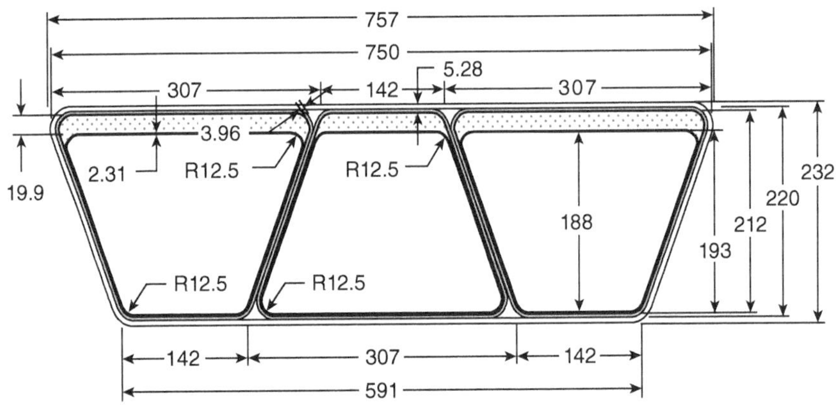

10.6 Cross-section of the test specimen (dimensions in mm).

installed in staggered positions on the top and bottom laminates with a required interval.

10.3.4 Experimental study

To verify the feasibility of the proposed hybrid FRP–concrete bridge superstructure by examining the performance under live loads, a series of quasi-static loading tests and a fatigue loading test were performed on a test specimen of the proposed bridge superstructure.

Test specimen

The test specimen is a one-fifth scale model of the 18.3 m hybrid FRP–concrete bridge superstructure shown in Fig. 10.3. Its cross-section is shown in Fig. 10.6. Each trapezoidal box section was fabricated individually by the hand layup process. Each section consists of two layers of laminates: the inner tube laminate and the outer tube laminate. The inner tube laminate was first constructed with a $[0^{\circ}_{7}]$ laminate construction, and the outer tube

Table 10.3 GFRP material properties

Test type	Direction	Modulus of elasticity (GPa)	Strength (MPa)	Ultimate strain
Tension	Fill	16.6	285	0.0218
	Warp	17.9	335	0.0235
	Average	17.3	310	0.0227
Compression	Fill	15.9	241	0.0177
	Warp	22.5	265	0.0158
	Average	19.2	253	0.0168
Shear	Fill	2.72	56.1	–
	Warp	2.45	63.8	–
	Average	2.59	60.0	–

laminate was constructed over the inner tube laminate with a $[(\pm 45)_4 0°_4]$ laminate construction. At the same time, a cavity for concrete was created at the top of each section. Then, three trapezoidal sections were assembled together. A layer of glass fiber chopped strand mat wetted with vinyl ester resin was applied between box sections as a bonding material. After being assembled, the three sections were wrapped with the outermost laminate, whose stacking sequence is $[0°_{16}]$.

Materials

E-glass woven fabric was chosen as reinforcement of FRP, and vinylester resin was chosen as the matrix. To determine the mechanical properties of an FRP laminate made of E-glass woven fabric and vinyl ester resin, tensile, compressive, and in-plane shear tests were conducted according to ASTM D3039-76, ASTM D3410-75, and ASTM D4255/D4255M-83, respectively. Tensile and shear coupons were cut from a 10-layer laminate, while compressive coupons were cut from a 20-layer laminate. These laminates were made by a hand layup process. Fiber volume fractions of the 10- and 20-layer laminates were 0.295 and 0.355, respectively.

The material properties based on testing are shown in Table 10.3. Modulus of elasticity and Poisson's ratio were determined for a strain range of 0.001 to 0.003, –0.005 to –0.007, and 0.001 to 0.005, for tensile, compressive, and shear tests, respectively. Stress–strain relationships of tension and compression were slightly nonlinear, while that of shear was highly nonlinear. Ultimate shear strain was not obtained because it was beyond the capacity of the strain gages used in the test.

As cavities for concrete in the test specimen were small, coarse aggregate was not used. The maximum aggregate size was 4.75 mm, and Type I Portland cement was used. The weight proportions of constituent materials in the concrete mix were: 9.4% water; 20% cement; 70% fine aggregate; 0.11% superplasticizer; and

0.39% shrinkage reducer. Four cylindrical specimens of 152.4 mm in diameter and 304.8 mm in length were prepared according to ASTM C192/C192M-98. Specimens were moist-cured until an age of 28 days after the mixing of concrete. Compressive test of cylindrical specimens was performed until an age of 28 days' cure according to ASTM C39-96. The obtained compressive Young's modulus and strength were 8.38 GPa and 37.9 MPa, respectively. The Young's modulus is about one-third that of normal con-crete, due to the lack of coarse aggregate.

Test set-up

Figure 10.7 shows the test set-up. Loads are applied vertically to the top surface of the test specimen by the actuator hanging from the top beam of a reaction frame. The specimen is instrumented with potentiometers and strain gages at various locations to measure displacement and strain, respectively. Spreader beams spread the load from the actuator to four contact points. The load configuration simulates the tandem load specified in the *AASHTO LRFD Bridge Design Specifications* (AASHTO, 1998). The design tandem load is a live load that has two axles of 110 kN, where one axle is 1200 mm away from the other. Each axle has two tires that are 1800 mm apart center-to-center, and each tire area is 510 mm long and 250 mm wide. For the one-fifth scale model, this design tandem load becomes two axles of 4.4 kN, 240 mm apart. Two tires of each axle are 360 mm apart and each tire area is 102 mm long and 50 mm wide.

The test specimen is supported by concrete blocks at its two ends. To protect the bottom surface of the specimen from damage and to allow rotation at the supports, elastomeric bearing pads are placed on the concrete blocks,

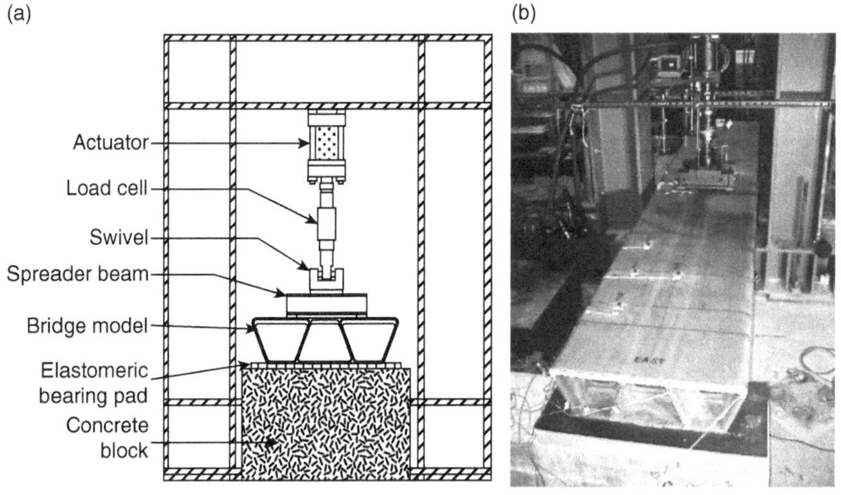

10.7 Test set-up. (a) Schematic drawing; (b) photo of test set-up.

and the specimen sits on the pads. Each elastomeric bearing pad is 13 mm thick and made of neoprene.

Test procedures

A series of non-destructive tests were first performed on the test specimen to obtain its structural characteristics in the elastic range under flexural loading. Then, the capacity of the specimen was investigated by destructive tests. Loading configuration for the static and fatigue loading is shown in Fig. 10.8. Displacement and strain measurement locations during the experiment are shown in Fig. 10.9.

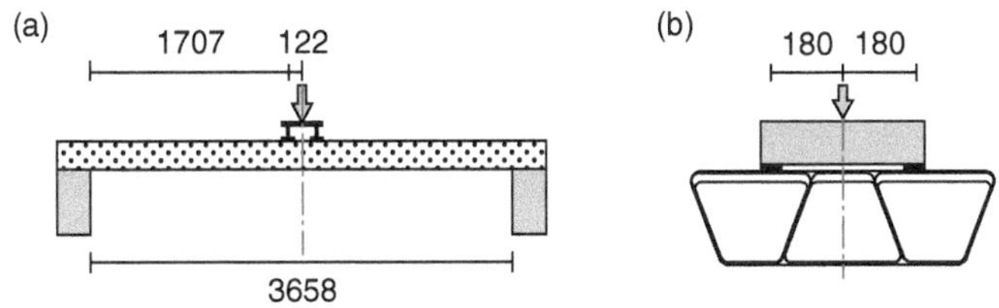

10.8 Loading configuration (dimensions in mm). (a) Elevation; (b) cross-section.

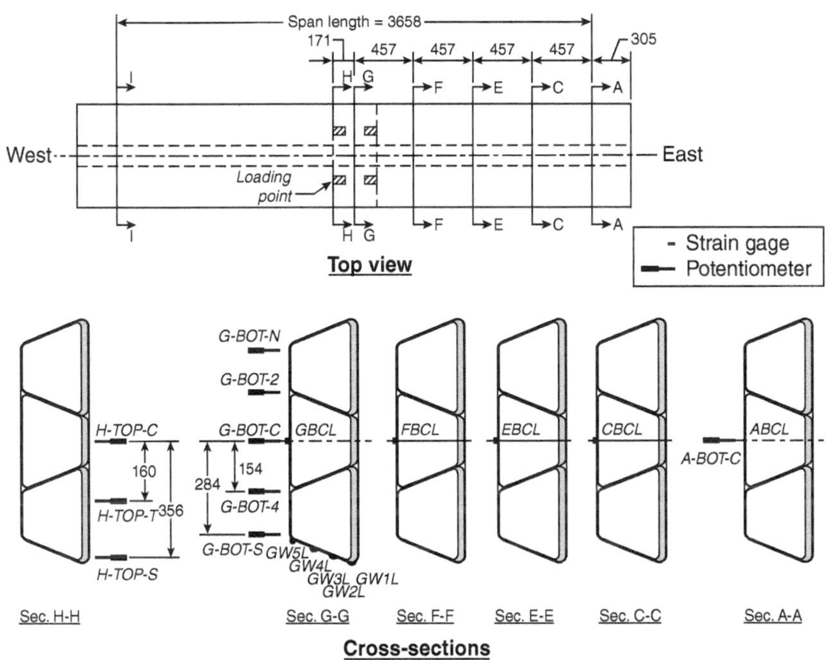

10.9 Instrumentation layout and displacement measurement (dimensions in mm).

Before concrete was cast into the specimen, the FRP-only specimen was tested in flexure in order to examine the behavior without concrete. The test was performed by displacement control, and the maximum displace-ment applied was $L/480$ (7.6 mm), where L is a span length. The hybrid FRP–concrete bridge specimen was then tested in flexure up to the maximum displacement of $L/480$. Then the test specimen was subjected to 2×10^6 load cycles in flexure. This fatigue test was performed by force control with a load range of 0–17.6 kN and a frequency of 3.0 Hz. The maximum load of 17.6 kN is equivalent to twice the design tandem load for this test specimen. The test specimen was subjected to a static flexural loading every 2×10^5 cycles to obtain its stiffness and examine stiffness degradation.

After the fatigue test, the bridge model was tested in flexure to failure to examine its residual strength and failure modes. This test was performed by displacement control, and it was divided into two steps. In the first step (Step I), a cyclic displacement profile was used with the amplitude gradually increased. For each cycle, the displacement ranged from 0 to a predetermined amplitude, and three cycles were applied for each amplitude.

Although it was planned to repeat this process until failure of the bridge model, the test was stopped at the maximum displacement of 77.7 mm ($17 \times L/800$) because the maximum force became close to the capacity of the load cell. By using a load cell with a higher capacity, the second step (Step II) of the test was performed. In this step, displacement was increased monotonically until the test specimen failed.

Test results

No sound of cracking of either the concrete or GFRP was heard during the non-destructive flexural loading where the applied maximum displacement is $L/480$. The obtained force–displacement responses were very much linear, as shown in Fig. 10.10. Stiffness of the hybrid bridge specimen was 19% higher than that of the FRP-only specimen, which shows the effectiveness of concrete. For the prototype bridge, the stiffness increase resulting from the inclusion of concrete is expected to be as much as 40%, because Young's modulus of normal concrete to be used in the prototype bridge will be about three times as high as that of the concrete used in the test specimen. Figure 10.11 shows the deformed shapes of the top and bottom surfaces, respectively. It can be observed from the figure that the three box shapes deformed as a unit and there was not significant local deformation, even near the loading points.

It is recommended in the *AASHTO LRFD Bridge Design Specifications* that the maximum deflection under live loads be smaller than $L/800$ (L = span length). The live load for this check is $(1 + \text{IM}) \times$ truck load, where IM = dynamic load allowance, and has a value of 0.33 in this case. In this

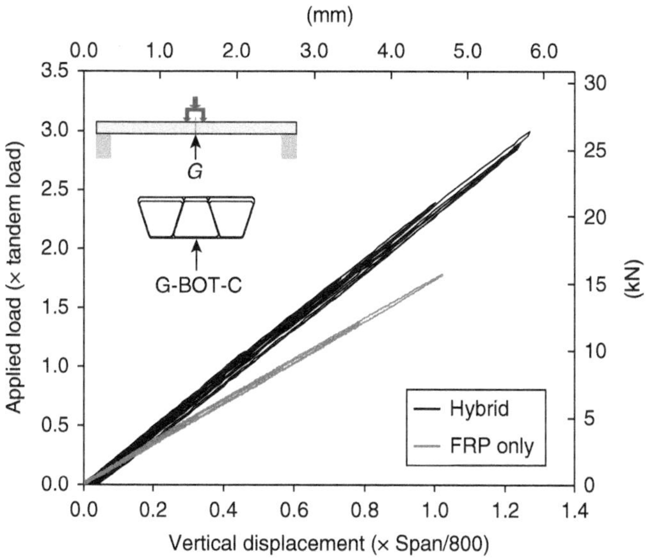

10.10 Force–displacement relationship at G-BOT-C.

study, the condition was checked with the tandem load instead of the truck load. As the span length of the test specimen is 3.66 m, the maximum deflection under (1 + IM) × tandem load has to be smaller than $L/800 = 4.6$ mm.

The obtained deflection of the hybrid specimen due to (1 + IM) × tandem load was 0.547 × $L/800$. The hybrid specimen well satisfied the AASHTO live-load deflection recommendation.

Variations of longitudinal strains over the height of the superstructure on the exterior web at the midspan of the specimen are shown in Fig. 10.12. Longitudinal strains on the web varied linearly, implying that the bending theory for small strain can apply, i.e., the plane section before deformation remains plane after deformation.

Figure 10.13 shows stiffness degradation over 2×10^6 cycles in the fatigue test. The stiffness degradation is defined as a ratio of the stiffness measured after a certain number of load cycles to the stiffness measured in the last cycle of the non-destructive flexural test. Although the trend of stiffness degradation is not very clear in this figure, it can be concluded that stiffness degradation was insignificant. The degradation was 5.9% after two million cycles.

Figure 10.14 shows force–displacement responses obtained at H-TOP-T and G-BOT-C during the destructive flexural test (Steps I and II). At the load of 19.1 × tandem load in Step I, there was a loud cracking sound from concrete under the loading points. This is a local failure mode. The failure load was 8.2 times the AASHTO requirement of 1.75 (1 + IM) × tandem

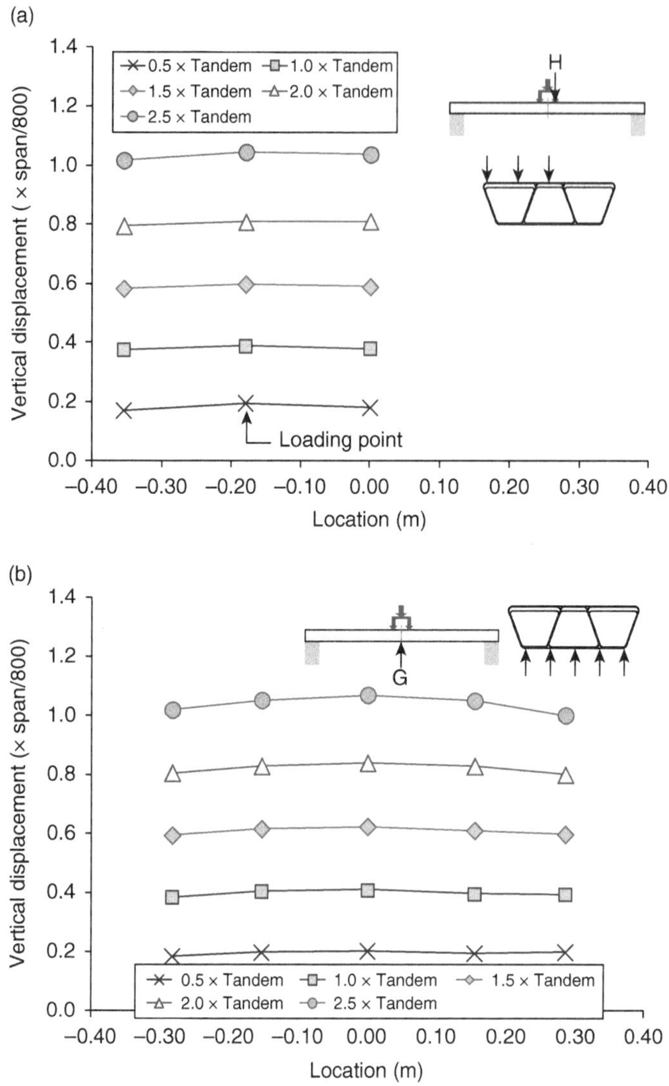

10.11 Deformed shapes of (a) top surface at Section H; (b) bottom surface at Section G.

load for live loads in the Strength Limit I State. After concrete cracked under the loading points, the bottom flanges of the spreader beams came in contact with the top surface of the specimen, and the loading configuration became two line loads instead of four-point loads. Therefore, further local deformation was not observed at H-TOP-T.

The load path from Step II shows that the reloading path is very much linear until the maximum load experienced before. A global failure occurred

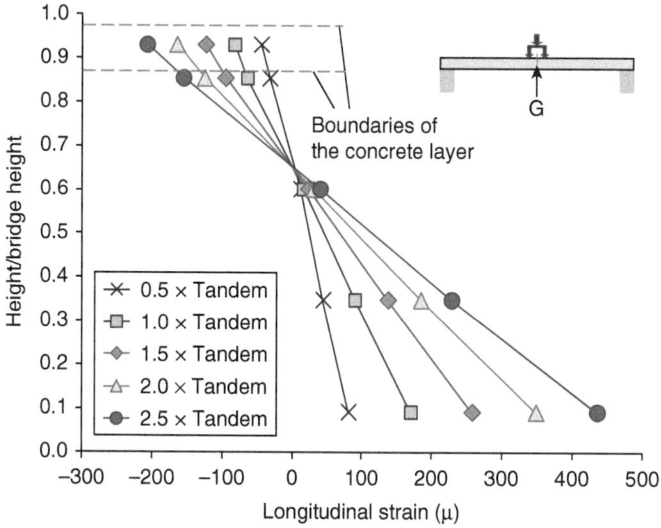

10.12 Longitudinal strain on the exterior web at Section G.

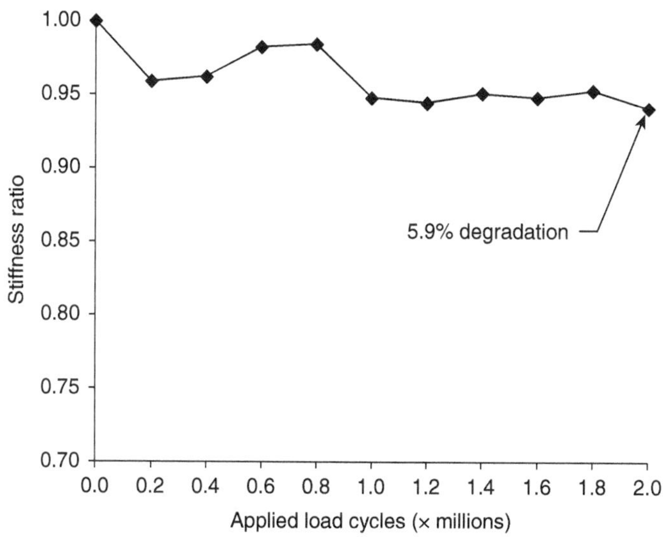

10.13 Stiffness degradation obtained from fatigue test.

at 35 × tandem load, which is 15 times the 1.75 (1 + IM) × tandem load. A sequence of failure leading to global failure can be described as follows. Concrete failed in compression first, then the GFRP flanges took over the compressive force that had originally been carried by concrete. GFRP compression flanges failed in compression when the compressive stress

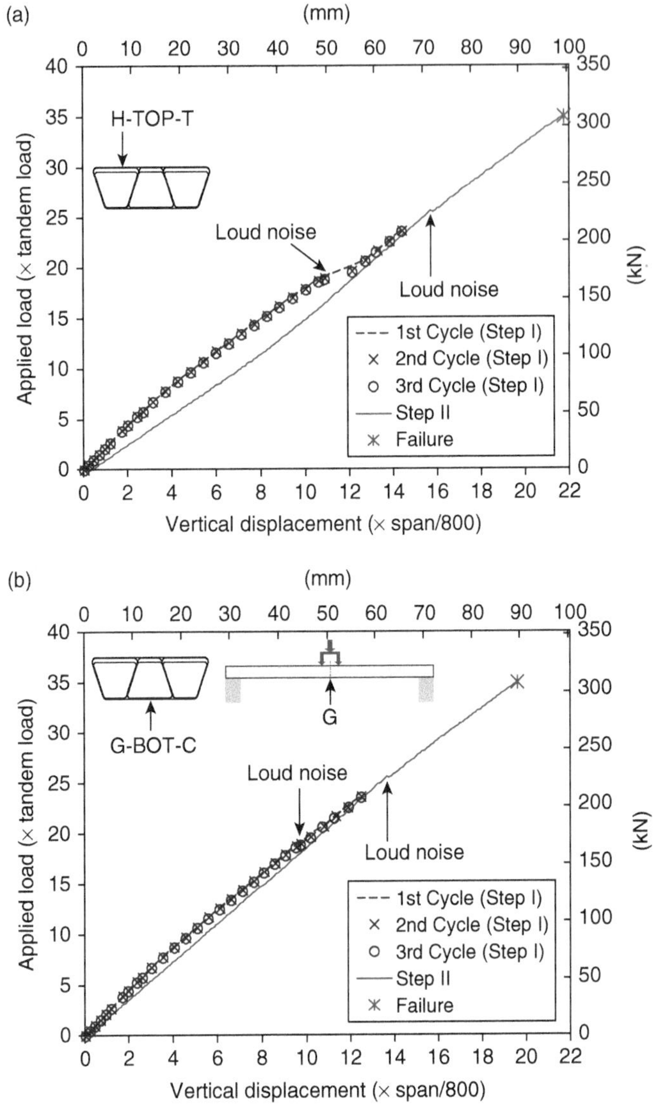

10.14 Force–displacement relationships in the destructive flexural test. (a) Top surface at Section H; (b) bottom surface at Section G.

reaches the compressive strength, followed by failure of the webs in compression. The bottom GFRP flange was found to be intact. Although the failure was sudden, due to the nature of GFRP composites, the obtained global failure mode can be considered to be favorable because it did not lead to collapse of the entire bridge. At the failure section, crushing of concrete, compressive failure of the GFRP top flange, buckling failure of the

GFRP interior flange, and significant delamination of the GFRP lami-nates were observed.

After the strength test, the test specimen was cut at a few sections that were away from the damaged section (Section H) to investigate the interface between the GFRP and concrete. Visual inspection did not find any trace of slippage between the GFRP and concrete.

The experimental study demonstrated excellent performance of the proposed hybrid FRP–concrete bridge superstructure. As is often the case with many GFRP composite bridges, stiffness governs the design of the proposed bridge superstructure. Clearly, the proposed design can be designed to meet the AASHTO live-load deflection recommendation. Also, the fatigue test revealed that the stiffness degradation was not significant over 2×10^6 cycles of the load equivalent to twice the tandem load. However, the test results showed that the proposed design has much higher strength than is required, i.e., the design was conservative. The depth of the bridge may be increased so that the same flexural rigidity can be achieved with less material and the strength can be used more efficiently.

10.3.5 Structural analysis

In this section, detailed finite element analysis (FEA) of the hybrid FRP–concrete bridge superstructure and simple methods of analysis are introduced, and their applicability for the proposed hybrid superstructure is discussed.

Linear finite element analysis

For the detailed finite element analyses of the hybrid FRP–concrete bridge superstructure, the general purpose commercial FEA software, ABAQUS ver. 6.1 (Hibbit, Karlsson & Sorensen, Inc., 2000), was used. A four-noded general shell element was used for GFRP laminates, while a general 3D eight-noded solid element was used for concrete. Figure 10.15 shows the finite element model for the one-lane hybrid FRP–concrete bridge superstructure. The longitudinal, transverse, and vertical directions are referred to as the x, y, and z directions, respectively. In this model, the total number of elements is 38°892 (22°764 shell elements and 16°128 solid elements), and the total number of nodes is 31°857.

The FEA model was supported by a line at each end. Boundary conditions were imposed to two lines of nodes of the bottom surface. Nodes at $y = 0$ and $z = 0$ were restrained in the y and z directions, and nodes at $y = L$ and $z = 0$ were restrained in the z direction. In addition, nodes at $x = 0$, $z = 0$, and $y = 0$ and L were restrained in the x direction, too.

Material properties of GFRP and concrete obtained from the experimental study presented in Section 10.3.4 were used in the FEA. Although the

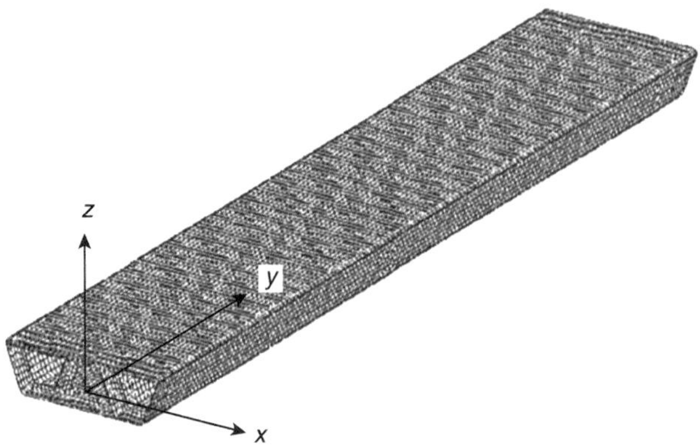

10.15 Finite element mesh.

obtained tensile and compressive properties are different to some degree, they are assumed to be the same in the analysis. The experimental study showed that GFRP almost behaves as a linear-elastic material until failure, and that concrete cracking or compressive failure occurs at a much higher load level than the design load level. Therefore, only the linearly elastic behavior of the hybrid bridge superstructure is examined in this section, and only elastic properties are used in the FEA. A perfect bonding between concrete and GFRP laminates was assumed in the analysis.

Simple methods of analysis

Because a detailed finite element analyses for every trial design is very time-consuming and not cost-effective to select the bridge design parameters in a preliminary design phase, it is necessary to have a simple and quick method to analyze the bridge behavior under various loading conditions. The beam and orthotropic plate analyses are presented here as the simple methods of analysis. It is assumed that the estimation of deflection will be the primary objective of the simplified analysis because the design of the hybrid bridge superstructure is controlled by deflection. To model the bridge superstructure as a beam or a plate, cross-sectional properties are the key parameters. The procedures to evaluate the cross-sectional properties are presented based on the classical lamination theory.

It is assumed in the beam analysis that the cross-section in the xz plane will not deform. The bridge superstructure is modeled as a beam with span length, L, effective flexural rigidity, EI_{eff}, and effective torsional rigidity, GJ_{eff}. The evaluation of EI_{eff} and GJ_{eff} is the main task of modeling the bridge superstructure as a beam. Equation [10.1] calculates the effective bending rigidity.

$$EI_{\text{eff}} = \int_{A_y} \bar{E}_y \bar{z}^2 \mathrm{d}A \qquad [10.1]$$

Table 10.4 Effective engineering properties of different laminates (units in GPa)

Laminate	$\bar{E}_x$	$\bar{E}_y$	$\bar{G}_{xy}$
Outermost	18.6	18.6	2.94
Outer tube	13.1	13.1	6.47
Inner tube	18.6	18.6	2.94

where $\bar{E}_y$ is the effective modulus for laminates in the y direction or Young's modulus for the concrete; $\bar{z}$ is the vertical coordinate taken from the location of the neutral axis; and A_y is the cross-sectional area in the xz plane.

The effective engineering properties of a laminate were obtained based on the classical lamination theory. The effective engineering properties of different laminates are given in Table 10.4, and names of different laminates are found in Fig. 10.4.

The effective torsional rigidity, GJ_{eff}, is evaluated by considering the shear flow around the box section. Here, the following assumptions are made:

- the net shear flow through interior webs is negligible and the shear flow through the top and bottom flanges and exterior webs is of prime significance;
- the magnitude of the shear flow does not change along the top and bottom flanges and the exterior webs;
- the shear flow goes through the median line of each member;
- the top flange consists of the outermost laminate, the outer tube laminate, the concrete layer, and the inner tube laminate; and the top flange extends over the bridge width without any interruptions by the interior webs;
- the bottom flange consists of the outermost laminate, the outer tube laminate, and the inner tube laminate; and the bottom flange extends over the bridge width without any interruptions by the interior webs.

Based on these assumptions, the effective torsional rigidity may be approximated by using Bredt's formula:

$$GJ_{\text{eff}} = \frac{4A_{\text{encl}}^2/1}{\int 1 \Big/ \sum_k \bar{G}_{xyk} t_k \, ds} \qquad [10.2]$$

where A_{encl} is the area enclosed by the median lines of the top and bottom flanges and the exterior webs; $\bar{G}_{xyk}$ is the effective shear modulus of the kth layer in a member; t_k is the thickness of the kth layer in a member; and s is the axis along the median line of each member.

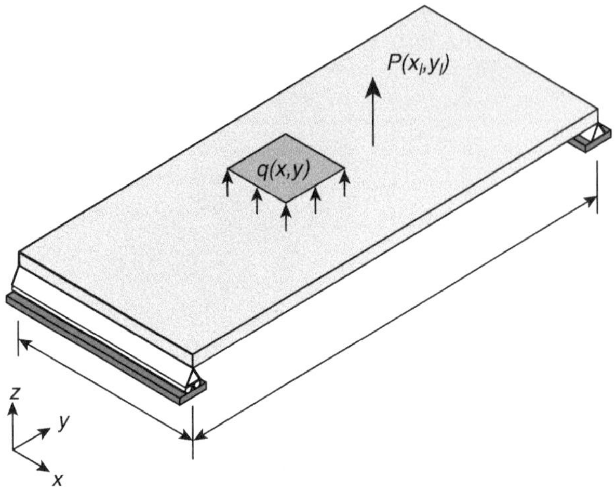

10.16 Bridge model for the orthotropic plate analysis.

In Equation [10.2], the summation is carried out over layers in a member; for instance, there are four layers in the top flange: the outermost laminate, the outer tube laminate, the concrete layer, and the inner tube laminate. The integration is carried out along the median line of each member.

In the orthotropic plate analysis, the hybrid bridge is modeled as an orthotropic plate with two opposite edges simply supported as shown in Fig. 10.16. By assuming some of the bending extension coupling stiffnesses are insignificant, the governing equations for the orthotropic plate model of the hybrid bridge superstructure can be expressed as follows (Aref *et al.*, 2005):

$$D_x \frac{\partial^4 w_o}{\partial x^4} + 2D_t \frac{\partial^4 w_o}{\partial x^2 \partial y^2} + D_y \frac{\partial^4 w_o}{\partial y^4} = q(x, y) \qquad [10.3]$$

where the following notations are used:

$$D_x = D_{11}^x, \quad D_y = D_{22}^x, \quad D_t = \frac{D_{xy} + D_{yx} + D_1 + D_2}{2},$$

$$D_1 = D_{12}^x, \quad D_2 = D_{12}^y, \quad D_{xy} = 2D_{66}^x \text{ and } D_{yx} = 2D_{66}^y \qquad [10.4]$$

and w_o is vertical deflection, and D_{ij}^x and D_{ij}^y are 3×3 bending stiffness matrices evaluated by the classical lamination theory for the transverse direction

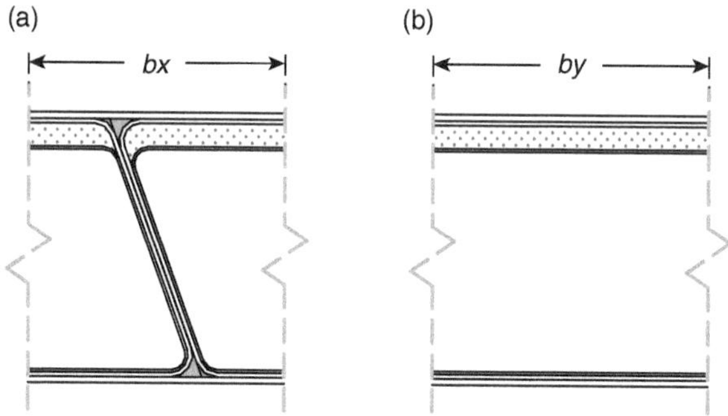

10.17 Representative units for the orthotropic plate analysis. (a) Longitudinal direction; (b) transverse direction. *bx* = width of the representative unit of the bridge in the longitudinal direction *by* = width of the representative unit of the bridge in the transverse direction

Table 10.5 Deflection at midspan under twice the tandem load

	Deflection		Difference from FEA
	(mm)	(*L*/800)	(%)
Linear FEA	3.89	0.850	−1.04
Beam analysis	3.85	0.842	0.687
Orthotropic plate analysis	3.92	0.856	0.706
Experiment	3.88	0.849	−0.135

and the longitudinal direction, respectively. These bending stiffness matrices can be evaluated by choosing representative units for the longitudinal and transverse directions as shown in Fig. 10.17. The solutions of Equation [10.3] can be found in Cusens and Pama (1975).

Table 10.5 compares deflections obtained from the three different analytical methods and the experiment at the center of the test specimen of the bridge superstructure subjected to the flexural loading of twice the tandem load. Figure 10.18 compares deformed shapes from the three different analytical methods and the experiment at twice the tandem load. As can be seen in the table and figure, the predictions by the beam and plate analyses are very close to those by the detailed FEA with the maximum difference being only about 1.0%. Therefore, in this particular case, either the beam or the orthotropic plate analysis can be a very good tool to estimate the overall bridge behavior under different loading conditions. However, when the bridge superstructure becomes wider, a variation of deflection in the transverse direction will get larger. To predict the maximum deflection of the hybrid bridge with more than three lanes, it is recommended to use the orthotropic plate analysis instead of the beam analysis.

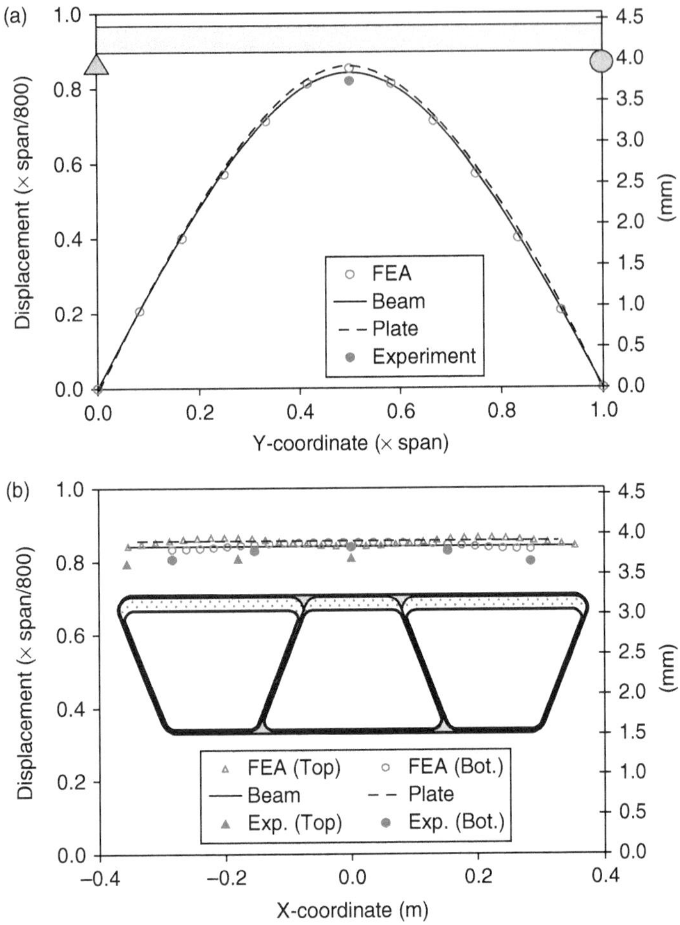

10.18 Deformed shapes from different analyses. (a) Longitudinal variation; (b) transverse variation.

10.3.6 Practical issues

Judging from results from the experimental and numerical study, the proposed hybrid FRP–concrete superstructure is highly feasible from a structural engineering point of view. To fully develop the superstructure and apply the concept in an actual bridge project, the following issues will have to be investigated in the future:

- automated fabrication process for trapezoidal sections,
- methods to expand traffic lanes,

- efficient methods to cast concrete in the field,
- design details to make better use of strength,
- design for the section to resist negative moments in a continuous girder,
- installation method of guard fences,
- response to thermal effects,
- long-term performance degradation and maintenance requirements, and
- life-cycle costs.

10.4 Conclusion

The concept of the hybrid design is a very prudent way to design a structure, because different materials can be used efficiently where they perform best. In this chapter, the hybrid FRP–concrete bridge superstructure is introduced, where the novel concept of the hybrid FRP–concrete design is applied. The trial design of the bridge superstructure was proposed, and its structural performance was investigated through the detailed FEA and a series of static and fatigue loading tests. In addition, simple methods of analysis for this hybrid bridge superstructure were also proposed. Results from the research to date have showed that the proposed hybrid bridge superstructure has excellent structural performance, and the hybrid FRP–concrete system for the bridge superstructure is highly feasible from a structural engineering point of view.

10.5 References

Alampalli, S. and Kunin, J. (2001), *Load Testing of an FRP Bridge Deck on a Truss Bridge*, Special Report 137, New York State Department of Transportation.

American Association of State Highway and Transportation Officials (AASHTO) (1998), *AASHTO LRFD Bridge Design Specifications*, 2nd edn, AASHTO, Washington, DC.

American Composites Manufacturers Association (2004), *Industry Overview: Overview of the FRP Composites Industry – A Historical Perspective*, http://www.mdacomposites.org/mda/overview.html (accessed 30 April 2012).

American Concrete Institute (2006), *Guide for the Design and Construction of Structural Concrete Reinforced with FRP Bars*, ACI 440.1R-06, American Concrete Institute, Farmington Hills, MI.

Aref, A. J. (1997), 'A novel fiber reinforced composite bridge structural system', PhD dissertation, the University of Illinois at Urbana-Champaign.

Aref, A. J., Kitane, Y. and Lee, G. C. (2005), 'Analysis of hybrid FRP–concrete multi-cell bridge superstructure', *Composite Structures*, **69**(3), 346–359.

Ashby, M. F. (1991), 'Overview No. 92 – materials and shape', *Acta Metallurgica et Materialia*, **39**(6), 1025–1039.

Bakeri, P. A. and Sunder, S. S. (1990), 'Concepts for hybrid FRP bridge deck systems', *Serviceability and Durability of Construction Materials*, Proceedings of the First Materials Engineering Congress, Denver, CO, 13–15 August, ASCE, Vol. 2, pp. 1006–1015.

Bakis, C. E., Bank, L. C., Brown, V. L., Consenza, E., Davalos, J. F., Lesko, J. J., Machida, A., Rizkalla, S. H. and Triantafillou, T. C. (2002), 'Fiber reinforced polymer composites for construction – state-of-the-art review', *Journal of Composites for Construction*, **6**(2), 73–87.

Bank, L. C. (2006), *Composites for Construction: Structural Design with FRP Materials*, John Wiley & Sons, New York.

Brailsford, B., Mikovich, S. M. and Hopwood, T. (1995), *Definition of Infrastructure Specific Markets for Composite Materials*, Technical Report, Project P93–121/A573, BIRL, Northwestern University, Evanston, IL.

Creative Pultrusions, Inc. (2002), *Superdeck Composite Bridge Deck Installation Profile – Laurel Lick Bridge*, http://www.creativepultrusions.com/library.html (accessed 30 April 2012).

Cusens, A. R. and Pama, R. P. (1975), *Bridge Deck Analysis*, John Wiley & Sons, London.

Deskovic, N., Triantafillou, T. C. and Meier, U. (1995a), 'Innovative design of FRP combined with concrete: short-term behavior', *Journal of Structural Engineering*, **121**(7), 1069–1078.

Deskovic, N., Meier, U. and Triantafillou, T. C. (1995b), 'Innovative design of FRP combined with concrete: long-term behavior', *Journal of Structural Engineering*, **121**(7), 1079–1089.

Dumlao, C., Lauraitis, K., Abrahamson, E., Hurlbut, B., Jacoby, M., Miller, A. and Thomas, A. (1996), 'Demonstration low-cost modular composite highway bridge', *Fiber Composites in Infrastructure*, Proceedings of the First International Conference on Composites in Infrastructure, ICCI'96, Tucson, AZ, 15–17 January, H. Saadatmanesh and M. R. Ehsani (eds.), pp. 1141–1155.

Dunker, K. F. and Rabbat, B. G. (1995), 'Assessing infrastructure deficiencies: the case of highway bridges', *Journal of Infrastructure Systems*, **1**(2), 100–119.

Fam, A. Z. and Rizkalla, S. H. (2002), 'Flexural behavior of concrete-filled fiber-reinforced polymer circular tubes', *Journal of Composites for Construction*, **6**(2), 123–132.

Farheyl, D. N. (2005), 'Long-term performance monitoring of the Tech 21 all-composite bridge', *Journal of Composites for Construction*, **9**(3), 255–262.

Hayes, M. D., Ohanehi, D., Lesko, J. J., Cousins, T. E. and Witcher, D. (2000), 'Performance of tube and plate fiberglass composite bridge deck', *Journal of Composites for Construction*, **4**(2), 48–55.

Hibbit, Karlsson & Sorensen, Inc. (2000), *ABAQUS/Standard User's Manual*, Version 6.1, Hibbit, Karlsson & Sorensen, Inc.

Hillman, J. R. and Murray, T. M. (1990), 'Innovative floor systems for steel framed buildings', *Mixed Structures, Including New Materials*, Proceedings of IABSE Symposium, Brussels, Belgium, Vol. 60, IABSE, Zurich, pp. 672–675.

Karbhari, V. M. and Zhao, L. (2000), 'Use of composites for 21st century civil infrastructure', *Computer Methods in Applied Mechanics and Engineering*, **185**, 433–454.

Mirmiran, A. (2001), 'Innovative combinations of FRP and traditional materials', *Proceedings of the International Conference on FRP Composites in Civil Engineering*, Vol. II, 12–15 December, Hong Kong, China, J. G. Teng (ed.), Elsevier Science, New York, pp. 1289–1298.

Moy, S. S. J. (ed.) (2001), 'FRP composites – life extension and strengthening of metallic structures', *ICE Design and Practice Guides*, The Institution of Civil Engineers, Thomas Telford Publishing, London.

Mufti, A. A., Labossière, P. and Neale, K. W. (2002), 'Recent bridge applications of FRPs in Canada', *Structural Engineering International*, **12**(2), 96–98.

Neely, W. D., Cousins, T. E. and Lesko, J. J. (2004), 'Evaluation of in-service performance of Tom's Creek Bridge fiber-reinforced polymer superstructure', *Journal of Performance of Constructed Facilities*, **18**(3), 147–158.

Ribeiro, M. C. S., Tavares, C. M. L., Ferreira, A. J. M. and Marques, A. T. (2001), 'Static flexural performance of GFRP–polymer concrete hybrid beams', *Proceedings of the International Conference on FRP Composites in Civil Engineering*, Vol. II, 12–15 December, Hong Kong, China, J. G. Teng (ed.), Elsevier Science, New York, pp. 1355–1362.

Saiidi, M., Gordaninejad, F. and Wehbe, N. (1994), 'Behavior of graphite/epoxy concrete composite beams', *Journal of Structural Engineering*, **120**(10), 2958–2976.

Seible, F., Burgueño, R., Abdallah, M. G. and Nuismer, R. (1995), 'Advanced composite carbon shell systems for bridge columns under seismic loads', *Proceedings of the National Seismic Conference on Bridges and Highways – Progress in Research and Practice*, San Diego, CA, 10–13 December, Section 10.

Seible, F., Karbhari, V. M., Burgueño, R. and Seaberg, E. (1998), 'Modular advanced composite bridge systems for short and medium span bridges', *Developments in Short and Medium Span Bridge Engineering '98*, The Fifth International Conference on Short and Medium Span Bridges, Calgary, 13–16 July, pp. 431–441.

Seible, F., Karbhari, V. M. and Burgueño, R. (1999), 'Kings Stormwater Channel and I-5/Guilman Bridges, USA', *Structural Engineering International*, **9**(4), 250–253.

Small, E. P. and Cooper, J. (1998), 'Condition of the nation's highway bridges – a look at the past, present, and future', *Transportation Research News*, No. 194, January–February, 3–8.

Tang, B. (1997), 'Fiber reinforced polymer composites applications in USA: DOT-Federal Highway Administration', *Proceedings of the First Korea/U.S.A. Road Workshop*, 28–29 January.

Tang, B. M. and Hooks, J. M. (2001), 'FRP composites technology is changing the American bridge building industry', *Proceedings of the International Conference on FRP Composites in Civil Engineering*, Vol. II, 12–15 December, Hong Kong, China, J. G. Teng (ed.), Elsevier Science, New York, pp. 1657–1663.

Tang, B. and Podolny, W. (1998), 'A successful beginning for fiber reinforced polymer (FRP) composite materials in bridge applications', *Proceedings of the International Conference on Corrosion and Rehabilitation of Reinforced Concrete Structures*, 7–11 December, Orlando, FL, Federal Highway Administration.

Telang, N. M., Dumlao, C., Mehrabi, A. B., Ciolko, A. T. and Gutierrez, J. (2006), *Field Inspection of In-Service FRP Bridge Decks*, NCHRP Report 564, Transportation Research Board.

US Department of Transportation, Federal Highway Administration, and Federal Transit Administration (2000), *1999 Status of the Nation's Highways,*

Bridges and Transit: Conditions and Performance – Report to Congress, http://www.fhwa.dot.gov/policy/1999cpr/report.htm (accessed 30 April 2012).

Van Erp, G. (2002), 'Road bridge benefits from hybrid beams', *Reinforced Plastics*, **46**(6), 4.

Van Erp, G., Cattell, C. and Ayers, S. (2005), 'The Australian approach to composites in civil engineering', *Reinforced Plastics*, **49**(6), 20–26.

Wu, H. F. (2005), *Composites in Civil Applications*, National Institute of Standards and Technology, http://www-15.nist.gov/atp/focus/99wp-ci.htm (accessed 30 April 2012).

Zureick, A. H., Shih, B. and Munley, E. (1995), 'Fiber-reinforced polymeric bridge decks', *Structural Engineering Review*, **7**(3), 257–266.

Index

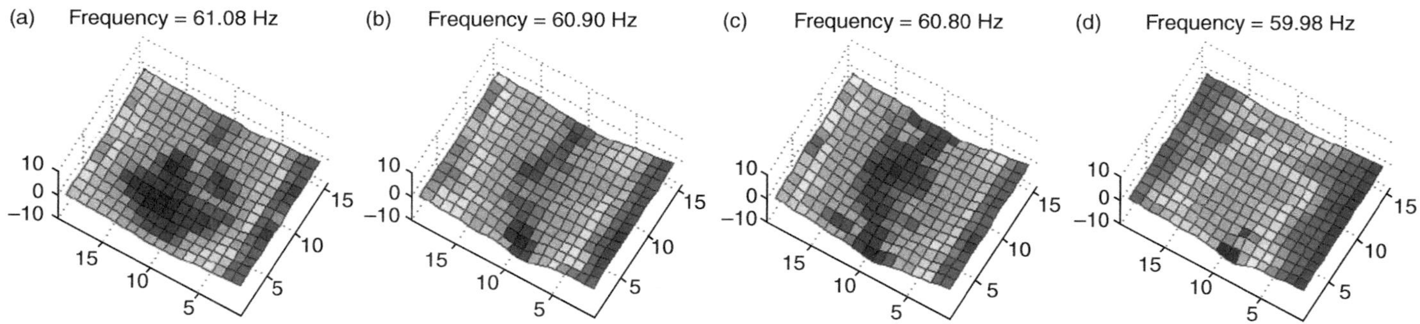

Plate I (Chapter 4) Longitudinal curvature mode shapes (κ_{yy}) of the 1st longitudinal bending mode from experiment. (a) Healthy; (b) 1st stage; (c) 2nd stage; (d) 3rd stage.

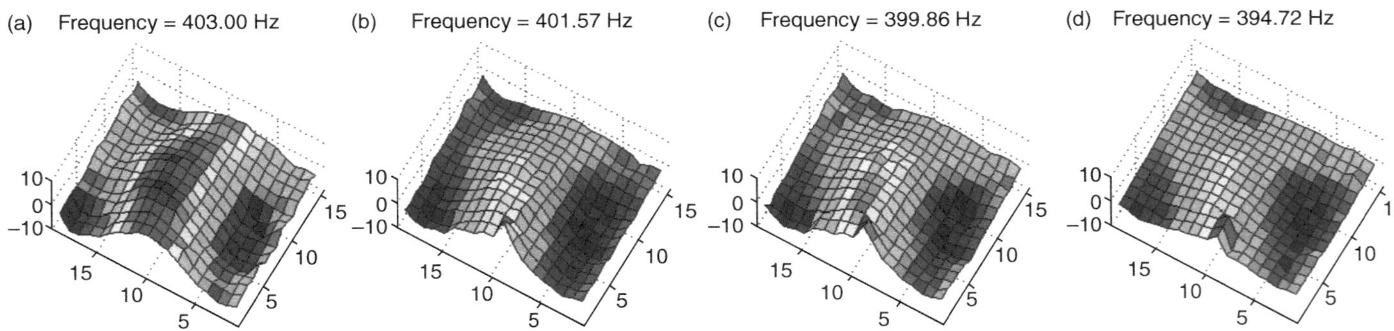

Plate II (Chapter 4) Longitudinal curvature mode shapes (κ_{yy}) of the 3rd longitudinal bending mode from experiment. (a) Healthy; (b) 1st stage; (c) 2nd stage; (d) 3rd stage.

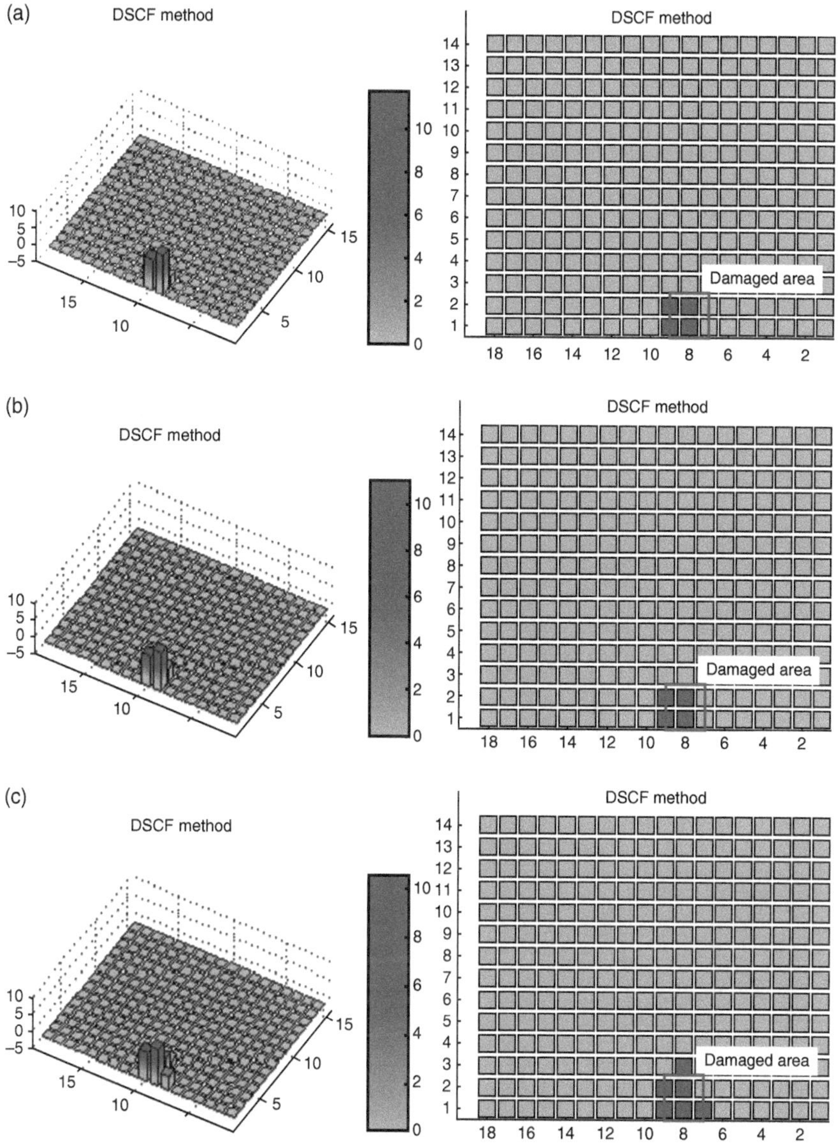

Plate III (Chapter 4) Damage localization of FRP sandwich plate at three damage stages: (a) Damage Stage 1; (b) Damage Stage 2; (c) Damage Stage 3.

(a) (b) (c)

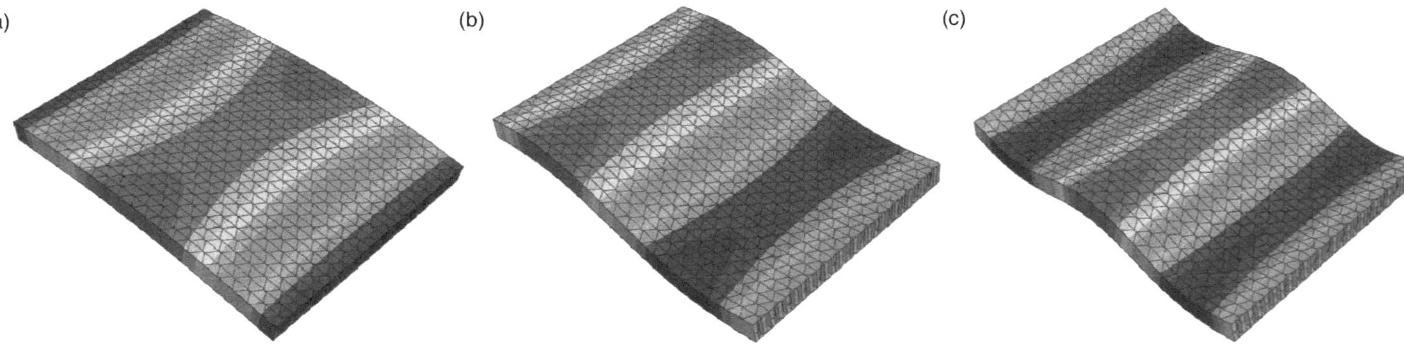

Plate IV (Chapter 4) The first three longitudinal bending mode shapes of the simply supported FRP sandwich deck panel from FEM (the contour of vertical displacement U3). (a) 1st mode; (b) 2nd mode; (c) 3rd mode.

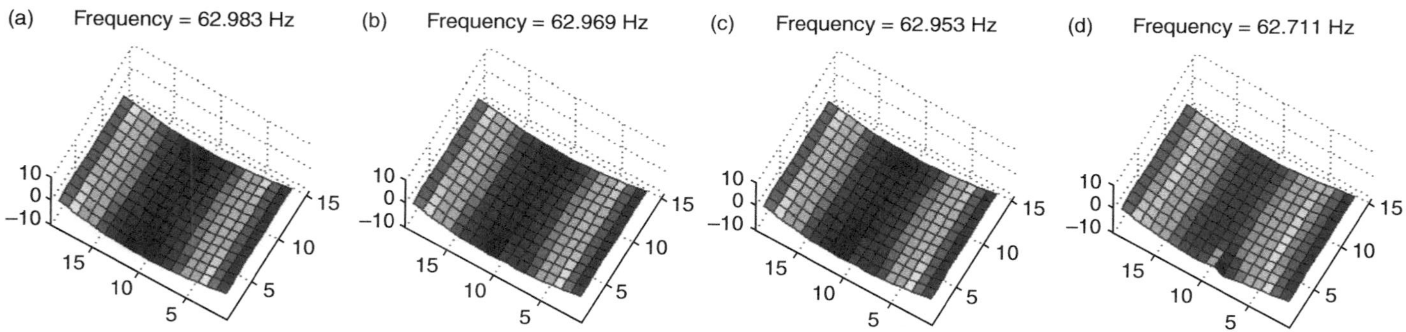

Plate V (Chapter 4) Longitudinal curvature mode shape (κ_{yy}) and modal frequency of 1st longitudinal bending mode from FEA. (a) Healthy; (b) 1st stage; (c) 2nd stage; (d) 3rd stage.

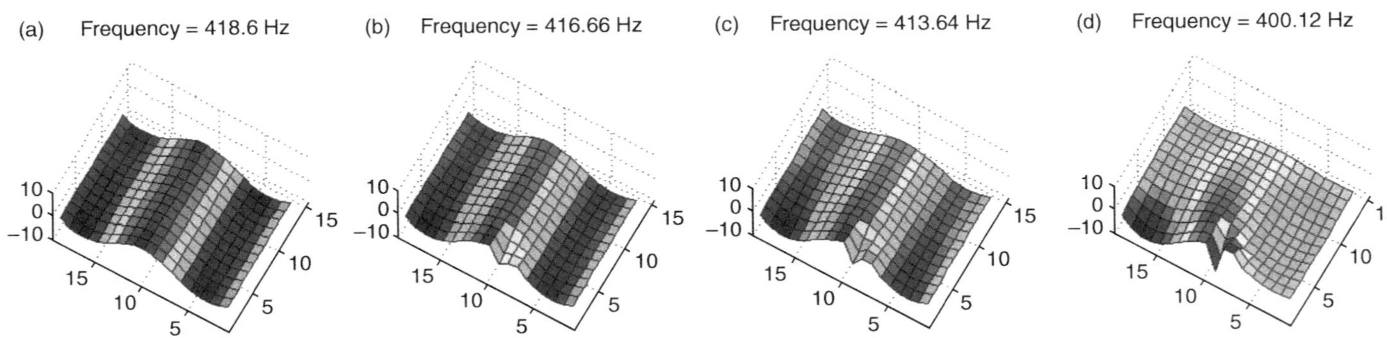

Plate VI (Chapter 4) Longitudinal curvature mode shape (κ_{yy}) and modal frequency of 3rd longitudinal bending mode from FEA. (a) 1st mode; (b) 2nd mode; (c) 3rd mode.

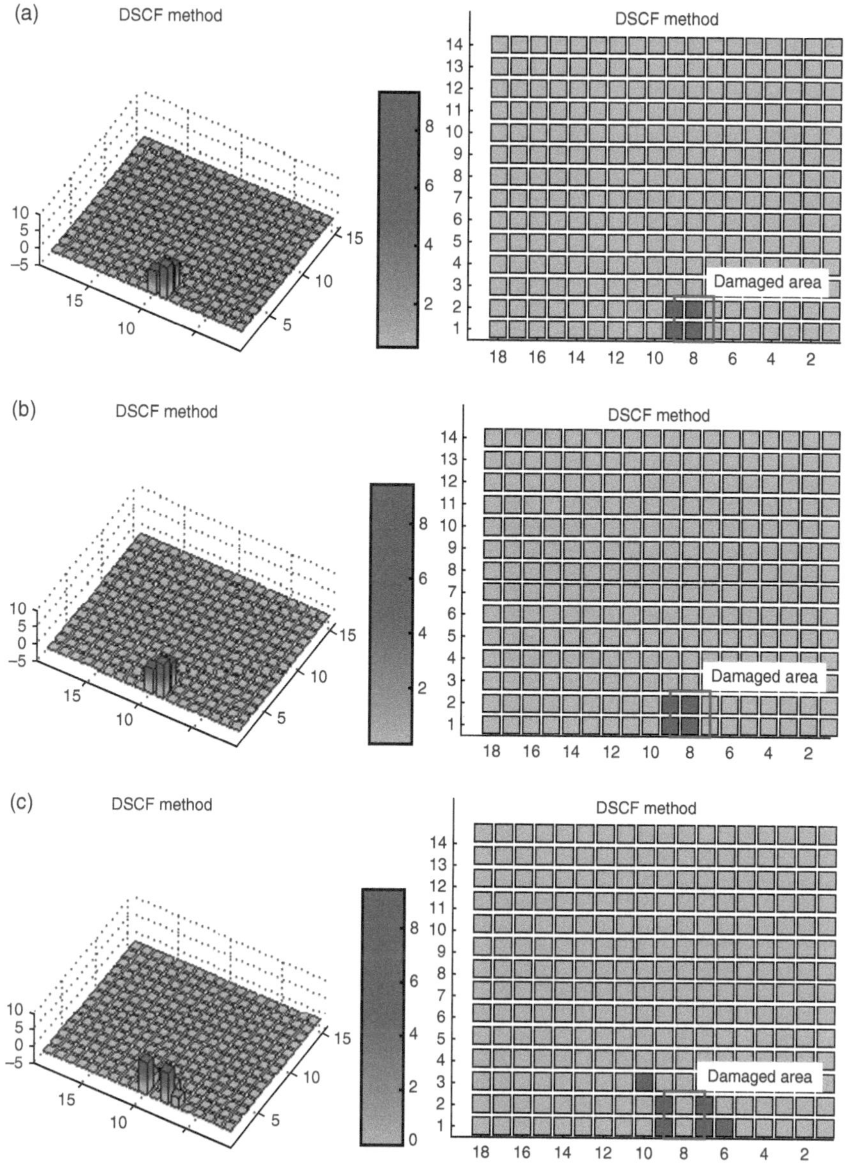

Plate VII (Chapter 4) Damage localization at three damage stages using FEA data: (a) Damage Stage 1; (b) Damage Stage 2; (c) Damage Stage 3.

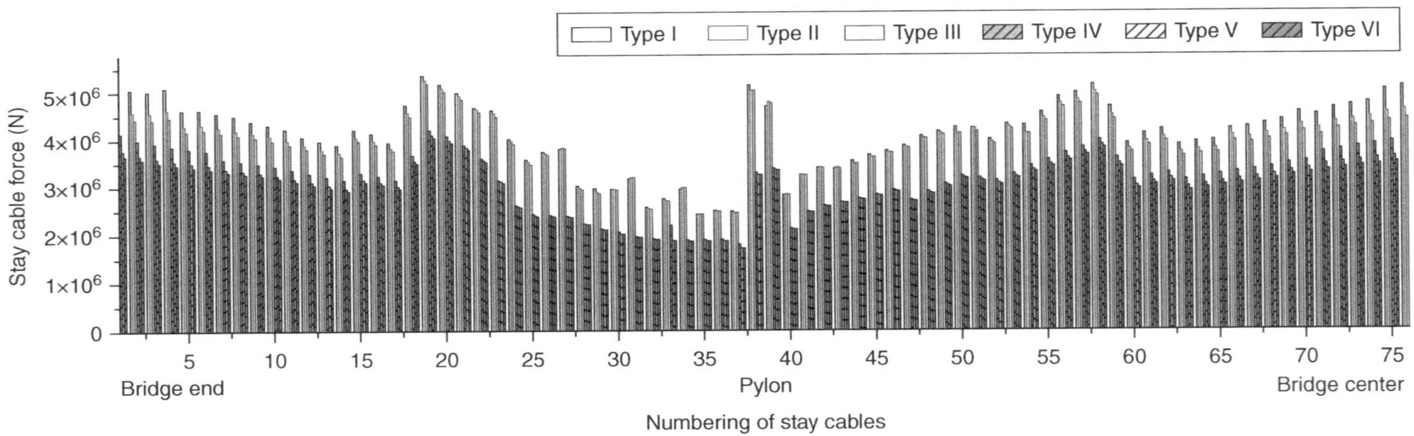

Plate VIII (Chapter 8) Reasonable stay cable forces for six bridges.

FSC
www.fsc.org
MIX
Papier | Fördert gute Waldnutzung
FSC® C083411

Kolibri 360 GmbH
Ferdinand-Jühlke-Straße 7
99095 Erfurt, Deutschland
produktsicherheit@kolibri360.de